Maritime Autonomous Surface Ships

Maritime Autonomous Surface Ships

Editors

Haitong Xu
Lúcia Moreira
Xianbo Xiang
Carlos Guedes Soares

MDPI

Basel • Beijing • Wuhan • Barcelona • Belgrade • Novi Sad • Cluj • Manchester

Haitong Xu
Centre for Marine Technology and Ocean Engineering (CENTEC)
University of Lisbon
Lisbon
Portugal

Lúcia Moreira
Centre for Marine Technology and Ocean Engineering (CENTEC)
University of Lisbon
Lisbon
Portugal

Xianbo Xiang
School of Naval Architecture and Ocean Engineering
Huazhong University of Science and Technology
Wuhan
China

Carlos Guedes Soares
Centre for Marine Technology and Ocean Engineering (CENTEC)
University of Lisbon
Lisbon
Portugal

Editorial Office
MDPI AG
Grosspeteranlage 5
4052 Basel, Switzerland

This is a reprint of the Special Issue, published open access by the journal *Journal of Marine Science and Engineering* (ISSN 2077-1312), freely accessible at: www.mdpi.com/journal/jmse/special_issues/668NU31H8V.

For citation purposes, cite each article independently as indicated on the article page online and using the guide below:

Lastname, A.A.; Lastname, B.B. Article Title. *Journal Name* **Year**, *Volume Number*, Page Range.

ISBN 978-3-7258-1532-6 (Hbk)
ISBN 978-3-7258-1531-9 (PDF)
https://doi.org/10.3390/books978-3-7258-1531-9

Cover image courtesy of Haitong Xu

Contents

About the Editors

Haitong Xu

Dr. Haitong Xu is an Adjunct Assistant Professor and researcher at the Centre for Marine Technology and Ocean Engineering (CENTEC), Instituto Superior Técnico, University of Lisbon. He carried out his PhD work at CENTEC, conducting research on the guidance and control of autonomous ships, and defended his PhD thesis in January 2021. He oversees research on autonomous ships and is currently in charge of two ongoing FCT projects at CENTEC. He has been involved in seven EU and FCT R&D projects and has published 60 original scientific research papers, consisting of 36 papers published in SCI journals. He has been invited to serve as the Guest Editor of four Special Issues on autonomous ships in the *Journal of Marine Science and Engineering*, was the Section-Chairman of the MARTECH2022 and 2024 conferences, is an IEEE Oceanic Engineering Society member, and is a Technical Committee member of *Marine Robotics, Autonomous Maritime Systems*. He oversees research on autonomous ships, sensors, and experimental instruments at CENTEC. The primary objective of his research is to design and test marine control systems for autonomous surface ships. This includes manoeuvring modelling, guidance and control systems, collision avoidance, data acquisition (DAQ) and developing prototypes.

Lúcia Moreira

Dr. Lúcia Moreira is an Adjunct Assistant Professor and researcher at the Centre for Marine Technology and Engineering (CENTEC), developing research mainly in the areas of the manoeuvring, guidance, and control of marine vehicles and artificial intelligence in maritime applications. From 2015 to 2018, she worked as a professor and coordinator of the mechanical engineering degree at the Instituto Superior Politécnico de Tecnologias e Ciências (ISPTEC) in Luanda, Angola, being responsible for the leadership, development, and strategic and operational management of the degree. She received a PhD in Naval Architecture and Marine Engineering from the Technical University of Lisbon (IST) in 2008, and from 2009 to 2014, she performed postdoc research at CENTEC. Previously, she received an engineering degree in naval architecture and marine engineering from IST in 1998 and an MPhil degree in marine technology from the University of Newcastle upon Tyne, U.K., in 2002. In 2002 and 2003, she pursued postgraduate studies in control systems at the Department of Electrical and Computers Engineering, IST, and in 2004, she continued her studies in the guidance and control of marine systems at the Norwegian University of Science and Technology (NTNU), Trondheim, Norway.

Xianbo Xiang

Prof. Xianbo Xiang received B.E. and M.E. degrees in automatic control and marine engineering from the Huazhong University of Science and Technology, Wuhan, China, in 2000 and 2003, respectively, and a Ph.D. degree in robotics from the University of Montpellier 2, Montpellier, France, in 2011. He is currently a Professor at the School of Naval Architecture and Ocean Engineering, Huazhong University of Science and Technology, Wuhan, China. In 2006, he was an EU Erasmus Mundus Visiting Scholar with the SpaceMaster Project for three months. From 2008 to 2011, he was with the European Project FreeSubNet as an EC Marie Curie ESR Fellow with the Laboratoire d' Informatique, de Robotique et de Microelectronique de Montpellier, CNRS UMR, Montpellier. His current research interests include robotics and marine vehicle systems. He has published more than 200 papers in refereed journals and conferences. He has been the general chair of conferences including IEEE USYS2018, IMET2022, and IMET2023.

He is currently serving as the associate editor for the *International Journal of Maritime Engineering* and as an editorial member for *Applied Ocean Research*, *China Ocean Engineering*, the *Journal of Marine Science and Engineering*, and *Brodogradnja/Shipbuilding*.

Carlos Guedes Soares

Prof. Carlos Guedes Soares is a Distinguished Professor of Instituto Superior Técnico, University of Lisbon. He is the Scientific Coordinator of the Centre for Marine Technology and Ocean Engineering (CENTEC), which is a research centre classified by the Portuguese Foundation for Science and Technology as Excellent in 2014 and 2018. He received MSc. and Ocean Engineer degrees from the Massachusetts Institute of Technology, USA, in 1976, a Ph.D. degree from the Norwegian Institute of Technology of the University of Trondheim in 1984, and a Doctor of Science degree from the Technical University of Lisbon, Portugal, in 1991. He is Doctor Honoris causa from the Technical University of Varna (2003) and University Dunarea de Jos in Galati (2004). He has supervised more than 55 PhD students, has co-authored more than 700 journal papers, and has more than 20,000 citations in Web of Science (29,000 in SCOPUS). He is one of the five Founding Members of the European Safety and Reliability Association (ESRA) and has been General Secretary, Vice-Chairman, and Chairman of ESRA. He has been Chairman of the International Maritime Association of the Mediterranean (IMAM) (2005–2011; 2017–2019) and of the ISSC - Ship and Offshore Structures Congress (2012–2015). He has been the Chair or Co-Chair of various conferences in the series of OMAE, ESREL, IMAM, MARSTRUCT, MARTECH, ICCGS, and RENEW Conferences. He has been coordinating the Symposium on Structures, Safety and Reliability of the Ocean, Offshore and Arctic Engineering (OMAE) Conference since 1989. He has been Editor and Editor-in-Chief of the *Reliability Engineering and System Safety Journal* since 1992, is Co-Editor in Chief of the *Journal for Marine Science and Application*, and belongs to the Editorial Board of more than 15 journals. He is a Fellow of SNAME, RINA, IMarEST, and ASME; "Membro Conselheiro" of Ordem dos Engenheiros; a member of ASCE, AGU, AMS, and SRA; and a member of the Portuguese Academy of Engineering.

Journal of
Marine Science and Engineering

MDPI

Editorial

Maritime Autonomous Surface Ships

Haitong Xu [1,*], Lúcia Moreira [1], Xianbo Xiang [2] and C. Guedes Soares [1]

[1] Centre for Marine Technology and Ocean Engineering (CENTEC), Instituto Superior Técnico, Universidade de Lisboa, Av. Rovisco Pais, 1049-001 Lisboa, Portugal; lucia.moreira@centec.tecnico.ulisboa.pt (L.M.); c.guedes.soares@centec.tecnico.ulisboa.pt (C.G.S.)

[2] School of Naval Architecture and Ocean Engineering, Huazhong University of Science and Technology, Wuhan 430074, China; xbxiang@hust.edu.cn

* Correspondence: haitong.xu@centec.tecnico.ulisboa.pt; Tel.: +351-218-417-607

The maritime industry faces many pressing challenges due to increasing environmental and safety regulations and crew safety concerns. In light of these challenges, autonomous ships can provide potential solutions for addressing smart shipping, fuel efficiency, and safety issues. The development of marine autonomy technology will significantly improve the situation and is expected to become a cost-efficient alternative to conventional ships. Currently, automated shipping technology is rapidly transitioning from theoretical to practical applications as the number and scope of autonomous ship prototypes increase around the globe. These prototypes are widely used in both navy and commercial applications, such as ocean observation, coast patrol, underwater monitoring, and underwater production system operation, to name just a few.

The main goal of this book is to address key challenges, thereby promoting research on Maritime Autonomous Surface Ships (MASS). Firstly, one review paper on developing Digital Twin (DT) technology in the maritime domain is provided [1]. The following topics on autonomous surface ships are included in this book: methods of ship control [2–5], collision avoidance [6,7], ship detection methods [8,9], and manoeuvring models [10].

Madusanka et al. [1] reviewed the development of Digital Twin (DT) technology and its applications within the maritime domain, mainly surface ships. In [2], a finite-time, active fault-tolerant control (AFTC) method was proposed for autonomous surface vehicles, and the framework was based on an integrated design of fault detection (FD), fault estimation (FE), and controller reconfiguration. Simulation tests using the CyberShip II were carried out to validate the proposed AFTC method. In [3], the authors studied collision avoidance in the formation-containment tracking control of multi-USVs with constrained velocity and propulsion forces. A multi-USV formation-containment tracking control strategy was designed based on a dual-layer control framework, and stability was validated using the Lyapunov method. In [4], the cooperative formation trajectory tracking problem for heterogeneous unmanned aerial vehicles (UAV) and multiple unmanned surface vessels (USV) was invested in this paper, and a simulation study is provided to show the efficacy of the proposed strategy. To provide a higher lift force and improve the seakeeping performance of a ship, a control method for the T-foil's swinging angle is established and optimized based on model testing [5]. The results obtained by model testing show that T-foils with pitch angular velocity control can decrease the vertical motion response in the resonance region of a ship's encounter frequency and increase the anti-bow acceleration effect.

Another important topic for autonomous surface ships is collision avoidance, since the ship must be able to avoid unexpected obstacles. Niu et al. [6] proposed a multi-ship autonomous collision avoidance decision-making algorithm using a data-driven method and adopted the Multi-agent Deep Reinforcement Learning (MADRL) framework. The 40 encounter scenarios were designed to verify the proposed algorithm, and the results

Citation: Xu, H.; Moreira, L.; Xiang, X.; Guedes Soares, C. Maritime Autonomous Surface Ships. *J. Mar. Sci. Eng.* **2024**, *12*, 957. https://doi.org/10.3390/jmse12060957

Received: 23 May 2024
Accepted: 28 May 2024
Published: 7 June 2024

show that this algorithm can efficiently make a ship collision avoidance decision in compliance with COLREGs. Hwang et al. [7] proposed a method for analyzing collision risk situations extracted from AIS data through graph-based modeling and establishing validation scenarios. Yasir et al. [8] presented a survey of AI- and ML-based techniques for ship detection in SAR images that provide a more effective and reliable way to detect and classify ships in various weather conditions, both onshore and offshore. In [9], a new attitude-estimation framework was proposed to extract the geometric features using point clouds from shipborne LiDAR and compute the attitude of the target ships. The experimental results demonstrated the filtering ability and practical applicability of the proposed method in real water-pool experiments under real environmental noises. In [10], a data-driven method, the truncated LS-SVM, was proposed for estimating the nondimensional hydrodynamic coefficients of a maneuvering model. The results demonstrate that the truncated LS-SVM method effectively models the hydrodynamic force prediction problems with an extensive training set, reducing parameter uncertainty and yielding more convincing results.

Author Contributions: Conceptualization, H.X., L.M., X.X. and C.G.S.; writing—original draft preparation, H.X.; writing—review and editing, H.X., L.M., X.X. and C.G.S. All authors have read and agreed to the published version of the manuscript.

Funding: This work contributes to the Strategic Research Plan of the Centre for Marine Technology and Ocean Engineering, financed by the Portuguese Foundation for Science and Technology (Fundação para a Ciência e Tecnologia—FCT) under contract UIDB/UIDP/00134/2020.

Conflicts of Interest: The authors declare no conflicts of interest.

References

1. Madusanka, N.S.; Fan, Y.; Yang, S.; Xiang, X. Digital Twin in the Maritime Domain: A Review and Emerging Trends. *J. Mar. Sci. Eng.* **2023**, *11*, 1021. [CrossRef]
2. Wang, X.; Ouyang, Y.; Wang, X.; Wang, Q. A Novel, Finite-Time, Active Fault-Tolerant Control Framework for Autonomous Surface Vehicle with Guaranteed Performance. *J. Mar. Sci. Eng.* **2024**, *12*, 347. [CrossRef]
3. Wang, J.; Shan, Q.; Li, T.; Xiao, G.; Xu, Q. Collision-Free Formation-Containment Tracking of Multi-USV Systems with Constrained Velocity and Driving Force. *J. Mar. Sci. Eng.* **2024**, *12*, 304. [CrossRef]
4. Huang, Y.; Li, W.; Ning, J.; Li, Z. Formation Control for UAV-USVs Heterogeneous System with Collision Avoidance Performance. *J. Mar. Sci. Eng.* **2023**, *11*, 2332. [CrossRef]
5. Sun, Y.; Zhang, D.; Wang, Y.; Zong, Z.; Wu, Z. Model Experimental Study on a T-Foil Control Method with Anti-Vertical Motion Optimization of the Mono Hull. *J. Mar. Sci. Eng.* **2023**, *11*, 1551. [CrossRef]
6. Niu, Y.; Zhu, F.; Wei, M.; Du, Y.; Zhai, P. A Multi-Ship Collision Avoidance Algorithm Using Data-Driven Multi-Agent Deep Reinforcement Learning. *J. Mar. Sci. Eng.* **2023**, *11*, 2101. [CrossRef]
7. Hwang, T.; Youn, I.-H. Development of a Graph-Based Collision Risk Situation Model for Validation of Autonomous Ships' Collision Avoidance Systems. *J. Mar. Sci. Eng.* **2023**, *11*, 2037. [CrossRef]
8. Yasir, M.; Niang, A.J.; Hossain, M.S.; Islam, Q.U.; Yang, Q.; Yin, Y. Ranking Ship Detection Methods Using SAR Images Based on Machine Learning and Artificial Intelligence. *J. Mar. Sci. Eng.* **2023**, *11*, 1916. [CrossRef]
9. Wei, S.; Xiao, Y.; Yang, X.; Wang, H. Attitude Estimation Method for Target Ships Based on LiDAR Point Clouds via An Improved RANSAC. *J. Mar. Sci. Eng.* **2023**, *11*, 1755. [CrossRef]
10. Xu, H.; Guedes Soares, C. Data-Driven Parameter Estimation of Nonlinear Ship Manoeuvring Model in Shallow Water Using Truncated Least Squares Support Vector Machines. *J. Mar. Sci. Eng.* **2023**, *11*, 1865. [CrossRef]

Journal of
Marine Science and Engineering

MDPI

Review

Digital Twin in the Maritime Domain: A Review and Emerging Trends

Nuwan Sri Madusanka, Yijie Fan, Shaolong Yang * and Xianbo Xiang *

School of Naval Architecture and Ocean Engineering, Huazhong University of Science and Technology, 1037 Luoyu Road, Wuhan 430074, China; nuwan_sl@hust.edu.cn (N.S.M.); m202172178@hust.edu.cn (Y.F.)
* Correspondence: yangsl@hust.edu.cn (S.Y.); xbxiang@hust.edu.cn (X.X.)

Abstract: This paper highlights the development of Digital Twin (DT) technology and its admittance to a variety of applications within the maritime domain in general and surface ships in particular. The conceptual theory behind the evolution of DT is highlighted along with the development of the technology and current progress in practical applications with an exploration of the key milestones in the extension from the electrification of the shipping sector towards the realization of a definitive DT-based system. Existing DT-based applications within the maritime sector are surveyed along with the comprehension of ongoing research work. The development strategy for a formidable DT architecture is discussed, culminating in a proposal of a four-layered DT framework. Considering the importance of DT, an extensive and methodical literature survey has also been carried out, along with a comprehensive scientometric analysis to unveil the methodical footprint of DT in the marine sector, thus leading the way for future work on the design, development and operation of surface vessels using DT applications.

Keywords: digital twin; digitalization; smart shipping; autonomous surface vehicles; citespace; scientometric analysis

check for updates

Citation: Madusanka, N.S.; Fan, Y.; Yang, S.; Xiang, X. Digital Twin in the Maritime Domain: A Review and Emerging Trends. *J. Mar. Sci. Eng.* **2023**, *11*, 1021. https://doi.org/10.3390/jmse11051021

Academic Editor: Rafael Morales

Received: 29 March 2023
Revised: 21 April 2023
Accepted: 4 May 2023
Published: 10 May 2023

1. Introduction

The technological development of the world is ever incrementing, and humankind is searching for innovative pathways to perform the tasks demanded by industries or services with improved efficiency and effectiveness. The digitalization of the world boomed with the evolution of Cloud computing, Internet of Things (IoT) [1,2], Big Data analytics, Virtual Reality (VR), Augmented Reality (AR), Artificial Intelligence (AI) [3] with Machine Learning, Deep Learning and Neural Networks, etc. The same has led the way to the development of the Cyber-Physical System (CPS), which is hailed as the epitome of the manufacturing and consumer service sector [4]. It is an automated system with a connection of physical reality featured with computing structures with smart networking tools. The technologies of 5G and Tactile Internet [5], with the provision of ultra-reliable ultra-low-delay, have enhanced communication and feedback between entities across oceans. To create the fusion between cyberspace and physical space, Digital Twin (DT) was introduced as a rational solution in associating the two extents. The concept of DT has become a real application within a limited time, and some disciplines are surpassing the expectations and predictions made during the early conceptual inception of DT.

In the present world, simulation has become an integral part of system development for every engineering discipline. Starting from solving design problems using numerical algorithms in the 1960s, simulation has taken us to the digital era, where simulation is integrated into the life cycle of the particular product including design, testing, manufacturing, commissioning, operating, and decommissioning [6]. This process of evolution has opened the way to DT, which is more versatile and dynamic than the physical twin concept. Ever since, the concept of DT has become a reality, expanding its outreach to various disciplines around the world including the maritime domain.

This survey is an endeavor to review and understand the DT technology and the meticulous footprint it has made within the maritime domain to comprehend the future compatibility to meet the upcoming high-tech requirements. Past technological milestones in DT evolution in the marine sector are highlighted in this article, which includes an evaluation of ongoing research and applications around the globe. Further, scientometric analysis has also been performed to identify the key research points such as author contributions, author impact research collaborations, demographic research impact, trending keywords, etc. The main contributions of this publication can be condensed as follows.

- A comprehensive literature review on DT from the initial notion up to the year 2022 with a focus on the maritime domain.
- Proposed a generic four-layer DT framework for a marine vessel.
- Scientometric analysis of the published research related to marine DT within the duration of 20 years including demographic collaboration and trending keywords.
- The progress of development and implementation of DT technology in the maritime sector, thus providing insight for researchers with key challenges and gaps within the marine DT domain.

This paper is structured in the following manner. Followed by the Introduction in Section 1, Section 2 describes the origination of the DT concept with nomenclature clarification. In the Section 3, an overview of the implication pathway of DT technology in the marine industry is discussed. In Section 4, various DT-based applications in different sectors in the maritime domain are discussed, while Section 5 describes the prerequisites for a DT-based system along with the overall DT framework for the marine sector. In Section 6, a comprehensive bibliometric study is carried out using the scientometric analysis approach. Section 7 describes the key challenges and future trends. Finally, the conclusion of the entire effort is provided in Section 8.

2. Inception of the Digital Twin Concept

The initial conception of having a "Twin" was first implemented by NASA in their "Apollo" program, which used two identically built satellites in their missions [7]. One would go on the actual voyage in outer space while the other physical twin will remain in the laboratory in a controlled space, thus allowing the mission scientists to analyze the conditions of the launched vehicle by comparing the conditions with the ground twin. The same helped them in parameter monitoring and fault analysis with a minimum data transfer between the two vehicles. A similar concept was utilized in the "Iron Bird" (Figure 1) ground-based aircraft system testing platform by Airbus Industries, where a hardware simulator twin was used in system evaluation, design and testing while a simulated cockpit was utilized by the test pilots [8].

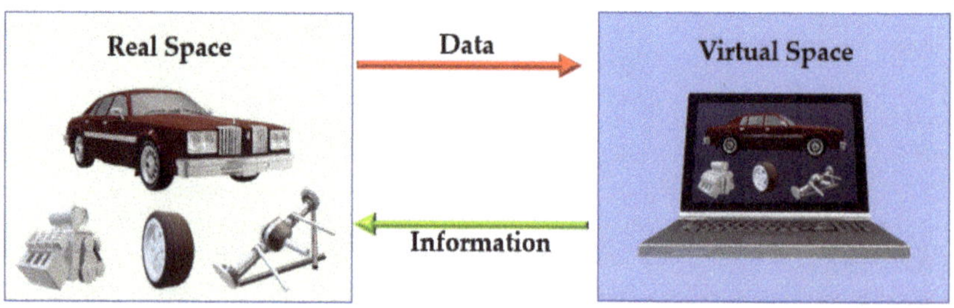

Figure 1. Reflection of the Original Conceptual Idea of DT by Dr M. Grieves in 2002.

The combined usage of the two terms of "Digital + Twin" first came up in a white paper related to the design for 3D arterial phantoms in coronary arteries published in the *Radiology* journal of RSNA (Radiological Society of North America) [9]. Authors have developed a realistic model of arteries using stereolithography using a computer-based model and they have referred to the computer-based design as a "digital twin".

The novel concept of the DT was first proposed in the year 2002 by Dr Michael Grieves, a leading scientist in the field of advanced manufacturing at the Florida Institute of Technology at a manufacturing engineers conference in Troy, Michigan. His idea was based on constructing digital information on the physical model on his own (Figure 1). This digital information is a replica of the data embedded in the physical system, which will be connected with the physical system in the complete life cycle of a particular system/component [10].

The concept of DT was also brought up by NASA under their "Modeling, Simulation, Information Technology & Processing Roadmap at Technology Area 11", where a DT is introduced as a multiscale simulation of a vehicle or system with its own incorporated physics by optimally utilizing its physical data, sensor data, historical data, etc., in the effort of obtaining a real-time image related to the life of its corresponding physical twin in outer space. Since the concept of DT is practical and more realistic in nature, application for a single vehicle or even for interdependent multiple assets can be performed robustly. The anomalies that occurred during the manufacturing stage that could imperil the space missions can also be foreseen during its manipulation. DT can act as the backbone of any high-fidelity physical model, supporting the integrated vehicle health management system in evaluating historical data. With the input dataset, DT will enhance the mission success credibility of the craft with continuous monitoring and evaluation of the operational condition and the remaining running hours ahead. A robust DT model will assist damage mitigation or degradation by real-time data analysis/forecasting and suggest required changes in an ongoing space mission to enhance the life span leading to mission success [11]. Later, USAF came up with the concept of Digital Thread, where each USAF aircraft enters the fleet with its own DT. The DT will help in Structural Health Monitoring (SHM), and maintenance, thus allowing calibration of the craft in its operation state by comparing the sensor readings of flight and DT. It also acts as a digital surrogate to plan the design, production, and acquisition phases of the project [12]. Though it was a clear definition for DT to be made considering the representation fidelity, model simulation capabilities, synchronization techniques, data collection, exchange attributes, etc., it is also stated as a virtual representation of the real-world asset with an exchange of information in a predetermined frequency [13]. It is a digital representation of the active product service system, which comprises preset characteristics, conditions, behaviors and properties [14].

Along with this concept, the Internet of Vehicles (IoVs) has also been developed with the integration of Vehicular Ad hoc Networks (VANETs) and the Internet of Things (IoTs) [15,16]. These were more focused on land-based locomotives, but the notion is widely accepted by many industrial sectors. IoT has supported the real-time data handshake component of DT, which has been the basement of existing topology. These also incorporate the diagnostics and prognostics embedded within the system architecture along with optimization of the process [17]. Risk analysis and accident prevention with DT support is another frontier that has already been adopted by the aviation industry. The system verification of the Boeing 737 Max after initial failures is a great example [18]. It ultimately promotes the concept of Smart Products (SP) in various phases of the product life cycle, which benefits the stakeholders in the user groups of the product or asset [19].

In the industry, DT is used under different names such as cyber objects, digital avatars, etc. The data flow between the DT and the physical unit will depend upon the purpose of the DT in the industry, which uses cyber objects or digital avatars. Accordingly, the nomenclature of DT has been proposed based on the two-directional data link between the DT and the actual unit. Kritzinger W. et al. have categorized the digital counterparts of a physical object based on the data flow into three types: Digital Model, Digital Shadow, and DT [20] (Figure 2).

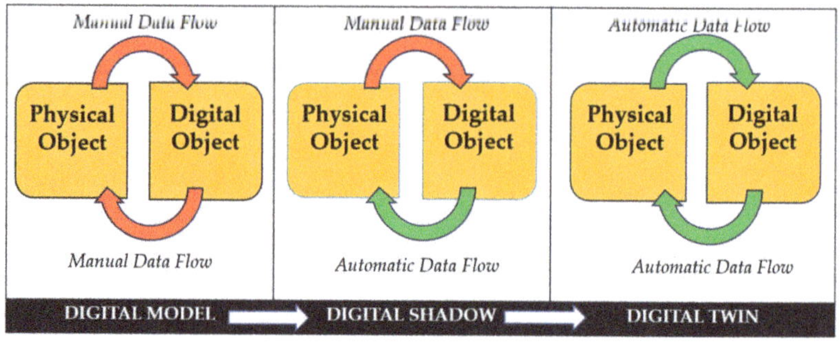

Figure 2. Difference of data topologies among Digital Model, Digital Shadow, and DT.

- Digital Model: There is no automated data exchange between the two units. Data integration is done manually to synchronize.
- Digital Shadow: There is only one-way data flow from the physical object towards the digital counterpart. The data arriving from the physical asset will update the digital object, but not vice versa. This requires offline actions with Human-in-the-loop (HITL) interaction between the physical object and the digital model.
- Digital Twin: Fully integrated data flow is available where a two-way automated data link is established. Both units are in a real-time synchronized state and the physical unit can be influenced by the digital object automatically.

3. Adoption of Digital Twin Technology for the Marine Sector

The induction of DT in marine vessels is already underway and Figure 3 explains the summarized trail of actions for said induction. The seafaring sector has been developing its technologies, targeting improvements in efficiency, emission controls, and ergonomic operating systems. Traditionally, most marine vessels, specifically surface vessels, are built with two separate power systems. One is dedicated to the propulsion system and the second is used to cater for the electricity demands. In this context, most of the vessels were built using diesel engine-driven propulsion systems due to the ergonomic simple robust architecture and ease of maintenance. However, traditional systems are highly polluting the marine environment, which has been regulated by the IMO regulations. Shipbuilders are finding ways to suffice these new regulations in their designs. Electrification of the vessels has been an ideal solution to gratify the emission control guidelines [21].

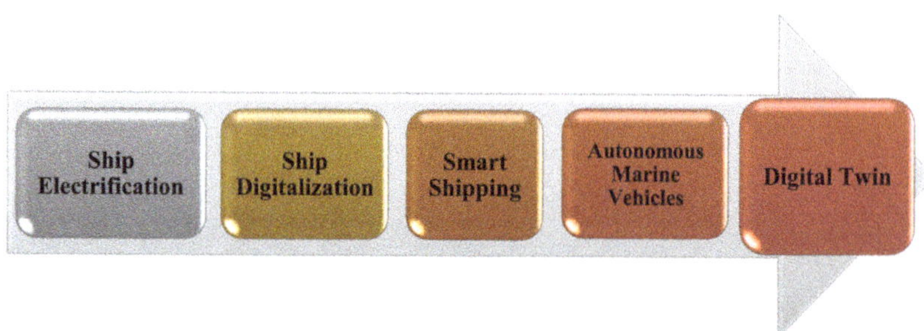

Figure 3. Pathway of Digital Twin Technology within the Shipping Industry.

3.1. Ship Electrification

The first electrical propulsion system was introduced by Russian scholar Boris Jacobi, which was a paddle boat driven by an electric motor. Due to the low battery capacity and efficiency, the maximum speed achieved was 1.5 knots, which discouraged the development of a system for a long time ahead [22]. In 1903, Swedish shipyard ASEA built the river tanker "Vandal" as the first vessel to carry a diesel-electric propulsion system [23]. At the end of WWII, more emphasis was placed on the development of electrical propellers and ships such as the Ice Tanker "Uikku" and Passenger Ferry Queen Mary 2. Uikku was the first ship to be installed with an Azipod propulsion system where the main electric motor was installed in a separate gondola that could direct the thrust with a 360° control. RMS Queen Mary 2 is a luxury cruise liner propelled by hybrid diesel turbo-electric propulsion with four 18 MW Azipods driven by two gas turbine generators and four diesel engine-driven generators. This propulsion topology has provided navigation flexibility along with economical gains in varying speed demands and different passage legs while complying with emission control protocols [24].

Integrated Full Electrical Propulsion (IFEP) is reducing fuel consumption heavily by avoiding the low load portion of fuel consumption curves in sharing the load of propulsion along with ships' service loads including the weapon systems [25]. It is far superior to the standard propulsion systems such as Combined Diesel Electric and Gas (CODLAG) systems or Combined Diesel Electric or Gas (CODLOG) [26]. The latest US Navy stealth Destroyer USS Zumwalt was commissioned in the year 2016 as the first full-electric warship equipped with an integrated power system [27]. Further, with the initial project of MS Medstraum, Norway has implemented the world's first battery-powered ferries to transfer vehicles and people with greener environmental insights [28].

With the development of IFEP, ship designers further developed the concept of All-Electric Ships (AES) [29]. The AES concept has provided high redundancy and mission capability even at a slight initial cost. At present, most of the high-end naval ship projects have opted for this technology due to the advantages posed by the electrical-based high-power navigation, communication, and weapon systems. Royal Navy Type 45 Destroyers and the aircraft carrier HMS Queen Elizabeth [30] are a few examples of AES-based power topology. Hybrid powering topology is also developed and implemented to achieve the redundancy and advantages of both mechanical and electrical drives.

Benefits of the Electrification in Marine Vessels [31]:

- Efficiency is improved as a huge prime mover will be replaced by an array of diesel alternators and the load can be catered at the highest efficient speed with the generated thrust by more efficient electric motors.
- The load can be managed easily to cater to the demand based on each mission.
- Improved maneuverability of the vessel and faster response where podded propellers will allow 360° steering along with dynamic positioning.
- Flexibility in placing the generators in the ship without considering the shafting arrangements.
- Allowing the introduction of cleaner and more efficient future power-generating technologies such as fuel cells, renewable energy, etc., to the existing vessel.
- Emission control goals can be easily achieved.

3.2. *Ship Digitalization*

Along with the electrification process, digitalization has also become a key innovative trend in every operation in the maritime sector due to its efficiency, effectiveness, superlative performance, etc. New tools in designing, performance evaluation, simulation, and information safety are being introduced and the same is being improved day by day, which further assists in producing more robust models. As per the digitalization service development framework proposed by Erikstad S., the process of digitalization of the maritime sector can be achieved through two different approaches [32].

- Service-driven Perspective: Considering the ship owners' specific requisites in operation and decision making, build is performed in implementation of hardware and software that suits the ship. It is operating in the need-space, where tactical-level improvements are more considered.
- Sensor-driven Perspective: It will consider the available inputs from the already installed sensors in a particular ship to design a digitalized framework to support its

operation. This method operates through the solution-space, which is the key doorway for future DT-based systems.

The marine industry will be able to gain the full merits of digitalization through the service-driven approach starting from the design phase of the vessel. However, existing platforms can be improved through the second approach to par with the present smart functions. Digitalization is further considered an effective, innovative, and optimized method enabling products and services as per Industry 4.0 [33]. Asset-intensive industries such as shipping, oil, gas, energy, etc., are looking for innovations that increase efficiency and reduce the cost while effectively managing operational risks and security. Main configurations such as the navigation system, power system and automation/control system are becoming fully digitalized, leaving the old analogue systems obsolete. Integrated Platform Management Systems (IPMS) available in present ships can link all the above system architectures into a centralized topology, thus providing a more ergonomic operation and increased domain awareness for all stakeholders.

3.3. Smart Shipping

With the availability of satellite communication, the onboard modules are easily linked with onshore supporting facilities, rather than opening the standalone data infrastructures onboard holistically to shore operators (Figure 4). As per the Guidance Notes of the American Bureau of Shipping on Smart Function Implementation, they have highlighted the importance of having a DT to support the data infrastructure in onshore processing during Big Data Transfer [34]. It will allow operators to continuously collect, transmit, manage, and analyze the data for real-time monitoring, increase awareness and decision making for both human-initiated decisions and autonomous command and control. It further enhances the ships' situational awareness, navigational safety, and reduction of crew fatigue and human error-related accidents.

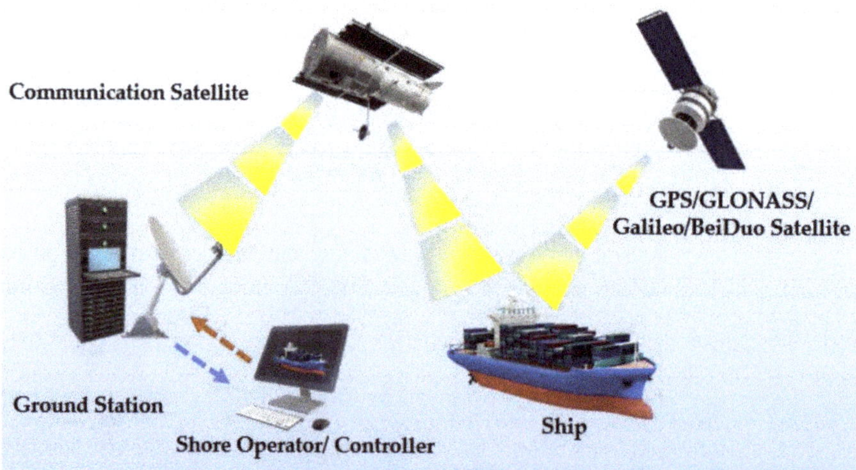

Figure 4. Realtime Satellite Data Link of Smart Ships.

The main requirement of a smart ship can be defined as the ability to perform the intended functions autonomously. The transformation from a traditional human-based system to full autonomy is achieved with the digitalized platforms in the electrified vessels. Smart ships highly depend upon the digitalized marine eco-system, which is focused on autonomous operation.

Furthermore, Wartsila and Rolls-Royce have announced the successful testing and demonstration of dock-to-dock navigating fully autonomous ferries with zero intervention of the master on-board. Wartsila, with their hybrid-powered ferry *Folgefonn*, made a three-stop route in Norway, while the Rolls-Royce ferry *Falco* made an unmanned voyage in Finland with autonomous collision avoidance and docking [35].

This evolution has opened the doorway for DT technology to enter the shipping industry by bringing all experts into a centralized topology, providing powerful analysis, understanding, and diagnostics that are crucial for decision-makers in every stage of the ship's life. This iteration has made the pathway for Ship DT (SDT), which incorporates the basic topology of the DT concept to develop and implement for various applications within the maritime domain. The limits are endless in this effort, but this DT can be primarily based on three main virtues, namely Asset Representation, Operational Behavior Model, and Parameter Measurement/Monitoring. The application of the technology can further be divided into the design phase, manufacturing phase, service phase and retire phase [36]. This encompasses every action that the ship must undergo during her life cycle. It will support the actions from preliminary conceptual design and optimization to the final validation of the model. The developed DT model will then assist the supply and production measures during her building at the shipyard till the final performance trials and commissioning. During the entire service period, DT will be the key tool to performing intended functions during her operational phase. Routine upkeeping of the vessel to Planned Preventive Maintenance (PPM) to Condition-Based Predictive Maintenance (CBPM) will be functioning through the DT model with all historical data accumulated in condition monitoring, prognostics, and diagnostics and even the simulated testing of the systems/components. However, the support it could render during the decommissioning phase has yet to be studied.

3.4. *Automation of Marine Vehicles*

Automation can be considered the final step of striding towards a fully DT-based environment within the maritime sector. Autonomous ships are under extensive research all over the world and are being implemented due to the versatility of unmanned operations. As per the International Maritime Organization (IMO) regulation enforced in 2021, Maritime Autonomous Surface Ships (MASS) are to be regulated under four separate degrees, which are categorized as per the level of autonomy incorporated into the vessel [37]. A ship may also have the interchange between the four states of autonomy during its operation based on the type of vessel and designated task. This allows flexibility in asset management onboard with crew augmentation as per the dynamics and constraints of specific operations of the vessel.

- Degree One: Ship with Automated Processes and Decision Support: Seafarers are present onboard; few functions may be operated onboard at times and operators are on standby and ready to take over control.
- Degree Two: Remotely Controlled Ship with Seafarers Onboard: The entire ship will be operated remotely from the shore station, but seafarers are available onboard to take over control at any time.
- Degree Three: Remotely Controlled Ship without Seafarers Onboard: The shore operator will control the functions of the entire ship remotely and have no humans onboard.
- Degree Four: Fully Autonomous Ship: The intelligent control system of the vessel will take actions based on the decisions made by itself with full autonomy.

As per Kaber, Human Automation Interaction (HAI) in complex systems will be governed by the Level of Automation (LOA), which is highly essential to be considered during the design stage of any autonomous system [38].

- LOA-0: No Automation: Full-time human operation with the assistance of parameter indications and warnings.
- LOA-1: Driver Assistance: Acceleration/deceleration/steering will be assisted by the operator by the inbuilt system but performed by the operator themself.

- LOA-2: Partial Automation: Steering/acceleration/deceleration would be performed by the automated system and all remaining decisions will be performed by the operator themself.
- LOA-3: Conditional Automation: Dynamic actions of the vessel will be performed by an automated system and the operator will intervene if required.

Hence, a fully autonomous intelligent surface vessel would be designed to be operated in the LOA under the fourth degree of IMO regulations for MASS. However, careful assessment of the usage is to be analyzed since ships are high-value cost-sensitive assets voyaging for weeks at times without seeing a shore. Thus, autonomous systems must be designed at a high-fidelity robust standard to be capable enough to address all complex situations during the operation of the vessel.

Autonomous Marine Vehicles (AMVs) are becoming popular both in commercial and military applications, as extensive research and development are underway all over the world [39]. It ranges from environmental surveys, cargo/passenger operations, hydrography, search and rescue, sea patrols, coastal protection, etc. They can be semi-submersible, conventional planing hulls, semi-planing hulls, and hydrofoils based on the sea surface, and these can be further augmented with the evitability of the limitations imposed by manning ergonomics [40,41]. LOA can vary by the application and the different scenarios she undergoes during the operations.

The applications of AMVs can be primarily divided into two parts: military and commercial. In military applications, DT can be incorporated into various missions in Anti-Submarine Warfare (ASW), Seabed Warfare (SBW), Mine countermeasures, Special operations forces support, Electronic Warfare (EW), Surveillance, Asymmetric Warfare, etc. Recently, in the Russia–Ukraine conflict, Unmanned Surface Vessels (USVs) have been used as self-destructing attack drones capable of destroying surface vessels or strategic land infrastructures such as piers, bridges, and port facilities, which has brought up a new dimension in the modern battlefield. These drones can also be swiftly deployed for the protection of high-valued assets including the major fleet units, port facilities, offshore platforms, coastal power generation stations, etc. The commercial applications can be stated as commercial shipping (cargo/passenger), marine research and survey, search and rescue (SAR), meteorology services, environmental protection, etc.

The functioning of these unmanned automated vessels including the integrated Guidance, Navigation and Control (GNC) will be monitored and governed by the shore control stations with the capacity to intervene where it is necessary depending on the situational shift based on the imposed level of autonomy protocol. In the case of a harbor, there will be various other drones, ROVs, and auxiliaries that will also be controlled by this station supporting the Smart Port functions. For example, the Norwegian University of Science and Technology (NTNU) Shore Control Lab has implemented a similar station to control the autonomous passenger transport ferries at Trondheim River [42]. A higher level of autonomy in the operation of any marine system will ensure a smooth and straightforward approach to the implementation of DT-induced functionalities in processes such as design, control, management, etc.

4. Present Digital Twin Manipulations in Marine Industry

Digital Twin Technology is still in the infancy stage when considering the maritime domain. Less research work has been conducted compared to other industrial sectors, but the researchers have identified the potential of this innovative concept and DTs are being developed for various maritime applications such as surface ships, underwater vehicles [43,44], offshore platforms [45,46], coastal electrical power stations [47], etc. Apart from the described examples, the limits to the application of DT technology are boundless, with numerous possible applications in the entire maritime domain. However, implementation is to be done by the marine industry with an opportunity-driven approach [48]. These DT applications are intended to improve the existing operations in the marine industry while many novel innovations can also be found with revolutionary performances.

Most research work published on DT technology so far is based on the development of new concepts of DT technology in manufacturing (55%), while only a few are based on practical case studies (28%). Out of the published work, the majority (49%) are focused on production planning and control (PPC) in the manufacturing sector [21], which is focused on Product Life Cycle Management (PLC) with the advent of DT-based solutions.

4.1. Embryonics in the Shipping Industry

DNV GL has tested a fully autonomous 60 m-long concept vessel, ReVolt, which is a battery-powered cargo carrier with zero crew [49]. DT is an integral part of the design and testing of this conceptual vessel and the same will be utilized during the operation of the vessel. Testing of the vital systems is being carried out using the DT of the concept vessel, which has saved time and money. The same institution also provides a cloud networking solution, Veracity [50], which can be used to design, access, and manipulate the data in DT. Similarly, Siemens has also launched the Mindsphere [51] venture, which has governance capabilities and data management under the concept of Maritime Data Space (MDS) [52], which is originally based on the concept of the Industrial Data Space (IDS) model. Meanwhile, it also allows for performing safety-critical tests and high-risk maneuvers for any number of attempts as desired in all varying conditions, which would have been a daunting task in a physical model. Using the historical data during the operational phase is used to predict the lifespan of components under varying conditions including position, dynamic environmental factors, speed, load conditions, etc. [53]. The South African polar supply and research vessel "SA Agulhas II", which operates in rural and risky arctic environments, has tested the feasibility of implementing a DT-based solution to operate with its benefits such as remote polar research capability, reduction of human impact on sensitive untouched environments, avail of real-time research data for scientists, etc. [54]. Further, the "Probabilistic Twin" risk-management tool used by DNV-GL also aims to forecast the probable accidents/damages based on analysis of past data and trends as an extension to the existing Blue Denmark DT Solution [55].

Having an unmanned vessel in connection to DT will deduct the human error factor in ship operation, and crew space can be spared for vital cargo space. Reduction of energy consumption and waiving of the extra maintenance expenses are other benefits of a DT-based system. Revolt itself is estimated to save a cost of 34 million USD in its 30 years of service life due to autonomous operation along with DT [56]. Emission control is also a huge benefit in such a system. Reducing the machinery and moving parts will further reduce the downtime for maintenance/repairs, reduce human fatalities and be cost-efficient. The vessel can also be designed into a more aerodynamic shape due to the absence of the huge superstructure used for crew accommodation and bridge.

Not only the direct merchant sector, but even the fisheries, ecological and maritime archaeologists have opted for DT-based solutions for underwater operations. In projects such as SUSHI DROP in Sweden, AUVs are developed and operated in DT-based systems to monitor/survey the marine environment investigating the fish population in a large sea space [57]. This DT-based module can further be incorporated into the ship's HVAC system to optimize the heat usage and distribution in waste heat recovery systems, charge air coolers, heat exchangers, etc. [58].

4.2. Vessel Life Cycle Management

Cost-saving is one main benefit of using the DT, as Condition-Based Predictive Maintenance (CBPM) is powered extensively with networked sensors, actuators, and control systems that are operated at their optimum condition. Prognostic Health Management (PHM) will be used with the projection of the state of the machinery with accurate data fidelity. It will further provide insight into failure modes, threshold settings, health indicators, and risks. With this analytic data, the operator will obtain the total domain awareness, which can be utilized to operate the physical unit at its prime operating condition or to make amends to optimize the performance. It further improves the predictability and management of real-time continuous quality control of marine diesel engine (MDE) critical parts with dynamic quality control starting from the machining stage of the components [59]. The historical data derived from the DT will be used to compute the Remaining Useful Life (RUL) of mission-critical components of the ship's systems. In DT-based CBPM solutions, data fusion [60] is highly important to implement formidable evidence-driven big data sets taken using Bayesian Inference, Demster–Shafer reasoning, etc. [61]. M. Xiang et al. proposed intelligent predictive maintenance using Implicit DT (IDT) formed with intelligent sensing, and reliable simulation data mapping with the historical operation data [62]. Critical components such as the main engine can be digitally modelled as a Cycle Mean Value (CMV) model, allowing the monitoring of the performance in both the steady state and transient response [63].

With the increasing visibility along with the real-time synching of high-fidelity models, monitoring of the physical system can be easily enhanced. Reduction of the time taken for system testing and development has led to the reduction of time to market by eliminating the risks of design failures. DT can provide a complete user experience even before the actual production. The entire operation of a ship within the total lifespan can be manipulated during the design phase with the support of its DT from a multi-user point of view (Passengers, Crew, Owner Company) [64]. Further, real-time connection ensures the optimum operation of the actual system with the ability to analyze the performance of the physical model [65]. It will further evaluate the existing ship's machinery with the support of machine learning with a quantitative assessment of the degradation of the components. This will result in the intelligent diagnosis by an inferential decision model leading to more accurate predictions [66]. These aspects will further support Structural Health Monitoring (SHM) and parameter trending of critical components of any machinery, aircraft, ship, plant, etc. [67]. This augments the Model-Based System Engineering (MBSE) [68,69] with available tools for simulation, rehearsals, machine learning applications, etc. Vessel Traffic Service (VTS) is also supported by MBSE-based DT solutions. For example, the Diamond model concept proposed by Boeing Defence Company [70] illustrates the DT-based MBSE.

4.3. Dockyard and Port Facility Augmentation

Not only the sea units, but also the shore establishments are predisposed to DT with advances in Building Information Modeling (BIM) [71] with uplifted interoperability, planning, construction, and maintenance. A vast array of sensors from the building will feed generic data input to the façade DT of the structure, which will be processed for intended output to control the operation, site environmental control and other ergonomics. Already implementations are performed in Smart Home concepts [72], which are highly successful in achieving the intended outcome. Further, Intelligent Manufacturing uses a knowledge-driven autonomous platform, which functions with dynamic knowledge bases along with a DT model [73]. For example, the "smartBRIDGE" project initiated by Hamburg Port Authority, Germany, has implemented BIM-based DT to monitor the functions of the infrastructure and the process of extending to quay walls, locks, and other port infrastructural assets [74]. These smart technologies will promote self-judgment, self-implementation, and self-development in autonomous manufacturing in the shipping industry.

Similarly, the factory design framework [75] is also possible with an increase in productivity along with optimized floor utilization with modular-based DT. In such an operation, it can be developed into a Warehouse Management System (WMS) with the ability to customize the process as a Decision-Support Tool (DST) [76]. It can connect the different value chains with eased-up system integration from the design, and production into the later stages of the lifecycle of a component. In the maritime section, dockyards around the world will be the main beneficiaries of such a DT-based solution.

Further, the DT solution will allow critical operations of a vessel operating at a far location to be handled by the experts in a shore base in an Augmented Reality-induced environment [77]. This can be a routine operation to critical maintenance or fabrication task, which might require the intervention of an expert operator from a distant location.

The Simulation and Remote Control Centre (SRCC) will perform a task beyond remote monitoring such as crane operation in various sea conditions [78,79].

4.4. Training and Testing Support

DT can be incorporated into the training of seafarers in Full Mission Simulators, Part Task Simulators or computer-based training systems. Part Task Simulators often use Hardware In the Loop (HIL) based solutions with real plants operating in parallel with a virtual platform [80]. These platforms can ensure the safety of the personnel (trainers/trainees) along with the engagement of critical operations without endangering expensive real-life platforms. DNV-GL is using HIL testing, integration, and optimization of new control systems with a DT-based simulation platform and later goes for actual production with complete runnable executable solutions [55]. Designers and equipment manufacturers will provide and share model parts with the shipbuilder where the complete outfit will be tested, leading to fewer errors and modifications during construction. In their "Nerves of Steel" venture, all hydrodynamic data were also simulated in varying sea conditions, leading to an efficient and safe hull design.

A DT model can be developed to address a specific problem, or it can be harnessed as an open data source ready to be manipulated by human, operational and environmental dimensions of a vessel or any other marine platform. It is a critical asset in enabling future technological advances in data-driven operations. Even producing a surrogate model is also possible using the original DT model, which can mimic the operation with less computational load, achieving the intended efficiency [81].

Increasing user engagement is another advantage of a DT-based marine system. During the product design, the DT model can be tested to virtually test the interaction between the operator and the physical model [65]. From that, data can be accumulated and processed to reflect the user habits, and the same can be incorporated to improve the physical model with naïve innovations. DT acts as a hub for technologies such as Big Data Analytics, IoT, machine learning, simulation, modelling, etc., providing a gateway to achieve much more complex analytical power to the operators and developers with the sensing data of the actual physical model contrasted with simulation data and Machine Learning algorithms.

4.5. Defence & Military Assets

DT is widely implemented on naval assets such as warships, submarines, maritime aircraft, etc. HIL testing is performed for the Naval Ship Combat Survivability Testbed (NSCST) [82], which is a power-generating system redundancy checking simulation platform in warships of the US Navy. These DT-based simulators will be subjected to Individual Subsystem Actuators Control Tests, Multi Subsystem Integration Tests, Main Supervisory Control Tests and User Control Tests [83]. Further, new naval air assets are demonstrated using their DT, which is used as a tool by prospective contractors to showcase their products to the Naval Air Command [84]. The "Spiral development project" at Pearl Harbor and Portsmouth naval shipyards in the USA has optimized vessel trials using a new DT-based approach [85]. New F110 class frigates developed by Navantia, Spain, used DT as an efficient life cycle management tool to address condition monitoring, maintenance, and casualty/damage/fault reaction [86].

BAE Systems use DT technology in their latest Archerfish Mine Neutralization System [87], which uses an AUV operated in a DT-based architecture to deploy mine-countermeasure action for underwater mooring and sea bottom mines with an unparallel Maritime Domain Awareness. It is further capable of developing a virtual 3D space with real ocean data such as currents, waves, and topography with real-time sensor data, which can lead to value-based reinforced learning followed by autonomous path planning [88]. With the support of DT technology, Maritime Domain Awareness (MDA), mission readiness, Fleet Management and sustainment of a single Naval Force or even multinational task forces can be easily strengthened and augmented.

5. Developing a Ship Digital Twin Architecture

The virtues of the existing functionality of any surface vessel require a clear understanding of the planned design specifications along with the precise functionality of the various components, systems, and departments of the vessel. The standard actions of any of the existing surface vessels during her life cycle can be encapsulated in the following diagram (Figure 5) [89]. These eleven general functions will vary from ship to ship depending on the type and role of the vessel. However, these standard functionalities can be assisted and boosted with much fidelity and robustness using a DT-based solution.

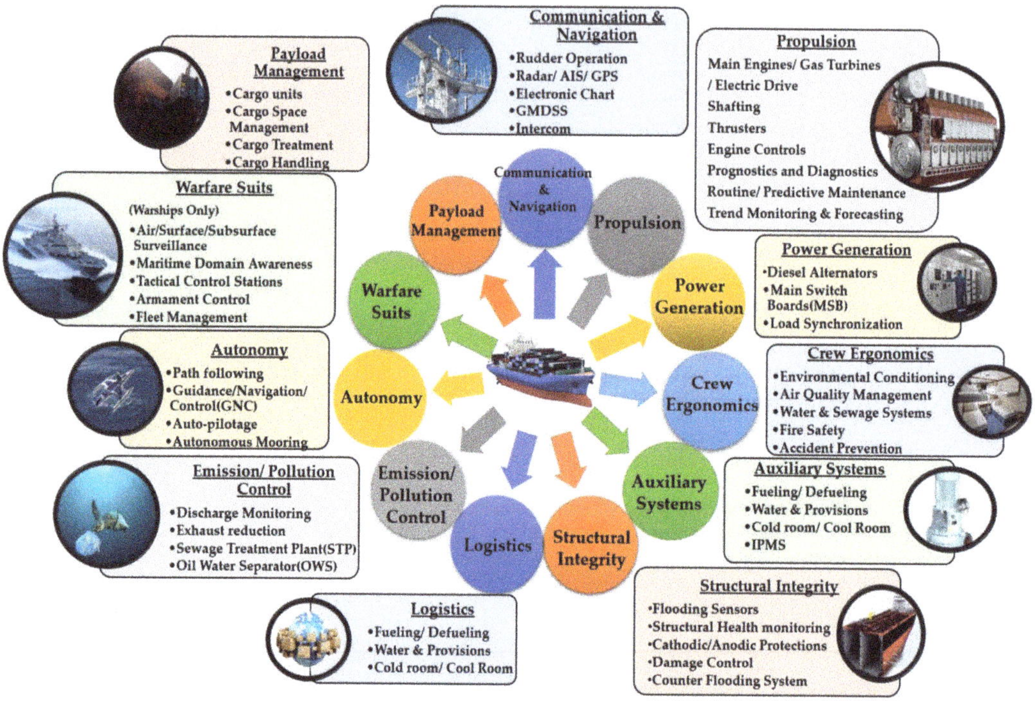

Figure 5. General Functions of a Surface Ship.

Developers of the DT are focused mainly on three user groups for the system, company, operators, and researchers. DT will continuously communicate with the physical ship to update its data as per the real world to support the above-mentioned stakeholders with different demands during the complete product life cycle. Starting from the marketing phase of the vessel, DT will be functioning till the decommissioning stage of the ship. Designing as per the end user requirement and validation of the design with a 3D representation of the entire ship layout can be used extensively before the building phase.

5.1. Functional Requirements of User Groups

The DT of a fully autonomous ship will act as a complete controller of the vessel while regular manned vessels will act as a Digital Model or Digital Shadow to maintain the required data structure. A DT solution should be capable of performing the following actions to provide service to the three main stakeholders: onboard crew (if present), shore operator and researcher/engineer (Figure 6).

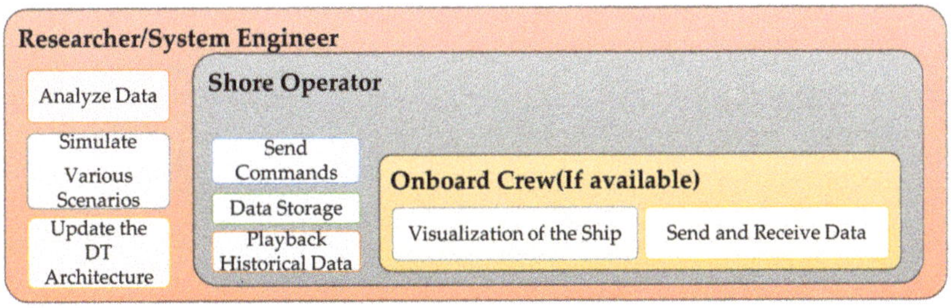

Figure 6. Direct Stakeholders of a Ship DT System.

A ship DT should be able to replicate all the Smart Functions [34] of a digitalized surface ship, including the following mandatory requirements:

- Structural Health Monitoring (SHM);
- Machinery Health Monitoring (MHM);
- Operational Performance Management (OPM);
- Asset Efficiency Monitoring (AEM);
- Crew Assistance and Augmentation (CAA).

Apart from the above functions, a robust communication link is to be established between the ship and the shore operator to obtain real-time situational awareness with complete control during the voyage. Digital images can act as a model, shadow or twin depending on the data handshake method. DT can be utilized to monitor and maintain the autonomous control of the vessel including self-navigation and collision avoidance. The formidable internal verification system generated due to the DT architecture will enhance the existing conceptual design optimization using virtual testing and intelligent process monitoring till the finalized model is developed [90].

Since the data handled in a fully integrated DT Process Model (DTPM) will be large and complex, therefore the model must be composed of Design Data (DD), Process Data (PD), Process Perception Data (PPD), Historical Running Data (HRD) and Simulation Data (SD) targeting the whole product life cycle [91]. Separate component schema for serialization of various components can be implemented covering the asset data, analysis and measured data of different segments of ships and can be shared among various stake holders [92] (Figure 7). The physical and control architecture of the future ships is also evolving each day with new concepts in improving the efficacy and efficiency of the vessels targeting economic and emission goals. Designing, testing, validating, and implementing a robust Digital Twin (DT) model for a surface ship will provide seamless support for the functioning of the ship/craft to its intended operation despite being an unmanned craft or a conventional crewed vessel.

5.2. Objectives of a Marine Digital Twin Framework

Overall, the DT framework should reflect the behavior of the vessel's response to the environmental inputs such as forces of waves, current, wind, temperature, etc. The input sensors will be limited as per the requirement only with relevant design patterns based on the target evaluations of the vessel. These design patterns for DT-based vessels are extensively discussed by S. Erikstad from NTNU, Norway [93], based on the original 23 design patterns derived by Gamma et al. in their classic book *Elements of Reusable Object-Oriented Software* in the year 1994 [94]. In this book, the following four design patterns for a ship DT have been proposed.

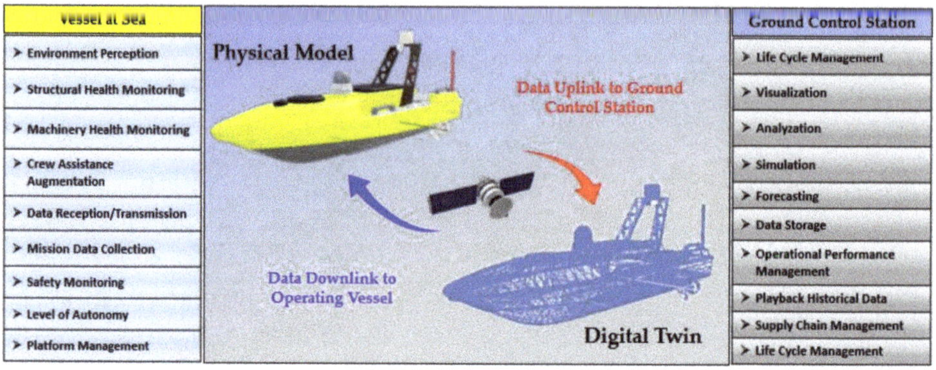

Figure 7. Overall Task Distribution of a DT-based Framework.

1. Structural Patterns: These are the main constituents of the DT package. It consists of the Product Data Model (PDM) of the DT that is formulated along with sensor-based big data of the vessel. The baseline is also a formulated source model derived from physics-based model behavior. Load-based sensor data are also acquired from the environment. The benchmark of the DT is to be designed to simulate the DT in parallel to the ship to compare anomalies. Finally, a Machine Learning Proxy layer can also be incorporated for DT to assist its behavior through ML rather than depending on the standard physics.
2. Creational Patterns: This is the realization of the DT as a proxy simulation tool. It will be a dynamic hybrid system with the ability to switch intelligently between ML and physics as per the situation. This can be either one vessel or an entire unmanned vessel fleet.
3. Insight Patterns: It is the layer of utilizing the DT that is operated with the ship. Anomaly patterns will be monitoring the deviation from normal operation and reverse analyzing to find the root cause for any abnormality in the system.
4. Computational Pattern: Designing the DT to perform efficiently with smart storage by a high-fidelity controlled regression to avoid data misconceptions. Offloading of data will also be performed with controlled dynamic regression offloading in physics analysis.

5.3. *Appreciating the Future Dynamics*

With the recent developments pertaining to software and hardware capabilities, IoT, Big Data Analytics, etc., the concept of DT has become a widely used technology in almost all sectors across the globe. DT is a key model portrayed in the main development strategies such as Industry 4.0 and Made in China 2025 (MIC 2025). The development of the Cyber-Physical Systems (CPS) architecture to transfigure the physical model into virtual space is a necessity to achieve the vision of the industrialists and researchers of the above development strategies [95]. The shipping industry is an ideal platform to utilize the positive effects of CPS in the form of DT. "Information Technology and Industry integration" is one of the key factors in MIC 2025 and "High-Tech ships" is one of the 10 key targeted sectors in the strategy [96]. In a similar example, the CyberFactory#1 project [97] powered by Industry 4.0 focuses on CPC modelling with human, social and economic dimensions. Following key technologies can profoundly uplift the system integration of DT-based solutions.

5.3.1. Smart Port Compatibility

To create better management solutions and increase the efficiency levels of port operations using DT techniques, initiatives such Smart Port or Port 4.0 [98] have become promising concepts. The digitalized port ecosystem will expect its stakeholders to follow the prerequisites to obtain or deliver the services. Automation will be the key component with supportive technologies such as IoT, AI, Smart Logistics [99], Blockchain port operations [100], DT, etc. With a similar intention, the "European Commission Roadmap for Waterborne Transportation Development Action Plan" [101], which was formulated in the year 2019, strives towards the implementation of a digitalized smart port concept with the ultimate goals of improving the safety, security, efficiency, environmental soundness, transparency and improvement in workforce factors in present port systems. The United Kingdom, in their long-term strategy Maritime-2050, has also emphasized the ambitions of digitalization of ports in the process of moving towards Smart Ports [102].

As the terminology itself portrays the correlation, Smart Ports will always be required to accommodate smart ships. Both systems will function hand in hand in energy management, navigation, communication, resource sharing, safety and security, marine environment conservation, intelligent awareness of each side's strengths/limitations, etc. Necessary historical data and real-time data are to be shared at both regional and global levels. DTs of both ship and port will be able to function as the bridge between two smart systems.

The Port of Rotterdam in the Netherlands has already taken the initial step in adopting DT technology by establishing a sensor array using IoT technology and obtaining real-time mission-critical hydro and weather data for port operations. It obtains tidal data, salinity, wind speed, and direction from a combination of 44 sensors and analyzes them along with prediction models to provide complete digital visibility to the port control station [103] and upbringing data to support the port governance and inter-port collaboration scheme [104]. The Port of Antwerp in Belgium is also adding digitalized concepts for its operations. Apart from the functioning the level DT model, digitalized radar infrastructure port control is assisted by an autonomous network of drones to assert domain awareness, spill/waste detection, firefighting tasks, etc. [105]. Further, Adelaide Port in Australia has optimized its container handling capacity by using AI-powered smart functions in the planning and utilization of assets in the terminal [106]. The TwinSim project at Hamburg port, Germany, is also underway to achieve 3D visualization of the port functions in collaboration with sensor data and AI [107]. However, the safety, physical security and cyber security of the smart port system [108] will be an utmost concern and a challenge, where a single deviation/disruption can halt the entire operational flow.

With the current trend of port digital extension, future ships may be mandated to operate with smart functions along with a DT to maintain compatibility with the DT of the smart ports. The transformation from regular to digital will require successful meta-level organizational alignment with strategic coalitions of stakeholders [109].

5.3.2. Incorporating Augmented Reality and Virtual Reality

Augmented Reality (AR) based applications are playing a pivotal role in the digitalization process, where the number of operations can be enhanced with AR-based solutions. It can be used as a presentation tool where interactive interfaces are utilized to visualize three-dimensional geometries and function as a supporting tool in executing the work while minimizing the errors incorporated in the manual process [110]. Additional information along with the Standard Operating Procedure (SOP) can be superimposed to provide more guidance to the operator to perform the designated task. The author can specify the linkage between the geometries of the component and metadata including the drawings, plans, documents, and other service instructions supporting the execution of the relevant process. In this methodology, a camera tracks the base object while feeding the user interface with virtual images, objects, and texts of relevant data [111]. AR-based service instructions can assist heavily in the application of CAA in the functioning of the smart functions onboard. At present, simulations play a vital role in the design decisions and validating of components/systems in the design, engineering, operation, and service phases of a product. DT-based AR functions realize all the aspirations of a seamless development of any critical product.

Virtual Reality (VR) is also a formidable domain that has stormed the digitalization process with its superior merits to the traditional 3D-based systems in providing more versatile information and interaction. It has already been utilized extensively in the shipbuilding industry with loaded CAD models in virtual environments with the facility to interact with them. A high-fidelity DT model will further allow virtual prototyping [112] in reducing the cost incurred during the validation and testing phases of any new vessel. Assigned controller buttons are used to navigate within the virtual environment while manipulating the model with incorporated control modes. Traditional Marine Engineering Systems (MES) with 2D interfaces are replaced by these naïve simulation/user interfaces with much-improved geometric modelling procedures with model integration with optimizations, general arrangement, equipment specifications, etc. [113]. Head-Mounted Displays (HMD) with room-scale tracking will allow the user to control the model within the 3D space along with motion-tracking handheld controllers. Eye tracking is used with interactive 3D screens in smaller applications while Cave Automatic Virtual Environment (CAVE) creates an immersive VR experience using projectors set to the entire room. These are used for training purposes due to cost concerns and the inability to produce high-definition output for real-time applications.

5.3.3. Artificial Intelligence

Artificial Intelligence (AI) will play a huge role in the development of DT in marine applications. Machine Learning (ML) has become a key component in AI-based systems in industrial applications. In a DT-based solution, the processing of large data sets can be incorporated with ML to reduce user participation in the training of the model. It can further conduit AI with Human Intelligence with effective integration processes in aspects such as remote sensing, social sensing, crowdsourcing, and analytics of the DT paradigm [114]. The positivity of AI algorithms can be easily integrated into the DT model with Deep Learning (DL) techniques. Diagnostics and predictive forecasting can be incorporated into the DT with its neural network with a deep network from the acquisition of data to the training of the final algorithm [115].

At present, applications of AI along with DT include Aerospace Industry, Driverless Automobiles, Intelligent Manufacturing, Smart City, Smart Home technology, etc. For example, real-time decision-making for pilots and Air Traffic Controllers (ATC) can be assisted by the AI-infused DT model in multiple scenarios with the aid of automation of various actions involved. It further supports collaborative decision-making in demanding situations with the ultimate aim of reducing the human factor in aviation operations [116]. To increase the operational readiness, availability and safety of UAV operations, an Internal AI-based meta-analysis system is incorporated with a communication data link to a ground-based DT model [117]. All these approaches are relevant to maritime applications, which demand similar navigation, control, communication, and operational requirements.

Self-driving cars with AI algorithms are already being used to minimize traffic accidents, increase fuel efficiency and ease driving including disabled personnel. For this, DT is essential to minimize failures and improve the ergonomics such as the Advanced Driver Assistance System (ADAS) in the Tesla Model-Y autopilot system [118]. Trust in these automated systems will improve with the system development process adopted by Tesla using the AI-based models implemented by the company. These individual DTs will form a city-level DT network, which can be incorporated for various environmental, traffic, legal and any other deemed administrative process. In a space engineering breakthrough, autonomous cognition in spacecraft is achieved using a DT-based on-orbit data analysis system. It enables the satellite to perceive its self-state in real-time with changing environment [119]. Similarly, marine applications also can be powered by AI within the designated DT architecture for autonomous operation with self-navigation, diagnostic and forecasting requirements, etc.

Similarly, a study by Heuijee Y. developed a simulation technique of the object detection, avoidance and lane detection algorithms in these AI-based vehicles using the online game GTA5 as a low-cost self-driving replication platform [120]. For these smart vehicles, highway traffic mapping and modelling can be done through 3D GIS-assisted DT with multi-domain scale modelling [121]. A similar approach can be manipulated using a ship simulator as a base platform to generate an AI-based self-navigation solution within the DT.

5.4. 4-Layer Digital Twin Framework for a Marine Vessel

From the initial contract of a ship until the final scrapping after decommissioning of the vessel, she will undergo various lifecycle phases with different actions in each stage. It will be governed by the functional requirements of the different stakeholders and Ship DT can provide valuable insights to support each activity in her design and development, validation, production, operations, and decommissioning stages. During the total lifespan of a ship, it will surpass the actions of marketing actions, the realization of the customer's intended design, optimization of the designed model using various physical/simulated methodologies, monitoring the entire functionality during her operational phase with correct control of all the actions, prognostics, diagnostics, predictive maintenance, etc. The DT of a marine vessel can deliver and support all the actions with the merits of its inherent advantages. Hence, the following 4-layer DT framework (Figure 8) is proposed in this paper with abstracted segments of Business, Tactical, Application and Operational Layers.

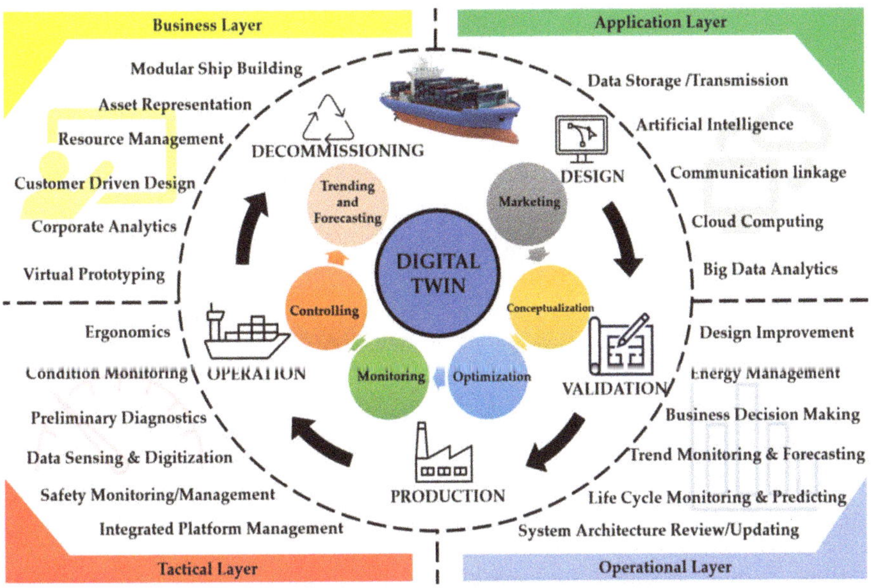

Figure 8. Proposed 4-Layer Ship DT Framework.

1. Business Layer: Initial specifications and ship owners' requirements will be met with the design, marketing and business analytics performed with the support of the DT. Simulation and validation of the prototype will be efficient and effective with the incorporated DT model and all information will be shared with the project team of the shipbuilder, allowing easy project management through a modular shipbuilding approach.
2. Tactical Layer: This will be mainly manipulated by the onboard crew if present or by the shore operator if unmanned during the operational phase of the vessel. With Integrated Platform Management System in the DT with real-time health monitoring,

navigation, communication, traffic control, load control and other ship functions will be governed with the support of DT with full integration of the ship's systems.

3. Application Layer: This is the primary segment of handling the collected data within the ship, shore command center, and ocean environment. Big data handling and cloud-based operations will heavily rely on the communication data uplink and downlinks. Artificial Intelligence will play a major part in future unmanned autonomous vessels, thus providing the advantages generated by higher levels of autonomy in the operational cycle. Machine Learning/Deep Learning can be incorporated as sub-layers in this process.
4. Operational Layer: This is used at the functionality level of the shore operators and system engineers to support their intended operational plans and strategies. The acquired data through the DT can be deeply analyzed and manipulated in decision-making related to every life cycle phase of the vessel. Supply chain management, CBPM and system development will be key benefits that will be availed through the ship DT in this layer.

The above framework is proposed as a common generic architecture and it can be easily modified, transformed, and adopted as per the user requirements. DT will be available for every patron in the hierarchy of ships operation with dedicated access to the intended data and information. A DT of any ship or offshore structure will be designed to render its present state and behavior with real-time updates in a continuous manner. This update will not only be limited to the data handshake, but also the continuous addition of parallel components to support different functions introduced in different stages of the vessel lifecycle. It will be identical to the physical asset cardinality covering all functions representing the engineering model with all possible metadata in the real-time state.

6. Scientometric Analysis of Digital Twin in the Maritime Domain

In this study, a scientometric analysis is performed with complete insight into the DT technology and to obtain a better approach to the past research work. This critical analysis and reflection will formulate recommendations with a complete mapping of available literature to visualize the pathways of previous work.

6.1. Methodology and Tools

It is thoroughly synthesized to obtain the exact outcome of the dataset derived from the large database obtained from Web of Science (WoS) [122] and refined the scientometric indicators to obtain the intended research data. In this study, the WoS inbuilt analytical tools were also used to provide a basic overview of the selected database of past research with a graph-based visualization. To obtain the visual analytics of the data set, computation tools of CiteSpace and VOSviewer are used in understanding how DT has evolved over the past passing intellectual turning points that are evident along the critical path, and the topics have attracted attention, thus synthesizing the development of DT closely and extensively [123,124].

Using the two analytical software, the database obtained from WoS is analyzed to generate the outcomes such as top authors, articles, journals with emerging literature, development, the impact of collaboration and intellectual dynamics to understand the vast subject area of DT in the maritime domain (Figure 9). In this study, research publications available in WoS Core Collection were considered from the years 2002 to 2022 (20 Years) as the DT concept was initially brought out in the year 2002. This comprehensive database includes all dominant journals and publications including IEEE Explore, MDPI, Elsevier, Springer, Emerald, Wiley online library, Taylor & Francis, etc.

Keywords were selected based on the target data set using required Booleans as stated below. The search was performed in WoS Core collection targeting the documents with the search equation of

$$TS = \{(DigitalTwinORDigitalizationORDigitalThreadORDigitalImage)AND (MarineORMaritimeORShippingOROffshoreORSeaOROceanOR PortORHarbour)\}$$

Search results provided 955 multilingual publications related to the above-narrowed research area targeting the exact timespan (Table 1).

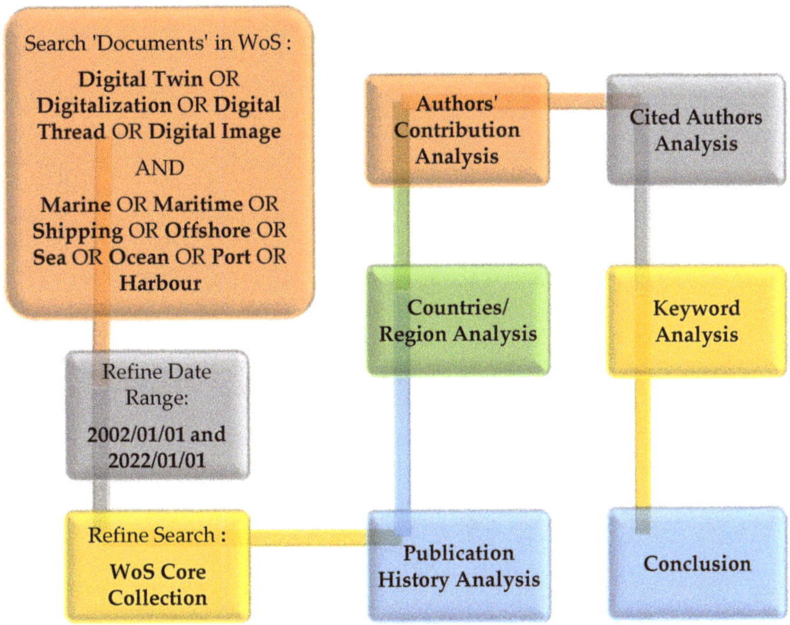

Figure 9. The Flowchart Demonstrating the Adopted Scientometric Analytical Process.

Table 1. Documents Extracted in Search Results of WoS Dataset.

Sr No	Document Type	Record Count	Percentage %
1	Articles	650	79.075
2	Meeting	234	28.467
3	Other	32	3.893
4	Reviews	26	3.163
5	Early Access	13	1.582

Based on the search results, visualizations of (1) Publication History Analysis, (2) Countries/Region Analysis, (3) Authors Contribution Analysis, (4) Cited Authors Analysis, and (5) Keyword Analysis were carried out using the analytical tools available in WoS, VOSviewer and CiteSpace. Respective analysis data and logical discussion against each step of the scientometric analysis process are described below.

6.2. *Analysis of the Dataset*

6.2.1. Publication History Analysis

The 955 publications found in the WoS dataset were analyzed and visualized based on the published year of the article chronologically. Starting with 14 publications in the year 2002, the number of articles published in the year is 231 records in 2022 with a percentage value of 24.2% of the bulk. It is 137% more than the previous year (2021), and the trend is set towards more increments in the coming years (Figure 10).

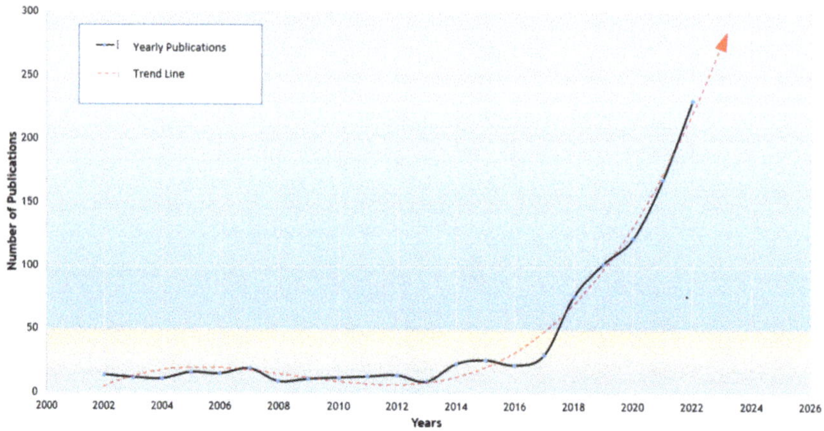

Figure 10. Publication Years-DT in Maritime Applications in WoS Analysis.

When we critically examine the holistic picture, we can observe the rapid increase in publications during the last five years, which indicates the blooming of DT technology within the maritime sector. From the year 2002 until the year 2017, an almost similar annual number of research articles have been published, but the same year has marked the takeoff year for the current trend. In the early years, very few publications originated, and these statistics may have been affected by the deficiencies of technology and knowledge gaps in early developments. Even the COVID-19 pandemic period has not affected the steep increment of the number of publications which have been launched from the year 2018. A similar trend line can be observed in similar studies performed by researchers on the DT applications and surrounding technologies in different disciplines such as in the manufacturing domain [125], business analytics [126], etc. Hence, the existing trend suggests more research work related to maritime DT concepts is underway and being published at present.

6.2.2. Analysis of Countries/Regions in Research Contribution

Figure 11 indicates that China has been the key stakeholder of research in DT towards the maritime industry with a 26% share of publications, whereas the USA is ranked second with 133 publications. Further, Europe has contributed a considerable impact on the scientific research of DT. With the ongoing research trend and industry innovations, many other regions of the world will also expect to produce more impact on the fast-growing technology. The concept of DT is still evolving with the latest technological inductions in the maritime domain, which has opened opportunities mostly for developed nations to conduct research and development of DT technology in the marine sector and produce good results over the past two decades.

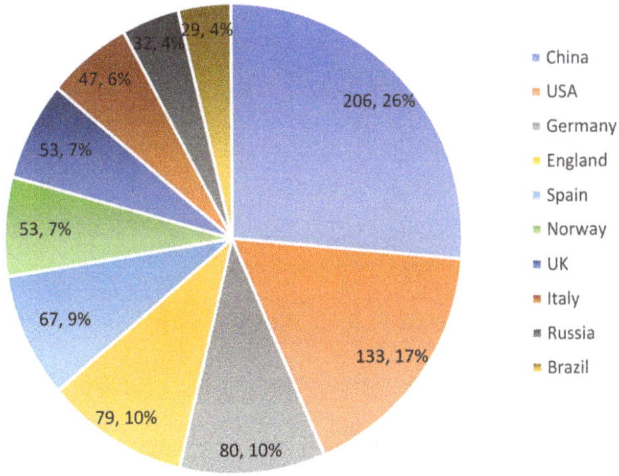

Figure 11. Contribution of Countries Towards DT in Maritime Applications in WoS Analysis.

6.2.3. Authors' Contribution Analysis

A comparison of the authors was performed considering the number of records in the database of the specific field with an insight into their corresponding profile (Figure 12). It is comprehended that "Rodriguez-tovar FJ" with 24 publications is the leading author, and the second author, "Dorador Jr", with a strong link with the No. 1 author in the same cluster, has published 17 records. "Corigliano P" (5 articles), and "Gonzalez-cancelas N" (5 articles) are in the 3rd and 4th position, respectively. Following the recent trend in DT adoption, more contributions are expected at a rapid rate in the following years from the research community.

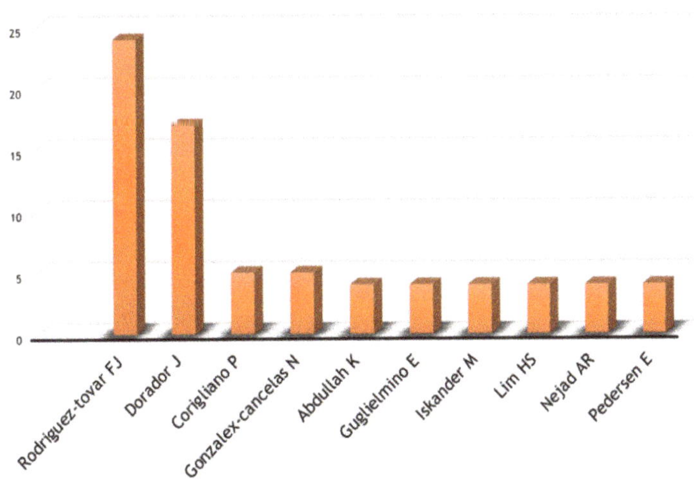

Figure 12. Authors' Contribution on DT in Maritime Applications.

Very little collaborative research is found among leading authors in the selected dataset as illustrated in Figure 13, where Rodriguez-Tovar and Francisco J. have the most collaborative publications. Since DT is a new-born concept for the maritime domain, it is observed that less cooperation is visible among the leading researchers. The major bloom started after the year 2017 and was succeeded soon by the COVID-19 pandemic. Due to the inflicted health risks, facility outreaches, travel restrictions and association barriers, the collaboration between the researchers must have had a negative impact on undergoing combined and collaborative research works. As we are now witnessing the probable end of the pandemic, more collaboration is expected among the research community towards the development of DT technology in the global maritime sphere.

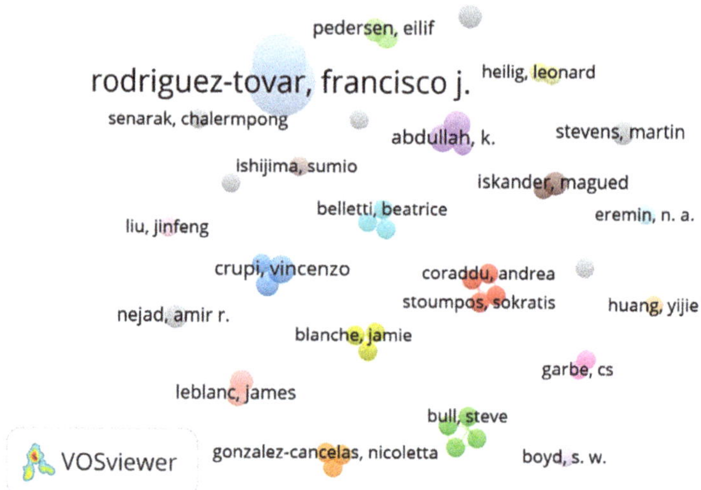

Figure 13. Authors' Collaboration on DT in Maritime Applications.

6.2.4. Cited Authors Analysis

The impact of the respective authors and the collaboration efforts is studied using the citation data (number of citations, citation numbers, citation bursts, centrality, etc.) derived using the CiteSpace analytics. The achieved data are depicted in Figure 14. There are 7 different clusters with varying silhouette values, and all can be considered reasonable depending on the value. Van D is found to be the highest-cited author with 76 citations for his publication, while Doradore J remains in second place with 24 citations.

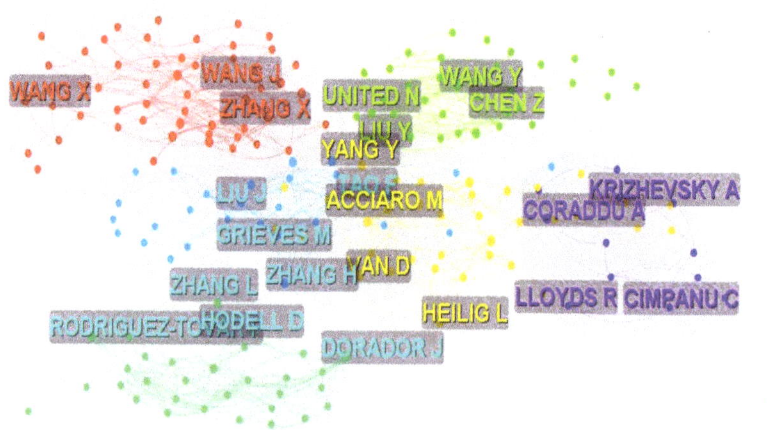

Figure 14. Visualization of Highest Cited Authors in Respective Clusters with a Threshold of 5.

Further, five references have achieved the strongest citation burst in their respective years, while Hodell D achieved the highest rate (strength 3.81) for his record published in 2013 and the burst commenced in 2014 and ended in 2016 (Table 2). The co-citation reference network visualized by VOSviewer indicates the development of the concept over time and 6 major clusters of collaborative research work within the timeframe from 2014 to 2022 (Figure 15).

Table 2. CiteSpace Analysis Output for Publications with Strongest Citation Bursts.

References	Year	Strength	Begin	End
Hodell D, 2013, SCI DRILL, v16, P13, DOI 10.5194/sd-16-13-2013 [127]	2013	3.81	2014	2016
Dorador J, 2014, MAR GEOL, V349, P55, DOI 10.1016/j.margeo.2014.01.003 [128]	2014	3.28	2015	2018
Sutton M,2009, IMAGE CORRELATION FOR SHAPE, V0, P0 [129]	2009	3.93	2009	2020
Rodriguez-Tovar F, 2015, ANN SOC GEOL POL, V85, P465 [130]	2015	3.2	2015	2020
Sanchez-Gonzalez P, 2019, SENSORS-BASEL, V19,P0,DOI 10.3390/s19040926 [131]	2019	3.56	2019	2022

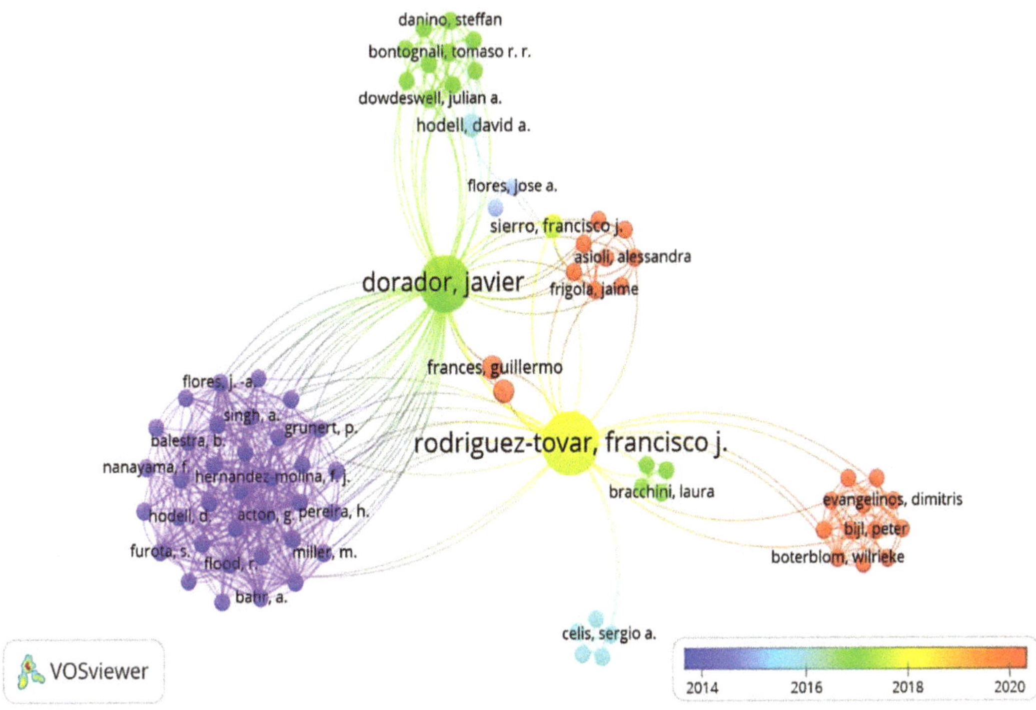

Figure 15. Co-citation Reference Network in the Timespan of 2014 to 2022 with 6 Main Clusters.

6.2.5. Keywords Analysis

Research subjects are often generated and influenced by the coexistence of the keywords amongst various publications making a deliberate trend for parallel research bases. CiteSpace keyword analysis was performed on the exported WoS database and the following visualization with highlighting of the prominent nodes indicating the highest counter and centrality (Figure 16).

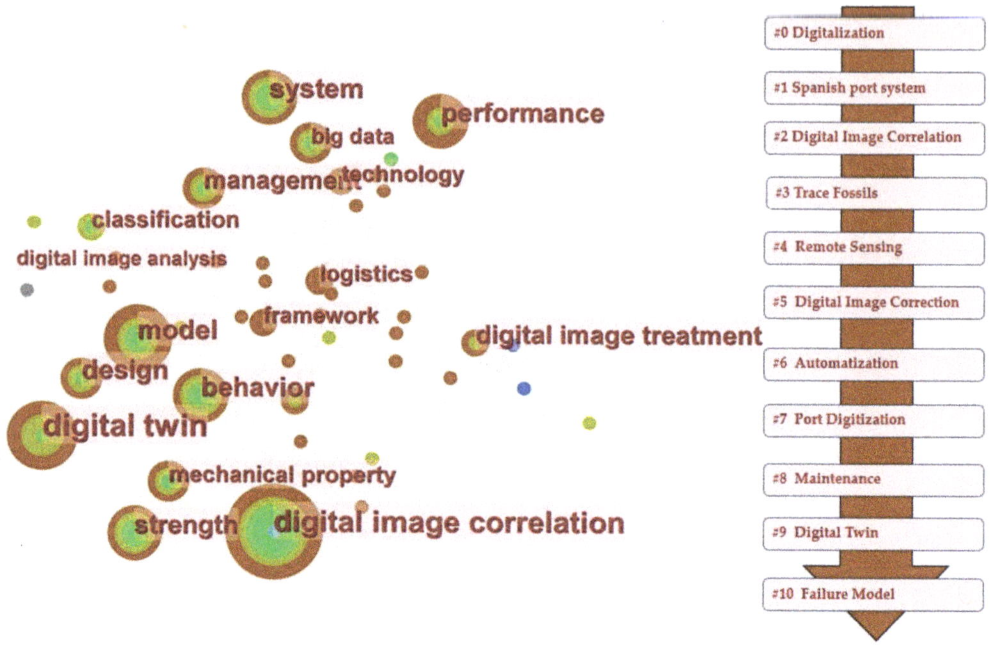

Figure 16. CiteSpace Top-50 Keyword Visualization with Top-10 Timeline Nodes.

It is observed that the frequency burst of most of these keywords occurred from 2016 to date, indicating the recent and ongoing research prominence among the research community. These keywords can provide an overview of the current trending research work within the maritime domain. Further, the available timeline view in CiteSpace brought up the progression of DT in the marine sector with the initial step of "Digitalization" (#0-First Node) leading the pathway passing cluster titles such as "digital image correlation", "remote sensing", "automatization", "port digitization" and finally to "Digital Twin" (#9-Ninth Node). This further provides substantial validation to the pathway of implementing DT technology within the maritime domain as discussed in the articles from initial digitalization, smart ports towards the final implementation of DT technology.

6.2.6. Results and Discussion

It is an evident fact that DT has had a slower beginning from its inception in the year 2002 up to the year 2017, but it has been thriving with major applications/achievements among researchers around the world. The present trend in publications related to DT technology in the maritime sector is following an exponential increment, which is also a promising factor in generating innovative avenues for the future industry. Further, the published data originated from almost all regions of the world, which emphasizes the attention received from the research communities around the world. The leading researchers are highlighted in the analysis along with the trending keywords within the highlighted domain, which will give insight to new researchers for their follow-up actions. Although cross-border and cross-discipline research work has occurred less in recent years, more collaborative research work is expected along with the easing of COVID-19 pandemic-imposed restrictions, which will generate a more positive impact on the progression of technology.

7. Challenges and Future Developments in Marine Digital Twin Technology

DT is at the infancy stage when it comes to implementation in the shipping industry. Limited research is conducted and implemented in projects compared to other industries such as aerospace, manufacturing, agriculture, construction, city planning, etc. The effectiveness of the Cyber-Physical Fusion [132] will be another key encounter. The following salient points can be prominently identified as the bottlenecks posed to the progression and implementation of this naïve technology within the maritime sector.

- Lack of necessary infrastructure within the existing maritime industry.
- Lack of software and hardware skills among various levels of stakeholders.
- Limited connectivity due to data transmission restrictions at sea.
- Communication barrier between the underwater vehicles (Submarines and UAVs) with the shore-operated DT.
- The reluctance to transition to digital systems from long-used manual hardware platforms.
- Knowledge gap of the DT technology and benefits reaching the industry as mostly academia is engaged in the research and development stage.
- Lack of research collaboration among key researchers/institutes.
- Technology readiness is lagging while concept development is at full throttle.
- Data security complications in dealing with confidentiality, non-repudiation, and authentication.
- Large data handling barriers in data retention, accessibility, and visualization.
- Very limited research has been conducted on the functioning of DT in the retiring/decommissioning phase.

With the known limitations highlighted above, DT has recently emerged as one of the most powerful tools with the aid of high-tech computational resources, AI functions, Big Data Analytics, etc. Thus, the same is being vastly adopted by academia, industry, and research organizations. In the maritime domain, one of the most critical aspects is communication, which is a mandatory prerequisite for any DT-based system. The data link between the onshore, offshore, or underwater physical model with the DT is to be established with continuous synchronization. With the development of satellite links and shore-based networks for coastal navigating vessels, DT is becoming much more practical for real-time implementations. Comprehensive data exchange and a model-based review process are to be inducted into existing manual operations. It will enable achieving a full Model-Based Definition (MBD) [133], which is mandatory in the DT implication process. However, a Digital Shadow or a Digital Model is a possibility by updating the digital image manually or whenever the communication link is established. A collaborative data handshake is mandatory with a cohesive data network accessed by all stakeholders [94]. Missing data or faulty data due to sensor failures can create an entire system failure in a DT-based system. This data should be coherent without any disparity to function in DT modules with the ability to unify the data among the cooperating partners. Further, most industries pose scattered ownership of data and the same are maintained in different formats, interfaces and software [134]. Unification and standardization of these mandatory data policies in DT can be a challenge.

Furthermore, with the development of more robust underwater communication technologies, in the future, DT solutions will come into play with automated submarines and other underwater applications. DT-based solutions will be able to overcome the communication gaps in AUV operations incorporating continuous research and development will be the key to imposing a complete transition from existing obsolete platforms to DT-based suits within the maritime industry.

It has been evident from the literature survey that DT technology has tremendous potential across various segments in the maritime domain and despite being in its early stages of implementation, worldwide research and development efforts are underway.

Currently, DT technology is primarily used for monitoring and simulation in the maritime domain, but its usage is expected to expand to other areas such as predictive maintenance, asset management, and decision-making. As the maritime industry shifts towards interconnectivity and collaboration, it is believed that DTs will play a crucial role in supporting and enhancing these efforts. In particular, the integration of DT technology with ASVs/AUVs is poised to play a significant role in connecting to future smart functions utilized by various stakeholders in the industry. AI is expected to be a key component in the DT functionality, where the least human intervention will be made possible in operating complex scenarios. Thus, the future of DT technology in the maritime domain is very bright, with continued technological advancements and increasing adoption expected in the coming years.

8. Conclusions

In this study, a comprehensive literature review has been performed to explore the applications of DT technology within the maritime domain. Since its introduction in the early 2000s, DT technology has progressed at a slower pace due to the absence of necessary components such as hardware, software, remote sensing techniques, cloud computing, big data analytics, digital manufacturing architecture, and satellite communication facilities. Despite the limited implementation of DTs in real-life projects during the early years of the millennium, the recognition of its capabilities and potential has led to its increasing adoption in the manufacturing and service sectors of the maritime industry. With the abundance of computational resources and recent emergent trends in AI-based technologies, DT is likely to become an effective solution for validating state-of-the-art designs in a virtual environment, resulting in significant time and cost savings. Many projects are already underway in many parts of the world to acquaint the existing operations with DT-based solutions in designing, monitoring, manufacturing, training, testing, simulations, etc. Based on the bibliometric study performed using scientometric analysis, it can be deduced that there is a growing trend of incorporating DTs into maritime operations as the industry becomes increasingly digitized and electrified. The proposed four-layer ship DT framework in this study has compiled and deduced these necessities to be addressed within the maritime sector and can be utilized as a base model in designing a comprehensive DT architecture covering the life cycle of a ship. The majority of the DT projects in the maritime sector are still in the developing stage with limited research work/data available on the subject. However, by observing the current bloom in technology and research trends on DT-based systems, it can be concluded that the potential of DT will revolutionize the maritime sector, especially the shipping and offshore engineering sectors with expected major outcomes in achieving high-tech-based design, development, manufacturing, and operations.

Author Contributions: Conceptualization, X.X. and S.Y.; methodology, S.Y. and X.X.; investigation, N.S.M., Y.F., S.Y. and X.X.; formal analysis, N.S.M. and S.Y.; resources, X.X. and S.Y.; writing—original draft preparation, N.S.M.; writing—review and editing, N.S.M., S.Y. and X.X. All authors have read and agreed to the published version of the manuscript.

Funding: This work is partially supported by the National Natural Science Foundation of China, grant number 52131101 and the Hubei Provincial Natural Science Foundation for Innovation Groups, grant number 2021CFA026.

Institutional Review Board Statement: Not applicable.

Informed Consent Statement: Not applicable.

Conflicts of Interest: The authors declare no conflict of interest. The funders had no role in the design of the study; in the collection, analyses, or interpretation of data; in the writing of the manuscript; or in the decision to publish the results.

References

1. Li, S.; Xu, L.D.; Zhao, S. The Internet of Things: A Survey. *Inf. Syst. Front.* **2015**, *17*, 243–259. [CrossRef]
2. Asthon, K. That 'Internet of Things' Thing. *RFiD J.* **2010**, *22*, 97–114.
3. Lv, Z.; Xie, S. Artificial Intelligence in the Digital Twins: State of the Art, Challenges, and Future Research Topics. *Digit. Twin* **2021**, *1*, 12. [CrossRef]

4. Enyedi, S.; Miclea, L. IEEE International Conference on Automation, Quality and Testing, Robotics (AQTR). *Proc.-Int. Test Conf.* **2019**, *2019*, 2–4. [CrossRef]
5. El Saddik, A. Digital Twins: The Convergence of Multimedia Technologies. *IEEE Multimed.* **2018**, *25*, 87–92. [CrossRef]
6. Huggins, D. The Treatment of Mixed Helminthic Infections with Oxantel/Pyrantel. *Folha Med.* **1978**, *76*, 475–478.
7. Hehenberger, P.; Bradley, D. *Mechatronic Futures: Challenges and Solutions for Mechatronic Systems and Their Designers*; Springer International Publishing: Midtown Manhattan, NY, USA, 2016; ISBN 9783319321561.
8. Taking Flight with the Airbus "Iron Bird"—Commercial Aircraft—Airbus. Available online: https://www.airbus.com/newsroom/news/en/2017/05/taking-flight-with-the-airbus-iron-bird.html (accessed on 3 December 2019).
9. Renaudin, C.P.; Barbier, B.; Roriz, R.; Revel, D.; Amiel, M. Coronary Arteries: New Design for Three-Dimensional Arterial Phantoms. *Radiology* **1994**, *190*, 579–582. [CrossRef]
10. Grieves, M. Foreword. Digital Twin: Mitigating Unpredictable, Undesirable Emergent Behavior in Complex Systems. In *Transdisciplinary Perspectives on Complex Systems*; Springer: Berlin/Heidelberg, Germany, 2016; pp. 85–113. [CrossRef]
11. Shafto, M.; Conroy, M.; Doyle, R.; Glaessgen, E.; Kemp, C.; LeMoigne, J.; Wang, L. Modeling, Simulation, Information Technology & Processing Roadmap-NASA. *Natl. Aeronaut. Space Adm.* **2012**, *32*, 1–38.
12. Endsley, M.; Headquarters U.S. Air Force during 2010–2030: The Air Force is critically dependent on science & technology advances. In Proceedings of the Presentation at the USAF Science and Technology 2030 Conference, Dayton, OH, USA, 14–15 June 2010.
13. GitHub-Evanderhorn/DSS_DT_Public. Available online: https://github.com/evanderhorn/DSS_DT_Public (accessed on 29 December 2022).
14. Jones, D.; Snider, C.; Nassehi, A.; Yon, J.; Hicks, B. Characterising the Digital Twin: A Systematic Literature Review. *CIRP J. Manuf. Sci. Technol.* **2020**, *29*, 36–52. [CrossRef]
15. Alam, K.M.; Saini, M.; El Saddik, A. Toward Social Internet of Vehicles: Concept, Architecture, and Applications. *IEEE Access* **2015**, *3*, 343–357. [CrossRef]
16. Atzori, L.; Iera, A.; Morabito, G. The Internet of Things: A Survey. *Comput. Netw.* **2010**, *54*, 2787–2805. [CrossRef]
17. Karve, P.M.; Guo, Y.; Kapusuzoglu, B.; Mahadevan, S.; Haile, M.A. Digital Twin Approach for Damage-Tolerant Mission Planning under Uncertainty. *Eng. Fract. Mech.* **2020**, *225*, 106766. [CrossRef]
18. Ibrion, M.; Paltrinieri, N.; Nejad, A.R. On Risk of Digital Twin Implementation in Marine Industry: Learning from Aviation Industry. *J. Phys. Conf. Ser.* **2019**, *1357*, 012009. [CrossRef]
19. Abramovici, M.; Göbel, J.C.; Savarino, P. Reconfiguration of Smart Products during Their Use Phase Based on Virtual Product Twins. *CIRP Ann.-Manuf. Technol.* **2017**, *66*, 165–168. [CrossRef]
20. Kritzinger, W.; Karner, M.; Traar, G.; Henjes, J.; Sihn, W. Digital Twin in Manufacturing: A Categorical Literature Review and Classification. *IFAC-Pap.* **2018**, *51*, 1016–1022. [CrossRef]
21. Vissamsetty, S.; Updhyayula, R.S. Study of Alternate Engine Propulsion Systems to Reduce CO_2. In Proceedings of the ICSOT: Technological Innovations in Shipbuilding, Indian Maritime University, Visakhapatnam, India, 12–13 December 2013; pp. 127–142.
22. Bolvashenkov, I.; Herzog, H.G.; Rubinraut, A.; Romanovskiy, V. Possible Ways to Improve the Efficiency and Competitiveness of Modern Ships with Electric Propulsion Systems. In Proceedings of the 2014 IEEE Vehicle Power and Propulsion Conference (VPPC), Coimbra, Portugal, 27–30 October 2014. [CrossRef]
23. M/s Vandal—A Historical Ship—Nobel Brothers. Available online: https://www.branobelhistory.com/innovations/m-s-vandal-a-historical-ship/ (accessed on 16 April 2023).
24. Paul, D. A History of Electric Ship Propulsion Systems [History]. *IEEE Ind. Appl. Mag.* **2020**, *26*, 9–19. [CrossRef]
25. Parker, D.S.; Hodge, C.G. The Electric Warship. *Power Eng. J.* **1998**, *12*, 5–13. [CrossRef]
26. Propulsion Systems Used in Modern Naval Vessels—Naval Post—Naval News and Information. Available online: https://navalpost.com/propulsion-systems-navies-gas-diesel-electric/ (accessed on 16 April 2023).
27. Doerry, N.; Amy, J.; Krolick, C. History and the Status of Electric Ship Propulsion, Integrated Power Systems, and Future Trends in the U.S. Navy. *Proc. IEEE* **2015**, *103*, 2243–2251. [CrossRef]
28. World's First All-Electric Fast Ferry Named in Norway. Available online: https://www.offshore-energy.biz/worlds-first-all-electric-fast-ferry-named-in-norway (accessed on 12 February 2023).
29. USS Zumwalt (DDG-1000)—Modern Warships—World of Warships Official Forum. Available online: https://forum.worldofwarships.eu/topic/639-uss-zumwalt-ddg-1000/ (accessed on 29 May 2020).
30. Sulligoi, G.; Vicenzutti, A.; Menis, R. All-Electric Ship Design: From Electrical Propulsion to Integrated Electrical and Electronic Power Systems. *IEEE Trans. Transp. Electrif.* **2016**, *2*, 507–521. [CrossRef]
31. Gilligan, B.K. *Electric Ship Digital Twin: Framework for Cyber-Physical System Security*; Massachusetts Institute of Technology, Department of Mechanical Engineering: Cambridge, MA, USA, 2019.
32. Erikstad, S.O. Designing Ship Digital Services. In Proceedings of the Compit'19, Tullamore, Ireland, 3 April 2019; pp. 458–469.
33. Schumann, C.-A.; Baum, J.; Forkel, E.; Otto, F. Digital Transformation and Industry 4.0 as a Complex and Eclectic Change. *Futur. Technol. Conf.* **2017**, *2017*, 645–650.
34. American Bureau of Shipping. Guidance Notes on Smart Function Implementation. *Houston, TX, USA*. 2018. Available online: https://ww2.eagle.org/content/dam/eagle/rules-and-guides/current/other/guidance-notes-smart-function-implementation.pdf (accessed on 5 January 2023).

35. Rolls-Royce and Wartsila in Close Race with Autonomous Ferries. Available online: https://www.maritime-executive.com/article/rolls-royce-and-wartsila-in-close-race-with-autonomous-ferries (accessed on 5 December 2019).
36. Liu, M.; Fang, S.; Dong, H.; Xu, C. Review of Digital Twin about Concepts, Technologies, and Industrial Applications. *J. Manuf. Syst.* **2021**, *58*, 346–361. [CrossRef]
37. Autonomous Shipping. Available online: https://www.imo.org/en/MediaCentre/HotTopics/Pages/Autonomous-shipping.aspx (accessed on 22 February 2023).
38. Kaber, D.B. Issues in Human-Automation Interaction Modeling: Presumptive Aspects of Frameworks of Types and Levels of Automation. *J. Cogn. Eng. Decis. Mak.* **2018**, *12*, 7–24. [CrossRef]
39. Li, J.; Xiang, X.; Yang, S. Robust Adaptive Neural Network Control for Dynamic Positioning of Marine Vessels with Prescribed Performance under Model Uncertainties and Input Saturation. *Neurocomputing* **2022**, *484*, 1–12. [CrossRef]
40. Development and Missions of Unmanned Surface Vehicle. Available online: https://cetcusv.en.ecplaza.net/ (accessed on 27 February 2023).
41. Wang, Z.; Yang, S.; Xiang, X.; Vasilijević, A.; Mišković, N.; Nađ, Đ. Cloud-Based Mission Control of USV Fleet: Architecture, Implementation and Experiments. *Control. Eng. Pract.* **2021**, *106*, 104657. [CrossRef]
42. Alsos, O.A.; Veitch, E.; Pantelatos, L.; Vasstein, K.; Eide, E.; Petermann, F.M.; Breivik, M. NTNU Shore Control Lab: Designing Shore Control Centres in the Age of Autonomous Ships. In Proceedings of the International Maritime and Port Technology and Development Conference (MTEC) & The 4th International Conference on Maritime Autonomous Surface Ships, Singapore, Singapore, 5–7 April 2022; Volume 2311.
43. Ahmed, F.; Xiang, X.; Jiang, C.; Xiang, G.; Yang, S. Survey on Traditional and AI Based Estimation Techniques for Hydrodynamic Coefficients of Autonomous Underwater Vehicle. *Ocean Eng.* **2023**, *268*, 113300. [CrossRef]
44. Li, J.; Xiang, X.; Dong, D.; Yang, S. Prescribed Time Observer Based Trajectory Tracking Control of Autonomous Underwater Vehicle with Tracking Error Constraints. *Ocean Eng.* **2023**, *274*, 114018. [CrossRef]
45. Nassar, W.M.; Anaya-Lara, O.; Ahmed, K.H.; Campos-Gaona, D.; Elgenedy, M. Assessment of Multi-Use Offshore Platforms: Structure Classification and Design Challenges. *Sustainability* **2020**, *12*, 1860. [CrossRef]
46. Amaechi, C.V.; Reda, A.; Butler, H.O.; Ja'e, I.A.; An, C. Review on Fixed and Floating Offshore Structures. Part I: Types of Platforms with Some Applications. *J. Mar. Sci. Eng.* **2022**, *10*, 1074. [CrossRef]
47. Lee, K.H.; Kim, M.G.; Lee, J.I.; Lee, P.S. Recent Advances in Ocean Nuclear Power Plants. *Energies* **2015**, *8*, 11470–11492. [CrossRef]
48. Taylor, N.; Human, C.; Kruger, K.; Bekker, A.; Basson, A. Comparison of Digital Twin Development in Manufacturing and Maritime Domains Nicole. *Stud. Comput. Intell.* **2019**, *853*, 158–170. [CrossRef]
49. The ReVolt-DNV GL. Available online: https://www.dnv.com/research/maritime/publications.html (accessed on 5 February 2023).
50. Veracity by DNV. Available online: https://www.veracity.com/ (accessed on 1 January 2023).
51. Siemens | MindSphere. Available online: https://siemens.mindsphere.io/en (accessed on 1 January 2023).
52. Rødseth, Ø.J.; Berre, A.J. From Digital Twin to Maritime Data Space: Transparent Ownership and Use of Ship Information. In Proceedings of the 13th International Symposium on Integrated Ship's Information Systems & Marine Traffic Engineering Conference ISIS-MTE 2018, Berlin, Germany, 27–28 September 2018; pp. 1–8.
53. Danielsen-Haces, A. Digital Twin Development—Condition Monitoring and Simulation Comparison for the ReVolt Autonomous Model Ship. Master's Thesis, Norwegian University of Science and Technology, Trondheim, Norway, 2018.
54. Bekker, A. Exploring the Blue Skies Potential of Digital Twin Technology for a Polar Supply and Research Vessel. *Mar. Des. XIII* **2018**, *1*, 135–146.
55. Bruun, K. *Digital Twins for Blue Denmark*; 2018-0006; DNV GL: Batam City, Indonesia, 2018.
56. ReVolt—Next Generation Short Sea Shipping—DNV GL. Available online: https://www.dnv.com/ (accessed on 22 March 2019).
57. Lambertini, A.; Menghini, M.; Cimini, J.; Odetti, A.; Bruzzone, G.; Bibuli, M.; Mandanici, E.; Vittuari, L.; Castaldi, P.; Caccia, M.; et al. Underwater Drone Architecture for Marine Digital Twin: Lessons Learned from SUSHI DROP Project. *Sensors* **2022**, *22*, 744. [CrossRef]
58. Manngård, M.; Lund, W.; Björkqvist, J. Using Digital Twin Technology to Ensure Data Quality in Transport Systems. In Proceedings of the TRA2020 Rethinking transport, Helsinki, Finland, 12 March 2020; pp. 27–30.
59. Cheng, D.J.; Zhang, J.; Hu, Z.T.; Xu, S.H.; Fang, X.F. A Digital Twin-Driven Approach for On-Line Controlling Quality of Marine Diesel Engine Critical Parts. *Int. J. Precis. Eng. Manuf.* **2020**, *21*, 1821–1841. [CrossRef]
60. Castanedo, F. A Review of Data Fusion Techniques. *Sci. World J.* **2013**, *2013*, 704504. [CrossRef]
61. Liu, Z.; Meyendorf, N.; Mrad, N. The Role of Data Fusion in Predictive Maintenance Using Digital Twin. *AIP Conf. Proc.* **2018**, *1949*, 02002. [CrossRef]
62. Xiong, M.; Wang, H.; Fu, Q.; Xu, Y. Digital Twin–Driven Aero-Engine Intelligent Predictive Maintenance. *Int. J. Adv. Manuf. Technol.* **2021**, *114*, 3751–3761. [CrossRef]
63. Bondarenko, O.; Fukuda, T. Development of a Diesel Engine's Digital Twin for Predicting Propulsion System Dynamics. *Energy* **2020**, *196*, 117126. [CrossRef]
64. Arrichiello, V.; Gualeni, P. Systems Engineering and Digital Twin: A Vision for the Future of Cruise Ships Design, Production and Operations. *Int. J. Interact. Des. Manuf.* **2020**, *14*, 115–122. [CrossRef]

65. Tao, F.; Zhang, M.; Nee, A.Y.C. *Digital Twin Driven Smart Manufacturing*; Elsevier Academic Press: Cambridge, MA, USA, 2019; Volume 53, ISBN 9788578110796.
66. Helman, J.; Bernard, O.; Bellouard, Y. The Application and Challenge of Digital Twin Technology in Ship Equipment. In Proceedings of the 2021 International Conference on Mechanical Engineering, Intelligent Manufacturing and Automation Technology (MEMAT 2021), Guilin, China, 23–25 April 2021; Volume 1939, p. 012068. [CrossRef]
67. Li, C.; MahaDeVan, S.; Ling, Y.; Choze, S.; Wang, L. Dynamic Bayesian Network for Aircraft Wing Health Monitoring Digital Twin. *AIAA J.* **2017**, *55*, 930–941. [CrossRef]
68. Madni, A.M.; Madni, C.C.; Lucero, S.D. Leveraging Digital Twin Technology in Model-Based Systems Engineering. *Systems* **2019**, *7*, 7. [CrossRef]
69. ÜZÜMCÜ, S.; MERT, A.A.; ATAY, F. Usage of Digital Twin Technologies during System Modeling and Testing in Vessel Traffic Services System Project. In *INCOSE International Symposium*; John and Wiley and Sons: Hoboken, NJ, USA, 2019; Volume 29, pp. 189–202.
70. Tracing the Origins of Boeing's "Diamond" from Apollo to NMA—The Air Current. Available online: https://theaircurrent.com/aircraft-development/tracing-the-origins-of-boeings-diamond-from-apollo-to-nma/ (accessed on 17 April 2023).
71. Khajavi, S.H.; Motlagh, N.H.; Jaribion, A.; Werner, L.C.; Holmström, J. Building Information Modeling.Pdf. In *Digital Twin: Vision, Benefits, Boundaries, and Creation for Buildings*; IEEE: New York, NY, USA, 2019; p. 14.
72. Marikyan, D.; Papagiannidis, S.; Alamanos, E. A Systematic Review of the Smart Home Literature: A User Perspective. *Technol. Forecast. Soc. Chang.* **2019**, *138*, 139–154. [CrossRef]
73. Zhou, G.; Zhang, C.; Li, Z.; Ding, K.; Wang, C. Knowledge-Driven Digital Twin Manufacturing Cell towards Intelligent Manufacturing. *Int. J. Prod. Res.* **2020**, *58*, 1034–1051. [CrossRef]
74. Hamburg Port Authority—SmartBRIDGE Hamburg—World Port Sustainability Program. Available online: https://sustainableworldports.org/project/hamburg-port-authority-smartbridge/ (accessed on 19 April 2023).
75. Guo, J.; Zhao, N.; Sun, L.; Zhang, S. Modular Based Flexible Digital Twin for Factory Design. *J. Ambient Intell. Humaniz. Comput.* **2019**, *10*, 1189–1200. [CrossRef]
76. Baruffaldi, G.; Accorsi, R.; Manzini, R. Warehouse Management System Customization and Information Availability in 3pl Companies: A Decision-Support Tool. *Ind. Manag. Data Syst.* **2019**, *119*, 251–273. [CrossRef]
77. Zhang, Q.; Zhang, J.; Chemori, A.; Xiang, X. Virtual Submerged Floating Operational System for Robotic Manipulation. *Complexity* **2018**, *2018*, 9528313. [CrossRef]
78. Major, P.; Li, G.; Zhang, H.; Hildre, H.P. Real-Time Digital Twin of Research Vessel for Remote Monitoring. *Proc.-Eur. Counc. Model. Simul. ECMS* **2021**, *35*, 159–164. [CrossRef]
79. Norwegian University of Science and Technology—NTNU. Available online: https://www.ntnu.edu/ (accessed on 4 November 2022).
80. Mukha, M.; Karianskyi, S. Real-time Propulsion Plant Simulation and HIL Implementation. In Proceedings of the 13th International Conference on Engine Room Simulators, Odessa, Ukraine, 20–21 September 2017.
81. Asher, M.J.; Croke, B.F.W.; Jakeman, A.J.; Peeters, L.J.M. A Review of Surrogate Models and Their Application to Groundwater Modeling. *Water Resour. Res.* **2015**, *51*, 5957–5973. [CrossRef]
82. Sudhoff, S.D.; Pekarek, S.; Kuhn, B.; Glover, S.; Sauer, J.; Delisle, D. Naval Combat Survivability Testbeds for Investigation of Issues in Shipboard Power Electronics Based Power and Propulsion Systems. In Proceedings of the IEEE Power Engineering Society Transmission and Distribution Conference, Chicago, IL, USA, 21–25 July 2002; Volume 1, pp. 347–350.
83. Dufour, C.; Soghomonian, Z.; Li, W. Hardware-in-the-Loop Testing of Modern On-Board Power Systems Using Digital Twins. In Proceedings of the 2018 International Symposium on Power Electronics, Electrical Drives, Automation and Motion (SPEEDAM), Amalfi, Italy, 20–22 June 2018; pp. 118–123. [CrossRef]
84. Naval Air Systems Command (NAVAIR) Digital Twin Technologies to Improve Mission Readiness and Sustainment. Available online: https://www.navysbir.com/n20_2/N202-105.htm (accessed on 19 April 2023).
85. Navy Optimizing Shipyards with Digital Twin Technology. Available online: https://www.nationaldefensemagazine.org/articles/2021/3/1/navy-optimizing-shipyards-with-digital-twin-technology (accessed on 19 April 2023).
86. Silvera, J.I.; Luis, J.L.; Luquero, J.M.; Bustelo, M. Navantia's Digital Twin Implementation Perspective in Military Naval Platform Life Cycle Report. 2022, Volume 1, pp. 1–30. Available online: https://www.google.com/url?sa=t&rct=j&q=&esrc=s&source=web&cd=&cad=rja&uact=8&ved=2ahUKEwiurb3NkOr-AhU3yLsIHZ_ABMwQFnoECAoQAQ&url=https%3A%2F%2Fwww.sto.nato.int%2Fpublications%2FSTO%2520Meeting%2520Proceedings%2FSTO-MP-MSG-171%2FMP-MSG-171-P1.pdf&usg=AOvVaw2Kw4F5wGOlzt25Zs6bAkgJ (accessed on 19 April 2023).
87. Digital Twin-Testing Real World Problems. Available online: https://www.baesystems.com/en/feature/digital-twin-testing-real-world-performance-in-the-digital-world (accessed on 4 November 2022).
88. Yang, J.; Xi, M.; Wen, J.; Li, Y.; Song, H.H. A Digital Twins Enabled Underwater Intelligent Internet Vehicle Path Planning System via Reinforcement Learning and Edge Computing. *Digit. Commun. Netw.* **2022**, in press. [CrossRef]
89. Papanikolaou, A. Holistic Ship Design Optimization. *CAD Comput. Aided Des.* **2010**, *42*, 1028–1044. [CrossRef]
90. Zhou, S.; Wu, W.; Zhang, T.; Zhang, Q. Digital Twin Technical System for Marine Power Systems. *Chin. J. Sh. Res.* **2021**, *16*, 151–156. [CrossRef]

91. Zhou, H.; Liu, X. The Modeling and Using Strategy for the Digital Twin in Process Planning. *IEEE Access* **2020**, *8*, 41229–41245. [CrossRef]
92. Fonseca, Í.A.; Gaspar, H.M. Challenges When Creating a Cohesive Digital Twin Ship: A Data Modelling Perspective. *Sh. Technol. Res.* **2021**, *68*, 70–83. [CrossRef]
93. Erikstad, S.O. Design Patterns for Digital Twin Solutions in Marine Systems Design and Operations Holistic Risk-Based Design for Sustainable Arctic Sea Transport (RISKAT) View Project Design Patterns for Digital Twin Solutions in Marine Systems Design and Operations. In Proceedings of the COMPIT 2018, Pavone, Italy, 14–16 May 2018; p. 10.
94. Gamma, E.; Helm, R.; Johnson, R.; Vlissides, J. *Design Patters: Elements of Reusable Object-Oriented Software*; Addison-Wesley: Boston, MA, USA, 1995; ISBN 5272003551.
95. Negri, E.; Fumagalli, L.; Macchi, M. A Review of the Roles of Digital Twin in CPS-Based Production Systems. *Procedia Manuf.* **2017**, *11*, 939–948. [CrossRef]
96. Institute for Security & Development Policy. *Made In China 2025: Backgrounder Report*; 2018; pp. 1–9. Available online: https://isdp.eu/publication/made-in-china-2025-backgrounder-report (accessed on 5 January 2023).
97. Bécue, A.; Maia, E.; Feeken, L.; Borchers, P.; Praça, I. A New Concept of Digital Twin Supporting Optimization and Resilience of Factories of the Future. *Appl. Sci.* **2020**, *10*, 4482. [CrossRef]
98. Heikkilä, M.; Saarni, J.; Saurama, A. Innovation in Smart Ports: Future Directions of Digitalization in Container Ports. *J. Mar. Sci. Eng.* **2022**, *10*, 1925. [CrossRef]
99. Douaioui, K.; Fri, M.; Mabrouki, C.; Semma, E.A. Smart Port: Design and Perspectives. In Proceedings of the 2018 4th International Conference on Logistics Operations Management (GOL), Le Havre, France, 10–12 April 2018; pp. 1–6. [CrossRef]
100. Shipfinex How Blockchain Technology Is Revolutionizing Port Operations. Available online: https://www.shipfinex.com/blockchain-technology-in-port-operations (accessed on 19 April 2023).
101. Smart Mobility and Services (SMO) | TRIMIS. Available online: https://trimis.ec.europa.eu/roadmaps/smart-mobility-and-services-smo (accessed on 19 April 2023).
102. "Maritime 2050-Navigating the Future". Department for Transport, UK Government. 2019. Available online: https://www.gov.uk/government/publications/maritime-2050-navigating-the-future (accessed on 5 January 2023).
103. Smart Ports. Available online: https://www.porttechnology.org/editions/smart_ports (accessed on 3 January 2020).
104. Chandra, D.R.; van Hillegersberg, J. Governance of Inter-Organizational Systems: A Longitudinal Case Study of Rotterdam's Port Community System. *Int. J. Inf. Syst. Proj. Manag.* **2018**, *6*, 47–68. [CrossRef]
105. Smart Port | Port of Antwerp-Bruges. Available online: https://www.portofantwerpbruges.com/en/our-port/port-future/smart-port (accessed on 19 April 2023).
106. Ports Australia. Port Digital Twins Working Group Report. 2022. Available online: https://www.portsaustralia.com.au/wp-content/uploads/2022/03/Ports-Australia-Port-Digital-Twins-Working-Group-Report.pdf (accessed on 5 January 2023).
107. Bachale, J. Hamburg Joint Project TwinSim. Available online: https://www.bwl.uni-hamburg.de/en/iwi/ueber-das-institut/news/2021/ihatecprojekt.html (accessed on 5 January 2023).
108. Yau, K.L.A.; Peng, S.; Qadir, J.; Low, Y.C.; Ling, M.H. Towards Smart Port Infrastructures: Enhancing Port Activities Using Information and Communications Technology. *IEEE Access* **2020**, *8*, 83387–83404. [CrossRef]
109. Heilig, L.; Lalla-Ruiz, E.; Voß, S. Digital Transformation in Maritime Ports: Analysis and a Game Theoretic Framework. *NETNOMICS Econ. Res. Electron. Netw.* **2017**, *18*, 227–254. [CrossRef]
110. Haringer, M.; Regenbrecht, H.T. A Pragmatic Approach to Augmented Reality Authoring. *Proc.-Int. Symp. Mix. Augment. Real. ISMAR* **2002**, *2002*, 237–246. [CrossRef]
111. Marcus, A.; Abromowitz, S.; Abulkhair, M.F. (Eds.) Design, User Experience, and Usability. In Proceedings of the 5th International Conference, DUXU 2016 Held as Part of HCI International 2016, Toronto, ON, Canada, 17–22 July 2016; Springer: Berlin/Heidelberg, Germany, 2016; Volume III, pp. 414–424.
112. Digital Twins and Digital Twin Technology in an Industrial Context. Available online: https://www.i-scoop.eu/internet-of-things-iot/industrial-internet-things-iiot-saving-costs-innovation/digital-twins/ (accessed on 30 December 2022).
113. Shen, H.; Zhang, J.; Yang, B.; Jia, B. Development of an Educational Virtual Reality Training System for Marine Engineers. *Comput. Appl. Eng. Educ.* **2019**, *27*, 580–602. [CrossRef]
114. Fan, C.; Zhang, C.; Yahja, A.; Mostafavi, A. Disaster City Digital Twin: A Vision for Integrating Artificial and Human Intelligence for Disaster Management. *Int. J. Inf. Manag.* **2021**, *56*, 102049. [CrossRef]
115. Shrestha, A.; Mahmood, A. Review of Deep Learning Algorithms and Architectures. *IEEE Access* **2019**, *7*, 53040–53065. [CrossRef]
116. Yiu, C.Y.; Ng, K.K.H.; Lee, C.H.; Chow, C.T.; Chan, T.C.; Li, K.C.; Wong, K.Y. A Digital Twin-Based Platform towards Intelligent Automation with Virtual Counterparts of Flight and Air Traffic Control Operations. *Appl. Sci.* **2021**, *11*, 10923. [CrossRef]
117. Singh, K.K.; Frazier, A.E. A Meta-Analysis and Review of Unmanned Aircraft System (UAS) Imagery for Terrestrial Applications. *Int. J. Remote Sens.* **2018**, *39*, 5078–5098. [CrossRef]
118. Dikmen, M.; Burns, C. Trust in Autonomous Vehicles. In Proceedings of the 2017 IEEE International Conference on Systems, Man, and Cybernetics (SMC), Banff, AB, Canada, 5–8 October 2017; pp. 1093–1098. [CrossRef]
119. Yang, W.; Zheng, Y.; Li, S. Application Status and Prospect of Digital Twin for On-Orbit Spacecraft. *IEEE Access* **2021**, *9*, 106489–106500. [CrossRef]

120. Yun, H.; Park, D. Simulation of Self-Driving System by Implementing Digital Twin with GTA5. In Proceedings of the IEEE International Conference on Electronics, Information and Communications, Jeju, Republic of Korea, 31 January 2021–3 February 2021; pp. 5–6. [CrossRef]
121. Wang, S.; Zhang, F.; Qin, T. Research on the Construction of Highway Traffic Digital Twin System Based on 3D GIS Technology. In Proceedings of the 7th International Conference on Computer-Aided Design, Manufacturing, Modeling and Simulation (CDMMS 2020), Busan, Republic of Korea, 14–15 November 2020; Volume 1802.
122. Trusted Publisher-Independent Citation Database—Web of Science Group. Available online: https://clarivate.com/webofsciencegroup/solutions/web-of-science/ (accessed on 18 October 2021).
123. van Eck, N.J.; Waltman, L. Software Survey: VOSviewer, a Computer Program for Bibliometric Mapping. *Scientometrics* **2010**, *84*, 523–538. [CrossRef]
124. Chen, C.; Song, M. Visualizing a Field of Research: A Methodology of Systematic Scientometric Reviews. *PLoS ONE* **2019**, *14*, e0223994. [CrossRef]
125. Application of Digital Twin in Industrial Manufacturing—FutureBridge. Available online: https://www.futurebridge.com/industry/perspectives-mobility/application-of-digital-twin-in-industrial-manufacturing/ (accessed on 18 April 2023).
126. Lim, K.Y.H.; Zheng, P.; Chen, C.H. A State-of-the-Art Survey of Digital Twin: Techniques, Engineering Product Lifecycle Management and Business Innovation Perspectives. *J. Intell. Manuf.* **2020**, *31*, 1313–1337. [CrossRef]
127. Hodell, D.A.; Lourens, L.; Stow, D.A.V.; Hernández-Molina, J.; Alvarez Zarikian, C.A. The "Shackleton Site" (IODP Site U1385) on the Iberian Margin. *Sci. Drill.* **2013**, *16*, 13–19. [CrossRef]
128. Dorador, J.; Rodríguez-Tovar, F.J.; Hernández-Molina, F.J.; Stow, D.A.V.; Alvarez-Zarikian, C.; Acton, G.; Bahr, A.; Balestra, B.; Ducassou, E.; Flood, R.; et al. Quantitative Estimation of Bioturbation Based on Digital Image Analysis. *Mar. Geol.* **2014**, *349*, 55–60. [CrossRef]
129. Schreier, H.; Orteu, J.-J.; Sutton, M.A. *Image Correlation for Shape, Motion and Deformation Measurements*, 1st ed.; Springer: New York, NY, USA, 2009; ISBN 9788527729833.
130. Rodríguez-tovar, F.J.; Dorador, J. Ichnofabric Characterization in Cores: A Method of Digital Image Treatment. *Ann. Soc. Geol. Pol.* **2015**, *85*, 465–471. [CrossRef]
131. Sanchez-Gonzalez, P.L.; Díaz-Gutiérrez, D.; Leo, T.J.; Núñez-Rivas, L.R. Toward Digitalization of Maritime Transport? *Sensors* **2019**, *19*, 926. [CrossRef] [PubMed]
132. Tao, F.; Zhang, H.; Liu, A.; Nee, A.Y.C. Digital Twin in Industry: State-of-the-Art. *IEEE Trans. Ind. Inform.* **2019**, *15*, 2405–2415. [CrossRef]
133. Quintana, V.; Rivest, L.; Pellerin, R.; Venne, F.; Kheddouci, F. Will Model-Based Definition Replace Engineering Drawings throughout the Product Lifecycle? A Global Perspective from Aerospace Industry. *Comput. Ind.* **2010**, *61*, 497–508. [CrossRef]
134. Tao, F.; Qi, Q. Make More Digital Twins. *Nat. Mag.* **2019**, *573*, 490–491. [CrossRef] [PubMed]

Journal of
Marine Science and Engineering

MDPI

Article

A Novel, Finite-Time, Active Fault-Tolerant Control Framework for Autonomous Surface Vehicle with Guaranteed Performance

Xuerao Wang [1], Yuncheng Ouyang [1], Xiao Wang [2] and Qingling Wang [2,*]

1 School of Artificial Intelligence, The Engineering Research Center of Autonomous Unmanned System Technology, Anhui University, Hefei 230601, China; xrwang@ahu.edu.cn (X.W.); ouyangyc@ahu.edu.cn (Y.O.)
2 School of Automation, Southeast University, Nanjing 210096, China; xwang2020seu@163.com
* Correspondence: qlwang@seu.edu.cn

Abstract: In this paper, a finite-time, active fault-tolerant control (AFTC) scheme is proposed for a class of autonomous surface vehicles (ASVs) with component faults. The designed AFTC framework is based on an integrated design of fault detection (FD), fault estimation (FE), and controller reconfiguration. First, a nominal controller based on the Barrier Lyapunov function is presented, which guarantees that the tracking error converges to the predefined performance constraints within a settling time. Then, a performance-based monitoring function with low complexity is designed to supervise the tracking behaviors and detect the fault. Different from existing results where the fault is bounded by a known scalar, the FE in this study is implemented by a finite-time estimator without requiring any *prioir* information of fault. Furthermore, under the proposed finite-time AFTC scheme, both the transient and steady-state performance of the ASV can be guaranteed regardless of the occurrence of faults. Finally, a simulation example on CyberShip II is given to confirm the effectiveness of the proposed AFTC method.

Keywords: fault-tolerant control; guaranteed performance; model uncertainties; autonomous surface vehicle; active fault-tolerant control

check for updates

Citation: Wang, X.; Ouyang, Y.; Wang, X.; Wang, Q. A Novel, Finite-Time, Active Fault-Tolerant Control Framework for Autonomous Surface Vehicle with Guaranteed Performance. *J. Mar. Sci. Eng.* **2024**, *12*, 347. https://doi.org/10.3390/jmse12020347

Academic Editors: Xianbo Xiang, Haitong Xu, Lúcia Moreira and Carlos Guedes Soares

Received: 17 January 2024
Revised: 13 February 2024
Accepted: 14 February 2024
Published: 17 February 2024

1. Introduction

In recent years, significant progress has been made in the field of marine autopilots, which has attracted a great deal of attention. An important area of research in this field is the control of autonomous surface vehicles (ASVs). The ability of ASVs to operate in remote and hazardous areas, coupled with their advanced sensing and control capabilities, make them valuable assets for various applications in the marine, research, and exploration industries. Numerous successful results have been developed for the control of ASVs, such as [1–7]. The authors of [1,2] presented a comprehensive literature review of the recent progress in ASVs' development, and highlighted more general challenges and future directions of ASVs towards more practical guidance, navigation, and control capabilities. Common issues encountered in ASV control include trajectory tracking [3–5], formation control [6], and cooperative target tracking control [7]. These positive results have led to widespread applications of ASVs in marine environments, encompassing complicated tasks such as ocean forecasting, surface inspection, and pipeline tracking. However, The presence of unpredictable factors such as rough waves, strong currents, and changing weather conditions can adversely affect the performance and integrity of the ASV system. specifically, the intricate and dynamic nature of the surface environment poses significant challenges to the reliable operation of various components within the ASV system, including sensors, actuators, and controllers. This complexity substantially increases the susceptibility of these components to potential malfunctions [8]. Furthermore, the repair of these components during operation is impractical [9]. This introduces significant safety risks for ASVs, making safety control a primary concern in fulfilling the vehicle's mission.

It becomes crucial to develop robust and resilient designs that can deal with these environmental risks and ensure the continued functionality of ASVs in demanding marine conditions. Confronting this challenge, fault-tolerant control (FTC) methods have been proposed to enhance the safety and reliability of ASVs, maintaining stable operation and eliminating the effects of system malfunctions [10–14].

Fault-tolerant control schemes are classified as passive fault-tolerant control (PFTC) and active fault-tolerant control (AFTC), depending upon the utilization of fault detection and diagnosis module and the implementation of redundancies [15]. In PFTC methods, a fixed controller was designed that maintains the stability and performance of the control system during both normal and faulty operating situations [16–18]. The fixed controller is pre-designed with system redundancies which can be invoked, such as switching to backup components or adjusting operational parameters to compensate for the fault. PFTC approaches can ensure that the control system remains stable and performs well even in the presence of faults, without requiring any significant changes to the control structure. In an AFTC system, the controller reacts to malfunctions in system components through the controller reconfiguration, guided by detection information generated by the fault detection (FD) module. Once a fault is detected, the AFTC scheme determines the most effective strategy for maintaining system stability and performance. For the controller reconfiguration, the AFTC system dynamically adjusts the control parameters, modifies the control laws, or redistributes control tasks among redundant components to eliminate the effects of the fault. In comparison, PFTC methods are typically simpler to implement and require less computational resources than AFTC techniques, making them a practical solution for the control system. On the other hand, AFTC methods are more complex and computationally demanding compared to PFTC methods but can offer greater flexibility and adaptability in responding to faults. By actively reconfiguring the control, AFTC techniques can effectively overcome faults and maintain system functionality, making them suitable for applications where immediate fault response and system optimization are critical.

As a result, AFTC schemes have attracted significant attention in research and engineering applications due to their flexibility and accuracy [19–21]. In [19], the authors proposed a distributed AFTC approach for satellite formation flying attitude control, where sensor errors can be diagnosed by nonlinear observers and static approximators. A novel AFTC scheme was proposed in [20] for uncertain fully actuated systems using the integrated integration structure with observer and controller to reveal the model characteristics, which include faults and uncertainty. In [21], an observer-based AFTC algorithm was designed for spacecraft with full state constraints, and the fault diagnosis was implemented by a linear matrix inequality (LMI)-based robust fault observer. Nevertheless, despite the advantages offered by AFTC methods, there are some issues with the aforementioned studies that require further investigation: (1) The utilization of an ideal data model in the FD makes it difficult to adapt and implement in real systems, and (2) the convergence time of the fault observer has not been considered to ensure the accurate and efficient estimation. Consequently, there is a pressing need to develop an AFTC scheme integrating implementable FD and precise fault estimation (FE) to guarantee the reliable tracking control of ASVs.

As a critical component of AFTC systems, FD has garnered significant attention in recent years, and researchers have published various meaningful results [22–24]. The integration of FD mechanisms plays a crucial role in enhancing the reliability and robustness of control systems, especially in the presence of component faults. By accurately identifying faults, the control systems can effectively adapt their control strategies to mitigate potential disruptions and ensure safe operation in dynamic environments. In [24], the authors introduced a robust FE strategy that relies on residual generation and evaluation modules. This approach enables the identification of fault occurrence, characteristics, and severity by analyzing input and state information. When the residual evaluation function surpasses the predefined threshold level, a fault is detected, triggering the generation of an alarm signal. It is worth mentioning that disturbance observation (DO) algorithms can provide many ideas and references for FD because of the similar uncertainty characteristics between disturbance and component fault. Until now, the control of surface and underwater vehicles has shown a wider range of achievements with DO as opposed to FD [25–27]. In [26], a fast estimation method was developed to assess the real-time evolution of wave disturbances acting on a vehicle and verified by incorporating the predicted loads within a Model Predictive Controller. An integrated deterministic sea wave predictor was proposed for underwater vehicles in [27], demonstrating high potential to effectively mitigate disturbances and facilitate accurate tracking performance even in the presence of high wave loading. These results offer valuable insights for the development of FD design. For example, an interval observer was constructed in [28] to detect and isolate the faults in multi-agent systems by generating the residual signals and implying the thresholds. In [29], the faults were detected by an adaptive interval observer, and isolated by a set of interval observers. However, ASVs operate in complex marine environments and are inevitably subject to high operational risks, failure types cannot be identified, and certain bounds of faults cannot be given. Although many scholars have devoted themselves to design thresholds and estimate the faults, the fault detection for ASV systems is still an open research problem.

From a practical perspective, the primary responsibility of ASVs is to maintain tracking performance, and fulfill their designated tasks accurately, reliably, and adaptively. To address this, several advanced control techniques have been employed in recent studies. These studies, referenced as [30–34], have explored different control strategies to enhance the tracking capabilities of ASVs in terms of accuracy, stability, and adaptability, enabling them to fulfill their tracking responsibilities effectively. Transient (convergence rate, overshoot, and undershoot) and steady-state performances are important performance metrics that should be considered for control systems. Considering these performance metrics is essential in evaluating the effectiveness of control strategies. For this purpose, a novel control method known as prescribed performance control, introduced in [35], has achieved plenty of positive results when applied to multiple control systems [36–38]. Prescribed performance control focuses on achieving specific performance objectives while ensuring robustness against uncertainties and disturbances. Due to this property, prescribed performance control algorithms have been designed for surface vessels in [39,40] to achieve the assigned trajectory mission. Building upon the concept of [35], a novel concept known as finite-time performance function (FTPF) was presented in [41], which achieves finite-time convergence while ensuring the transient and steady-state performances. An FTPT-based fuzzy adaptive controller was developed in [42] for the trajectory tracking problem of multiple input multiple output nonlinear systems to ensure the tracking error has the predefined performance in finite time. In [43], the FTPF was utilized to design an air–ground cooperative consensus control scheme by integrating with the fixed-time scheme, which can guarantee the predefined time and given formation performance simultaneously. However, maintaining and restoring the guaranteed performance becomes more notably challenging when faults occur in ASVs. Therefore, it is significant to develop an AFTC scheme for ASV that can both detect faults and maintain predefined performance, while ensuring safety and reliability in the whole operating process. However, when faults occur in ASVs, maintaining and restoring the guaranteed performance becomes notably more challenging. Therefore, it is crucial to develop an AFTC scheme for ASVs that can detect faults and maintain predefined performance while ensuring safety and reliability throughout the entire operating process. By integrating the FTPF with the AFTC scheme, it is possible to achieve both predetermined performance objectives and fault tolerance capabilities in ASVs. This integration allows for effective tracking and control of ASVs, even in the presence of faults or disturbances.

Motivated by the above discussion and observation, in this paper, we aim to develop an AFTC scheme for ASV with a predefined finite-time tracking performance guaranteed. By incorporating the FTPF and Barrier Lyapunov function, a nominal controller is proposed to maintain the performance under normal conditions, and a fault monitoring function is obtained to achieve fault detection in time. Once the fault is detected, a reconfigured controller with a finite-time estimator is employed to ensure the predefined performance is guaranteed all the time. The main characteristics and contributions are summarized as follows:

(1) The paper makes the first attempt to develop an integrated FD, FE, and FTC framework for ASV. Through the utilization of transformed performance constraints, a monitoring function with low complexity is formulated to supervise system behavior and facilitate fault detection. This approach eliminates the need for intricate threshold calculations as seen in existing works such as [44–46].
(2) The concept of FTPF is first introduced to solve the fault-tolerant problem of ASVs. A nominal controller and a reconfigured controller are proposed by integrating the FTPF and Barrier Lyapunov functions. Using the proposed controllers, the tracking errors are guaranteed within a specified performance metric in a settling time.
(3) To enable efficient controller reconfiguration, a finite-time estimator is designed to accurately estimate uncertainties and faults. In comparison to previous works such as [20,21,44], the proposed estimator does not require a priori knowledge of the upper bound of the fault.

The remaining part of the paper is organized as follows. In Section 2, the system modeling and essential knowledge are introduced. The nominal controller and the reconfigured controller design process are given in Section 3. In Section 4, the simulation result is presented to illustrate the effectiveness of the designed controllers. The conclusion is clarified in Section 5.

Throughout this paper, the following notations are adopted. $\mathbb{R}$ is the set of all real numbers, and $\mathbb{R}^n$ represents the Euclidean space with dimension n. For a vector $x \in \mathbb{R}^n$, $x_i (i = 1, 2, \ldots, n)$ means the corresponding ith component of x, $\lambda_{\max}(x)$ and $\lambda_{\min}(x)$ mean the minimum and maximum eigenvalues, respectively. $|\cdot|$ denotes the absolute value of a scalar, $\|\cdot\|$ denotes the Euclidean norm of a vector, $\mathrm{diag}(\cdot)$ is a diagonal matrix. $I_{n\times n}$ denotes an identity matrix of dimension n.

2. Problem Formulation and Preliminaries

2.1. Problem Statement

The standard three degrees of freedom (DOF) model of the ASV under two right-hand coordinate systems is considered, as illustrated in Figure 1. According to the trajectory tracking mission of ASV, the nonlinear motion equation of the vehicle in the horizontal planes can be described as

$$\begin{aligned} \dot{\eta} &= R(\eta)\nu, \\ M\dot{\nu} + C(\nu)\nu + D(\nu)\nu + d(t) &= \tau + \tau_d, \end{aligned} \tag{1}$$

where $\eta = [x, y, \psi]^T \in \mathbb{R}^3$ describes the position and yaw angle of the vehicle represented in inertial coordinates, and $\nu = [u, w, r]^T \in \mathbb{R}^3$ is the surge, sway, and yaw velocities represented in body-fixed coordinates. The rotation matrix between two coordinates is expressed by

$$R(\eta) = \begin{bmatrix} \cos(\psi) & -\sin(\psi) & 0 \\ \sin(\psi) & \cos(\psi) & 0 \\ 0 & 0 & 1 \end{bmatrix}. \tag{2}$$

For simplicity, $R(\eta)$ is denoted as R in the following. It can be found that the determinant of (2) is positive, so R is invertible. The matrix $M = M^T \in \mathbb{R}^{3\times 3}$ denotes the inertial matrix, $C(\nu) \in \mathbb{R}^{3\times 3}$ describes the Coriolis and centripetal matrix, $D(\nu) \in \mathbb{R}^{3\times 3}$ represents the nonlinear damping matrix, $d(t) \in \mathbb{R}^3$ denotes the unmodeled dynamics, and $\tau_d \in \mathbb{R}^3$ is the unknown disturbance from wind, wave, and marine currents. The control forces and torque are given by $\tau \in \mathbb{R}^3$.

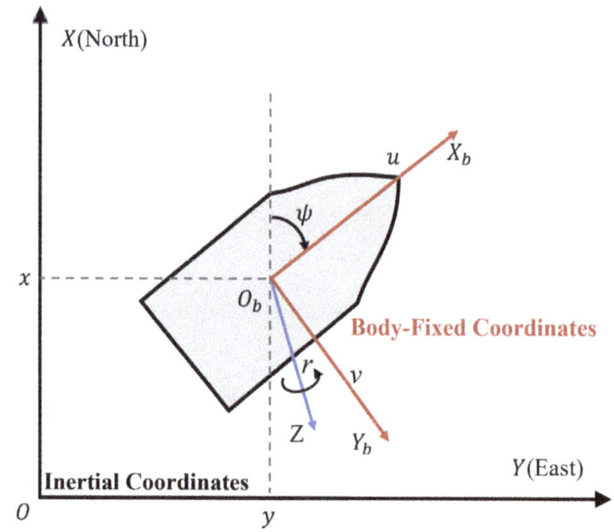

Figure 1. ASV model with two right-hand coordinate systems.

Using the conversion relation between η and ν, we have

$$\begin{aligned} \dot{\eta} &= R\nu \Leftrightarrow \nu = R^{-1}\dot{\eta} \\ \ddot{\eta} &= \dot{R}\nu + R\dot{\nu} \Leftrightarrow \dot{\nu} = R^{-1}(\ddot{\eta} - \dot{R}R^{-1}\dot{\eta}). \end{aligned} \tag{3}$$

Based on (3), the ASV model (1) under the fault-free condition is rewritten by

$$\ddot{\eta} = \dot{R}R^{-1}\dot{\eta} - RM^{-1}(C(\eta,\dot{\eta}) + D(\eta,\dot{\eta}))R^{-1}\dot{\eta} + RM^{-1}\tau + RM^{-1}(\tau_d - d). \tag{4}$$

The challenging operating conditions of ASVs increase the possibility of malfunctions in sensors, actuators, and controllers. In this paper, fault represents a state where a system or component does not meet its intended function or performance requirements. Specifically, a component fault refers to a failure or malfunction of an individual component within a control system, such as a sensor, actuator, controller, or any other hardware or software element involved in the control process. According to [20], the bias component faults can be modeled as $f_a \in \mathbb{R}^3$, satisfying $\sup_{t\in[0,\infty]} \|f_a\| < \infty$ and $\sup_{t\in[0,\infty]} \|\dot{f}_a\| < \infty$. In practice, it is challenging to determine the upper bounds of component faults due to the complex failure modes of ASVs. According to [47], the possible transition from the fault-free case to the fault case is unidirectional. Furthermore, we also assume that the fault occurred once during operation. Then, the faulty ASV model with the general component fault is considered as

$$\begin{aligned} &\dot{\eta} = R(\eta)\nu, \\ &M\dot{\nu} + C(\nu)\nu + D(\nu)\nu + f_a + d = \tau + \tau_d, \end{aligned} \tag{5}$$

The control objective of this paper is to develop an integrated finite-time AFTC framework for ASV so that the fault can be detected and estimated precisely, and the predefined tracking performance is ensured under both fault-free and faulty cases.

Assumption 1. *The desired trajectories η_d along with their time derivatives $\dot{\eta}_d$, $\ddot{\eta}_d$ are smooth and bounded.*

Assumption 2. *The external disturbance τ_d is bounded, i.e., there is a positive constant $\overline{\tau}_d$, such that $\|\tau_d\| \leq \overline{\tau}_d$.*

Assumption 3. *Under normal operation of ASV, the unmodeled dynamics $d(t)$ is bounded by $\|d(t)\| \le \bar{d}$ with $\bar{d}$ being a conservative constant.*

Lemma 1 ([41]). *Given a function $\varphi \ge 0$ and*

$$\frac{d\varphi(t)}{dt} = -\iota[\varphi(t)]^{\kappa} \tag{6}$$

holds, where $\iota > 0$ and $0 < \kappa < 1$ are the constants. Then (6) can be solved for

$$\varphi(t) = \begin{cases} \left((\varphi(0))^{1-\kappa} - (1-\kappa)\iota t\right)^{\frac{1}{1-\kappa}}, & t \in [0, T_0) \\ 0, & t \in [T_0, +\infty) \end{cases} \tag{7}$$

where $T_0 = \frac{(\varphi(0))^{1-\kappa}}{\iota(1-\kappa)}$.

2.2. Finite-Time Performance Function

In this subsection, we introduce the definition of FTPF, which aims at achieving two goals: first, it serves as a criteria for establishing a fault detection mechanism to identify component faults. Secondly, it ensures that tracking errors converge to the small specified residual sets within a settling time interval, even if fault occurs.

The definition of FTPF is as follows.

Definition 1 ([41]). *A function $\rho(t)$ is designated as the FTPF when it exhibits these properties:*

- $\rho(t) > 0$;
- $\dot{\rho}(t) \le 0$;
- $\lim_{t \to T_s} \rho(t) = \rho_{T_s} > 0$;
- $\rho(t) = \rho_{T_s}, \forall t \ge T_s$ *with T_s being the settling time.*

From Definition 1, it can be observed that $\rho(t)$ can converge to a specified set within the settling time T_s, indicating that $\rho(t)$ converges in finite time. According to Lemma 1, the FTPF employed in this paper is chosen as

$$\rho_i(t) = \begin{cases} (\rho_{i0}^{\varepsilon} - \iota\varepsilon t)^{\frac{1}{\varepsilon}} + \rho_{iT_s}, & t \in [0, T_s) \\ \rho_{iT_s}, & t \in [T_s, +\infty) \end{cases} \tag{8}$$

where $i = 1, 2, 3$, $\rho_{i0}, \rho_{iT_s}, \iota \in \mathbb{R}$ are positive constants to be chosen, $\varepsilon = \frac{\varepsilon_1}{\varepsilon_2} \in (0, 1]$ with ε_1, ε_2 are positive odd integers. Based on (8), the settling time T_s can be calculated by $T_s = \frac{\rho_0^{\varepsilon}}{\iota\varepsilon}$.

2.3. Error Transformation

In this subsection, the performance constraints for the tracking error is given first. Then, an error transformation is presented to transfer the time-varying constraints into an equivalent constant one to facilitate the design of AFTC scheme.

To achieve the control target, the ASV is requested to track the reference trajectories with guaranteed performance, which indicates that the tracking error $e = \eta - \eta_d$ should satisfy

$$\underline{\rho}_i(t) < e_i(t) < \bar{\rho}_i(t), \tag{9}$$

where $i = 1, 2, 3$, $\underline{\rho}_i$ and $\bar{\rho}_i$ represent the lower and upper constraints, respectively. Suppose $e_i(0)$ satisfies $\underline{\rho}_i < |e_i(0)| < \bar{\rho}_i$, depending on the sign of $e_i(0)$, the following should hold:

$$e_i(0) \ge 0 : \begin{cases} \underline{\rho}_i = -\sigma_i \rho_i(t) \\ \bar{\rho}_i = \rho_i(t) \end{cases}, \quad e_i(0) < 0 : \begin{cases} \underline{\rho}_i = -\rho_i(t) \\ \bar{\rho}_i = \sigma_i \rho_i(t) \end{cases}, \tag{10}$$

where $\rho_i(t)$ is the FTPF given in (8), σ_i is the design parameter.

Then, a sufficient necessary condition is deduced to ensure that the guaranteed performance described in (9) is achieved. The error transformation technology introduced and employed to convert the complex bounds (8) into more concise bounds as

$$z_i(t) = T_i(e_i(t), \underline{\rho}_i(t), \overline{\rho}_i(t)), i = 1,2,3. \tag{11}$$

where $z = [z_1, z_2, z_3]^T \in \mathbb{R}^3$ denotes the transformed error. According to [35], the given attributes should be present in the error transformation function $T_i(\cdot)$:

- $T_i(\cdot)$ is smooth and strictly increasing;
- $\lim_{e_i \to \overline{\rho}_i} T_i(e_i(t), \underline{\rho}_i(t), \overline{\rho}_i(t)) = 1$;
- $\lim_{e_i \to \underline{\rho}_i} T_i(e_i(t), \underline{\rho}_i(t), \overline{\rho}_i(t)) = -1$.

Following these considerations, the error transformation function is designed as

$$T_i(e_i(t), \underline{\rho}_i(t), \overline{\rho}_i(t)) = \frac{2e_i(t) - (\underline{\rho}_i(t) + \overline{\rho}_i(t))}{\overline{\rho}_i(t) - \underline{\rho}_i(t)}. \tag{12}$$

It follows from (12) that (9) is guaranteed if $|z_i(t)| < 1$. For simplicity, the independent variable t is omitted as the default time variable in the following. From (12), the original time-varying constraint is transformed into a constant one, which provides a simple solution for the design of the monitoring function and the AFTC scheme. Differentiating (11) yields

$$\dot{z}_i = \chi_i \dot{e}_i - \sigma_i(e_i, \overline{\rho}_i, \underline{\rho}_i), \tag{13}$$

where

$$\begin{aligned} \chi_i &= \frac{2}{\overline{\rho}_i - \underline{\rho}_i}, \\ \sigma_i(e_i, \overline{\rho}_i, \underline{\rho}_i) &= \frac{\dot{\overline{\rho}}_i + \dot{\underline{\rho}}_i}{\overline{\rho}_i - \underline{\rho}_i} + \frac{(2e_i - (\overline{\rho}_i + \underline{\rho}_i))(\dot{\overline{\rho}}_i - \dot{\underline{\rho}}_i)}{(\overline{\rho}_i - \underline{\rho}_i)^2} \end{aligned} \tag{14}$$

Thus, the transformed tracking error dynamics of (4) is given by

$$\begin{cases} z_i = T_i(e_i, \underline{\rho}_i, \overline{\rho}_i), \\ \dot{z}_i = \chi_i \dot{e}_i - \sigma_i(e_i, \overline{\rho}_i, \underline{\rho}_i), \\ \ddot{e} = \dot{R}R^{-1}\dot{\eta} - RM^{-1}(C(\eta, \dot{\eta}) + D(\eta, \dot{\eta}))R^{-1}\dot{\eta} + RM^{-1}\tau + RM^{-1}(\tau_d - d) - \ddot{\eta}_d. \end{cases} \tag{15}$$

Then, the sufficient necessary condition to guarantee performance bounds (9) can be derived.

Proposition 1. *Consider the ASV system (1) and its corresponding transformed tracking error dynamics (15). The performance bounds (9) can be guaranteed if and only if the transformed system (15) is stable, and the transformed error satisfies $|z_i(t)| < 1, i = 1,2,3$.*

Proof. If the performance bound (9) is guaranteed, then there exists an admissible continuous input τ, such that e_i is uniformly ultimately bounded (UUB). Employing (11)–(13), one has

$$\frac{2\underline{\rho}_i - (\underline{\rho}_i + \overline{\rho}_i)}{\overline{\rho}_i - \underline{\rho}_i} < \frac{2e_i - (\underline{\rho}_i + \overline{\rho}_i)}{\overline{\rho}_i - \underline{\rho}_i} < \frac{2\overline{\rho}_i - (\underline{\rho}_i + \overline{\rho}_i)}{\overline{\rho}_i - \underline{\rho}_i}, \tag{16}$$

resulting in

$$-1 < z_i < 1, i = 1,2,3. \tag{17}$$

Hence, the transformed error z_i is bounded. Thereby, the transformed system (15) is stable. Conversely, if the transformed error z_i satisfies $|z_i(t)| < 1$, one has

$$\underline{\rho}_i - \overline{\rho}_i < 2e_i - (\underline{\rho}_i + \overline{\rho}_i) < \overline{\rho}_i - \underline{\rho}_i. \tag{18}$$

It can be easily obtained that $\underline{\rho}_i < e_i(t) < \overline{\rho}_i$ holds. □

Remark 1. *Compared to the given error transformation function in [41], the transformation function in (11) has a simpler structure, which can potentially reduce computational complexity or implementation challenges. Moreover, by converting the performance constraints for tracking errors in (9) into a constant constraint, it becomes possible to establish a fixed threshold for fault detection and monitoring functions.*

3. Main Results

In this section, a nominal controller is presented first to ensure the tracking performance of ASV, and a performance-based monitoring function is given to monitor the control behavior and detect the fault. Upon detection of a fault, the reconfigured controller is constructed to maintain the system's stability.

3.1. Nominal Controller Design

From Proposition 1, it can be concluded that the predefined constraints can be ensured when the transformed tracking error z satisfies $|z_i| < 1$. The Barrier Lyapunov function proposed in [48] is utilized to construct the Lyapunov function as follows:

$$V_1 = \frac{1}{2}\sum_{i=1}^{3} \ln \frac{1}{1-z_i^2}. \tag{19}$$

Define the filtering error $s = \dot{\eta} - \alpha$. $\alpha \in \mathbb{R}^3$ is a virtual control signal to design. Taking the time derivative of V_1 yields

$$\dot{V}_1 = \sum_{i=1}^{3} \frac{z_i(\chi_i(s_i + \alpha_i - \dot{\eta}_{d,i}) - \sigma_i(e_i, \overline{\rho}_i, \underline{\rho}_i))}{1-z_i^2}. \tag{20}$$

Then, α can be designed as

$$\alpha = \dot{\eta}_d - k_1\overline{\chi}z + \overline{\chi}\sigma, \tag{21}$$

where $k_1 = \mathrm{diag}(k_{1,1}, k_{1,2}, k_{1,3})$ is the filtering gain matrix, $\overline{\chi} = \mathrm{diag}(1/\chi_1, 1/\chi_2, 1/\chi_3)$, and $\sigma = [\sigma_1(e_1, \overline{\rho}_1, \underline{\rho}_1), \sigma_2(e_2, \overline{\rho}_2, \underline{\rho}_2), \sigma_1(e_3, \overline{\rho}_3, \underline{\rho}_3)]^T$. Substituting (21) into (20) results in

$$\begin{aligned} \dot{V}_1 &= -\sum_{i=1}^{3} k_{1,i}\frac{z_i^2}{1-z_i^2} + \sum_{i=1}^{3}\frac{\chi_i z_i s_i}{1-z_i^2}, \\ &\leq -\lambda_{\min}(k_1)\sum_{i=1}^{3}\frac{z_i^2}{1-z_i^2} + \sum_{i=1}^{3}\frac{\chi_i z_i s_i}{1-z_i^2}. \end{aligned} \tag{22}$$

Define $\overline{M}(\eta) = MR^{-1}$, and the second Lyapunov function is considered as

$$V_2 = V_1 + \frac{1}{2}s^T\overline{M}(\eta)s. \tag{23}$$

Differentiating V_2 to time gives

$$\begin{aligned} \dot{V}_2 &= \dot{V}_1 + \frac{1}{2}s^T\dot{\overline{M}}(\eta)\dot{\eta}s + s^T\overline{M}(\eta)\dot{s}, \\ &= \dot{V}_1 + \frac{1}{2}s^T\dot{\overline{M}}(\eta)\dot{\eta}s + s^T g(\eta, \dot{\eta}) + s^T(\tau + \delta - \overline{M}(\eta)\dot{\alpha}), \end{aligned} \tag{24}$$

where $g(\eta, \dot{\eta}) = (\overline{M}(\eta)\dot{R} - C(\eta, \dot{\eta}) - D(\eta, \dot{\eta}))R^{-1}\dot{\eta}$, $\delta = \tau_d - d(t)$ denotes the lumped uncertainty. According to Assumptions 2 and 3, we have $\|\delta\| \leq \Delta$ for a bounded constant $0 < \Delta := \overline{\tau}_d + \overline{d}$. Recalling (4), the nominal controller is designed as

$$\tau_n = -k_2 s - g(\eta, \dot{\eta}) - \frac{1}{2}\dot{\overline{M}}(\eta)\dot{\eta}s + \overline{M}(\eta)\dot{\alpha} - \Sigma - \Phi, \tag{25}$$

where $k_2 = \mathrm{diag}(k_{2,1}, k_{2,2}, k_{2,3})$ is the control gain matrix, the auxiliary vector Σ is given as $\Sigma = [z_1\chi_1/(1-z_1^2), z_2\chi_2/(1-z_2^2), z_3\chi_3/(1-z_3^2)]^T$, and the uncertainty compensator Φ

is designed as $\Phi = (k_d + \Delta)\text{sgn}(s)$ with $k_d \in \mathbb{R} > 0$, $\text{sgn}(s) = [\text{sgn}(s_1), \text{sgn}(s_2), \text{sgn}(s_3)]^T$. Let $\tau = \tau_n$ and substituting (25) into (24) leads to

$$\begin{aligned} \dot{V}_2 \leq & -\lambda_{\min}(k_1)\sum_{i=1}^{3}\frac{z_i^2}{1-z_i^2} - k_2 s^T s - \sum_{i=1}^{3}(s_i\delta_i - (k_d+\Delta)\text{sgn}(s_i)s_i) \\ \leq & -\lambda_{\min}(k_1)\sum_{i=1}^{3}\frac{z_i^2}{1-z_i^2} - \lambda_{\min}(k_2)\|s\|^2 - \sum_{i=1}^{3}k_d|s_i|. \end{aligned} \tag{26}$$

The following theorem is proposed to point out the stability of the closed-loop system.

Theorem 1. *Consider the ASV system described in (1), and Assumptions 1–3 hold. Assuming the ASV system is fault-free on $[0, T_f)$, for $0 \leq t < T_f$, the proposed controller (25) is intended to ensure the following properties.*

(1) *The closed-loop control system is semi-globally stable, i.e., all signals are bounded. The tracking error converges to the origin within the predefined performance (9) at a settling time.*

(2) *The transformed tracking error provided by the error transformation (11) satisfies*

$$|z_i| < \gamma < 1, i = 1,2,3. \tag{27}$$

with

$$\begin{aligned} \gamma &= \sqrt{1 - e^{-\mu}}, \\ \mu &= \sum_{i=1}^{3}\ln\frac{1}{1-z_i^2(0)} + \lambda_{\max}(MR^{-1}(0))\|s(0)\|^2, \end{aligned} \tag{28}$$

where γ denotes a tighter bound for the guaranteed performance, μ is a constant depending on the initial state.

Proof. It can be concluded from (26) that if the control parameters are selected to satisfy $k_1, k_2, k_d > 0$, then $\dot{V}_2 \leq 0$. It can be further obtained that

$$\dot{V}_2 \leq -\lambda_{\min}(k_1)\sum_{i=1}^{3}\ln\frac{1}{1-z_i^2} - k_2\|s\|^2 - \sum_{i=1}^{3}k_d|s_i|. \tag{29}$$

Let $\zeta_n = \min\{2\lambda_{\min}(k_1), 2k_2/\lambda_{\max}(\overline{M}(\eta))\}$, we have

$$\dot{V}_2 \leq -\zeta_n V_2. \tag{30}$$

Integrating (30) from 0 to t yields

$$V_2(t) \leq V_2(0)e^{-\zeta_n t} \leq V_2(0). \tag{31}$$

It can be concluded from the above inequality that $\ln(1/1 - z_i^2)$ and s are bounded. Therefore, z_i remains in $z_i \in (-1, 1)$, and all signals in the closed-loop control system are bounded. Proposition 1 implies that tracking error e_i can converge within the predefined performance.

Furthermore, according to (23), it follows that

$$V_2(0) \leq \frac{1}{2}\sum_{i=1}^{3}\ln\frac{1}{1-z_i^2(0)} + \lambda_{\max}(MR^{-1}(0))\|s(0)\|^2. \tag{32}$$

Then, a tighter bound for z_i can be computed by (32), i.e., $z_i < \gamma = \sqrt{1 - e^{-\mu}}$, and it is clear that $\gamma < 1$ is valid. □

Remark 2. *To ensure that the tracking errors are kept within the predefined bound, a tighter monitoring bound is required. Through (11), the initial bound for e is transformed to a constant bound for z, so that the monitoring bound for z_i can also be set to a smaller constant. Compared with [44], the complex residual calculations are avoided. Different from the time-varying monitoring* *bound given in [45], we propose a more concise bound for subsequent fault detection. This simplified approach can make it easier to implement and analyze fault detection and monitoring strategies.*

3.2. Fault Detection and Reconfigured Controller Design

In Section 3.1, the nominal controller is presented for the ASV under fault-free condition, and the uncertainty is assumed to be bounded within a known region. However, it is crucial to consider the possibility of faults occurring at any time, denoted as T_f. Based on the designed nominal controller, monitoring functions are given to detect the component faults, and a reconfigured controller with a fault estimator are presented.

Theorem 1 indicates that when the fault occurs, the condition given in (27) is violated first before the predefined performance (9) is broken. As a result, by utilizing the tighter bounds presented in (27), we can derive the monitoring functions and identify the precise instant at which the fault is detected as

$$T_d := \inf\{t : |z_i| > \gamma, i = 1,2,3\}. \tag{33}$$

Once the fault is detected, fault estimation and compensation must be completed as quickly as possible to restore performance.

To guarantee the efficiency of fault estimation and performance restoration, a finite-time fault estimator is presented first. Define new state variables $\xi_1 = \eta$ and $\xi_2 = \dot{\eta}$, then the faulty ASV (4) can be described by the following equivalent dynamics:

$$\begin{cases} \dot{\xi}_1 = \xi_2, \\ \dot{\xi}_2 = -\kappa_1 \underline{M}(\xi_1)\xi_2 + f(\xi_1,\xi_2) + M(\xi_1)\tau + \underline{M}(\xi_1)\delta_f, \end{cases} \tag{34}$$

where $\kappa_1 \in \mathbb{R}$ is a positive constant to be chosen, $\underline{M}(\xi_1) = RM^{-1}$, $f(\xi_1,\xi_2) = \dot{R}R^{-1}\xi_2 - RM^{-1}((C(\xi_1,\xi_2) + D(\xi_1,\xi_2))R^{-1} - \kappa_1)\xi_2$, $\delta_f = f_a + \tau_d - d(t)$. It is noted that the upper bound of δ is unknown due to the component fault is unpredictable.

To obtain the estimation of δ_f, an auxiliary state variable $\xi_a \in \mathbb{R}^3$ is defined, and its dynamics is given as

$$\dot{\xi}_a = -\kappa_1 \underline{M}(\xi_1)\xi_a + f(\xi_1,\xi_2) + \underline{M}(\xi_1)\tau. \tag{35}$$

The difference between state variables of (34) and (35) is denoted by $\xi_e = \xi_2 - \xi_a$. Then, a modified two-order estimator is designed to precisely estimate the lumped uncertainty including fault as

$$\begin{cases} \hat{\delta}_f = \kappa_1 \hat{\xi}_e + MR^{-1}\dot{\hat{\xi}}_e, \\ \dot{\hat{\xi}}_e = -\kappa_2 \hat{\xi}_e + \dot{\xi}_e + \kappa_2 \xi_e + \kappa_3 \mathrm{sig}(\tilde{\xi}_e)^{\frac{r_1}{r_2}}, \end{cases} \tag{36}$$

where

$$\mathrm{sig}(\tilde{\xi}_e)^{\frac{r_1}{r_2}} = [\mathrm{sgn}(\tilde{\xi}_{e,1})|\tilde{\xi}_{e,1}|^{\frac{r_1}{r_2}}, \mathrm{sgn}(\tilde{\xi}_{e,2})|\tilde{\xi}_{e,2}|^{\frac{r_1}{r_2}}, \mathrm{sgn}(\tilde{\xi}_{e,3})|\tilde{\xi}_{e,3}|^{\frac{r_1}{r_2}}]^T, \tag{37}$$

$\kappa_2, \kappa_3 \in \mathbb{R}$ are positive constants to be chosen, $r_1, r_2 \in \mathbb{R}$ are positive odd integers and are selected to satisfy $r_1 < r_2$. Define the estimation errors of (36) as $\tilde{\delta}_f = \delta_f - \hat{\delta}_f$ and $\tilde{\xi}_e = \xi_e - \hat{\xi}_e$; the following Lemma is obtained.

Lemma 2. *Based on the modified two-order estimator designed in (36) for the ASV system (4) without component faults, and Assumptions 1–3 holding. Then, the estimation errors $\tilde{\delta}_f$ and $\tilde{\xi}_e$ can converge to zero in a finite time.*

Proof. According to the estimator given in Equation (36), the time derivative of $\tilde{\xi}_e$ can be calculated as

$$\begin{aligned} \dot{\tilde{\xi}}_e &= \dot{\xi}_e + \kappa_2 \hat{\xi}_e - \dot{\xi}_e - \kappa_2 \xi_e - \kappa_3 \mathrm{sig}(\tilde{\xi}_e)^{\frac{r_1}{r_2}} \\ &= -\kappa_2 \tilde{\xi}_e - \kappa_3 \mathrm{sig}(\tilde{\xi}_e)^{\frac{r_1}{r_2}}. \end{aligned} \tag{38}$$

It can be obtained from (34) and (35) that

$$\dot{\tilde{\xi}}_e = -\kappa_1 \underline{M}(\xi_1)\tilde{\xi}_e + \underline{M}(\xi_1)\delta_f, \tag{39}$$

and

$$\tilde{\delta}_f = \underline{M}^{-1}(\xi_1)\dot{\xi}_e + \kappa_1 \xi_e - \kappa_1 \hat{\xi}_e - MR^{-1}\dot{\xi}_e = \kappa_1 \tilde{\xi}_e. \tag{40}$$

Differentiating (40) with respect to time yields

$$\dot{\tilde{\delta}}_f = \kappa_1 \dot{\tilde{\xi}}_e = -\kappa_2 \tilde{\delta}_f - \kappa_3 \mathrm{sig}(\tilde{\delta}_f)^{\frac{r_1}{r_2}}. \tag{41}$$

Consider a Lyapunov function for the estimator given in Equation (36) as $V_d = \frac{1}{2}\tilde{\delta}_f^T \tilde{\delta}_f$, and the time derivative of V_d is given as

$$\begin{aligned} \dot{V}_d &= \tilde{\delta}_f^T (-\kappa_2 \tilde{\delta}_f - \kappa_3 \mathrm{sig}(\tilde{\delta}_f)^{\frac{r_1}{r_2}}) \\ &\le -\kappa_2 ||\tilde{\delta}_f||^2 - \kappa_3 ||\tilde{\delta}_f||^{\frac{r_1+r_2}{2r_2}} \\ &\le -\Gamma_1 V_d - \Gamma_2 V_d^r, \end{aligned} \tag{42}$$

where $\Gamma_1 = 2\kappa_2$, $r = \frac{r_1+r_2}{2r_2}$, $\Gamma_2 = 2^r\kappa_3$. It can be concluded from (40) and (42) that V_d is bounded, and the boundedness of $\tilde{\delta}_f$ and $\tilde{\xi}_e$ can be ensured. It follows from [49] that the estimation errors converge to zero in a finite time, and the convergence time can be obtained as

$$T_c \le \frac{1}{\Gamma_1(1-r)} \ln \frac{\Gamma_1 V^{1-r}(\tilde{\delta}_f(0)) + \Gamma_2}{\Gamma_2}. \tag{43}$$

□

Based on Lemma 2, the nominal controller can be reconfigured as

$$\tau_r = -k_2 s - g(\eta, \dot{\eta}) - \frac{1}{2}\dot{\overline{M}}(\eta)\dot{\eta} s + \overline{M}(\eta)\dot{\alpha} - \Sigma - \hat{\delta}_f, \tag{44}$$

Similar to the analysis in Theorem 1, let $\tau = \tau_r$ and substituting (44) into (24) leads to

$$\dot{V}_2 \le -\lambda_{\min}(k_1) \sum_{i=1}^{3} \frac{z_i^2}{1-z_i^2} - \lambda_{\min}(k_2) s^T s - s^T \tilde{\delta}_f. \tag{45}$$

Theorem 2. *Consider the ASV system described by (4) subject to component faults, and Assumptions 1–3 hold. Assume the component fault occurs at $t = T_f$ and is detected at $t = T_d$, for $t > T_d$, the proposed controller (44) with the estimator (36) can ensure the following:*

(1) *The closed-loop control system is semi-global stable, i.e., all signals are bounded.*
(2) *The transformed tracking error $z_i, i = 1,2,3$, is kept in in the compact set $(-1,1)$.*
(3) *The tracking error can converge to the origin within the predefined performance (9) at a settling time.*

Proof. Select the Lyapunov function as

$$V_n = V_2 + V_f. \tag{46}$$

Differentiating V_n and substituting (45) results in

$$\begin{aligned} \dot{V}_n \le &-\lambda_{\min}(k_1) \sum_{i=1}^{3} \frac{z_i^2}{1-z_i^2} - k_2 s^T s - s^T \tilde{\delta}_f + \dot{V}_f \\ \le &-\lambda_{\min}(k_1) \sum_{i=1}^{3} \frac{z_i^2}{1-z_i^2} - \frac{1}{2}\lambda_{\min}((2k_2 - I))||s||^2 - (\kappa_2 - \frac{1}{2})||\tilde{\delta}_f||^2 - \Gamma_2 ||\tilde{\delta}_f||^{2r}. \end{aligned} \tag{47}$$

If the control parameters are selected to satisfy $k_1 > 0$, $2k_2 - I > 0$, $\kappa_2 > \frac{1}{2}$, it can be obtained that $\dot{V}_n \le 0$. Similar to Theorem 1, we can obtain that z_i remains in $z_i \in (-1,1)$, and all signals in the closed-loop control system are bounded. Furthermore, Proposition 1 implies that tracking error e_i converges within the guaranteed performance. □

Remark 3. *MPC methods have been gradually applied for performance optimization in vehicle trajectory tracking scenarios [50–52]. In comparison to the MPC scheme, the advantages of the proposed FTPF + AFTC method are as follows: (1) Fault Recovery Time: the FTPF method imposes strict constraints on both reaction time and convergence range, which cannot be achieved by other methods currently available. (2) Fault Detection: the FTPF method not only ensures tracking performance within an ideal region but also provides a concise bound for designing the monitoring function. (3) Model Mismatch and Measurement Bias: in MPC control, performance can be sensitive to discrepancies between the prediction model and the actual system dynamics, potentially resulting in degraded control performance. If long-term stability is a critical requirement for ASV operations, the FTPF-based AFTC algorithm becomes a preferable choice.*

4. Simulation Study

To validate the feasibility of the designed AFTC scheme, a simulation example is carried out on CyberShip II model [53]. The parameters $M, C(\nu), D(\nu)$ in (1) are given as

$$M = \begin{bmatrix} m_{11} & 0 & 0 \\ 0 & m_{22} & m_{23} \\ 0 & m_{32} & m_{33} \end{bmatrix},$$
$$C(\nu) = \begin{bmatrix} 0 & 0 & c_{13}(\nu) \\ 0 & 0 & c_{23}(\nu) \\ -c_{13}(\nu) & -c_{23}(\nu) & 0 \end{bmatrix}, \tag{48}$$
$$D(\nu) = \begin{bmatrix} d_{11}(\nu) & 0 & 0 \\ 0 & d_{22}(\nu) & d_{23}(\nu) \\ 0 & d_{32}(\nu) & d_{33}(\nu) \end{bmatrix}.$$

where $m_{11} = m_0 - X_{\dot{u}}$, $m_{22} = m_0 - Y_{\dot{w}}$, $m_{23} = m_0 x_g - Y_{\dot{r}}$, $m_{32} = m_0 x_g - N_{\dot{w}}$, $m_{33} = I_z - N_{\dot{r}}$, $c_{13}(\nu) = -m_{11}w - m_{23}r$, $c_{23}(\nu) = m_{11}u$, $d_{11}(\nu) = -X_u - X_{|u|u}|u| - X_{uuu}u^2$, $d_{22}(\nu) = -Y_w - Y_{|w|w}|w|$, $d_{23}(\nu) = -Y_r - Y_{|w|r}|w| - Y_{|r|r}|r|$, $d_{32}(\nu) = -N_w - N_{|w|w}|w| - N_{|r|w}|r|$, and $d_{33}(\nu) = -N_r - N_{|w|r}|w| - N_{|r|r}|r|$. The system parameters are listed in Table 1.

Table 1. Main parameters for cybership II.

Factor	Value	Factor	Value	Factor	Value
m_0	23.8	Y_w	−0.8612	$X_{\dot{u}}$	−2
I_z	1.76	$Y_{\lvert w\rvert w}$	−36.2823	$Y_{\dot{w}}$	−10
x_g	0.046	Y_r	0.1079	$Y_{\dot{r}}$	0
X_u	−0.7225	N_w	0.1052	$N_{\dot{w}}$	0
$X_{\lvert u\rvert u}$	−1.3274	$N_{\lvert w\rvert w}$	5.0437	$N_{\dot{r}}$	−1
X_{uuu}	−5.8664				

In the simulation example, the control objective is to force the vehicle to track the desired trajectory as $\eta_d = [4\sin(0.02\pi t), -4\cos(0.02\pi t), 0.1t(1-e^{-t/5}]^T$. The initial states of ASV are set as $\eta(0) = [1,-2,1]^T$, $\nu(0) = [0,0,0]^T$. The parameters of performance function are chosen as $\rho_{i0} = 4.7, \rho_{iT_s} = 0.3, \iota = 1.2, \varepsilon = 0.3$, $i = 1,2,3$, and the settling time can be calculated by (6) as $T_s = 4.4190s$. For simplicity, the lumped uncertainty δ is described in a general term as time-varying forces/moment:

$$\delta = \begin{Bmatrix} 1.6 + 2\sin(0.01\pi t) \\ -0.9 + 1.5\sin(0.1\pi t - \pi/6) + 1.5\sin(0.01\pi t) \\ \sin(0.09\pi t + \pi/3) + \cos(0.01\pi t) \end{Bmatrix}. \tag{49}$$

To verify the fault-tolerant ability of the proposed AFTC scheme, the component faults are intentionally introduced into the ASV control system at a specific time

$t = T_f = 40$ s and $f_a = [6.2 + 5\sin(0.01\pi t - 10), 5.4 + 5\cos(0.01\pi t - 10), 7.2 + 0.3e^{-0.2t}]^T$. The control system's response to these faults can be analyzed and assessed to determine the effectiveness of the fault-tolerant control strategy. The control gains are selected as $k_1 = \text{diag}(23, 28, 2)$, $k_2 = \text{diag}(1.95, 1.7, 0.5)$, $k_d = 3.5$. According to the initial states, we can calculate that $\gamma = 0.6075$. The controller switches from nominal controller to reconfigured controller when $|z_i| > \gamma$ at $t = T_d$. The estimator gains are selected as $\kappa_1 = 0.001$, $\kappa_2 = 0.9$, $\kappa_3 = 15$, $r_1 = 99$, $r_2 = 101$.

4.1. Fault-Tolerant Ability Verification

To illustrate the effectiveness of the proposed nominal controller and the reconfigured controller, two experiments are conducted under fault-free and component fault conditions. In the first experiment, the fault-free condition is considered, where no component faults are present in the system. This experiment aims to demonstrate the performance of the control system under a nominal controller when there are no faults affecting the system's behavior. In the second experiment, the component fault condition is considered. Component faults are intentionally introduced into the system at a specific time, as mentioned earlier. This experiment aims to evaluate the performance of the control system in the presence of component faults. The control system switches from the nominal controller to the reconfigured controller when the fault detection threshold is exceeded. By comparing the results of these two experiments, the effectiveness of both the nominal controller and the reconfigured controller can be evaluated. The control system's ability to maintain stability, tracking performance, and fault tolerance can be assessed, providing insights into the overall performance of the proposed AFTC scheme.

The simulation results are shown in Figures 2–9. Among them, Figures 2–4 are the results of fault-free experiment and Figures 6–9 are the results of component fault experiment. It can be observed from Figures 2–4 that the ASV can follow the given trajectories within the designed nominal controller, and the predefined performance bounds are guaranteed. The trajectories of the ASV closely track the desired reference trajectories in Figure 2, indicating accurate tracking performance. The predefined performance bounds are observed to be satisfied in Figure 3. This ensures that the system operates within the desired performance criteria. The trajectories of transformed error are given in Figure 4, and it is obvious that the transformed errors are kept within the given sets. The control inputs are given in Figure 5.

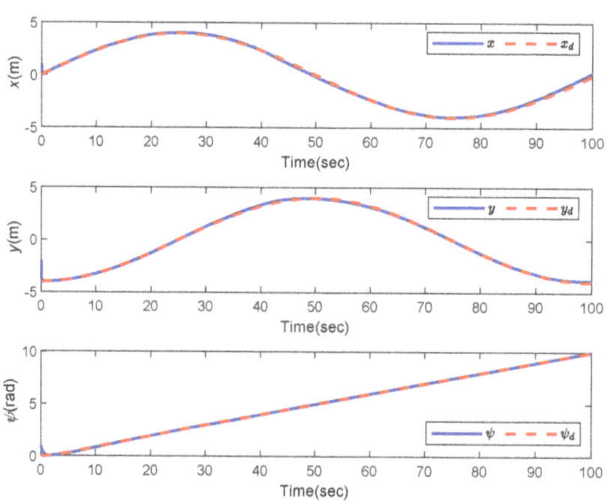

Figure 2. Fault−free experiments: trajectories under the nominal controller.

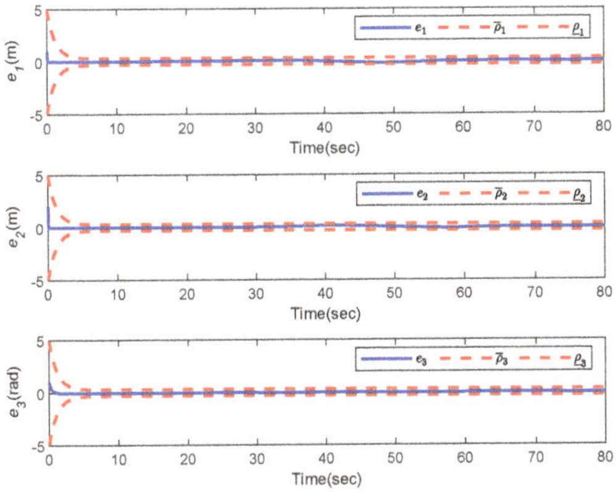

Figure 3. Fault-free experiments: tracking errors with performance bounds.

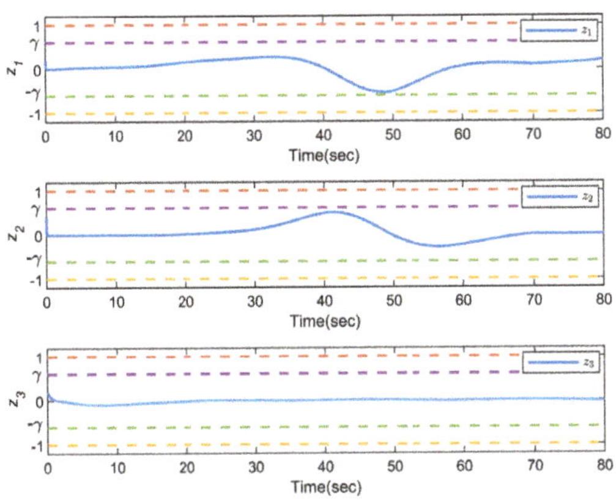

Figure 4. Fault-free experiments: transformed errors and monitoring functions.

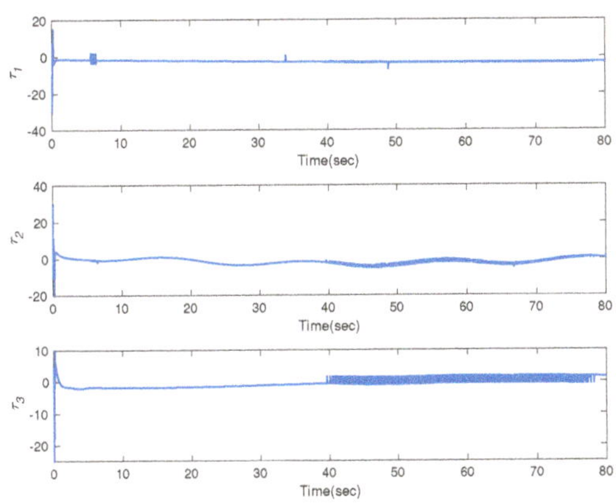

Figure 5. Fault-free experiments: control inputs.

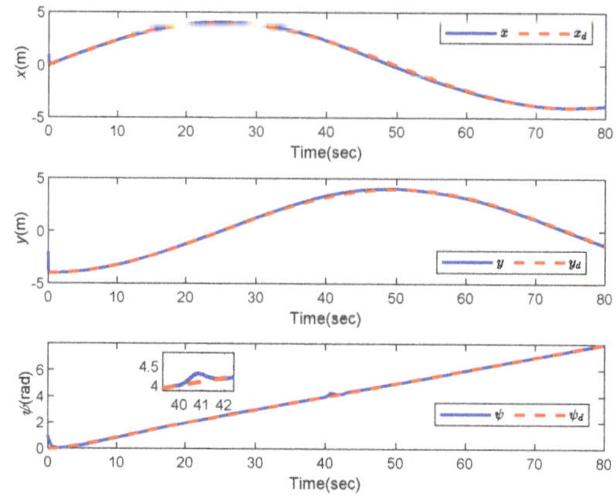

Figure 6. AFTC experiments: trajectories under the nominal and reconfigured controller.

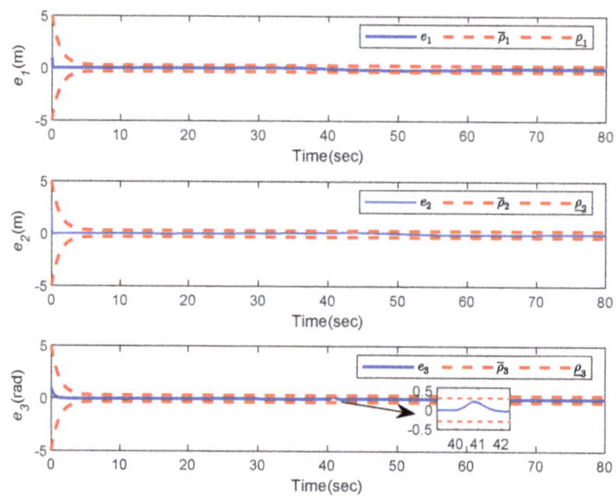

Figure 7. AFTC experiments: tracking errors with performance bounds.

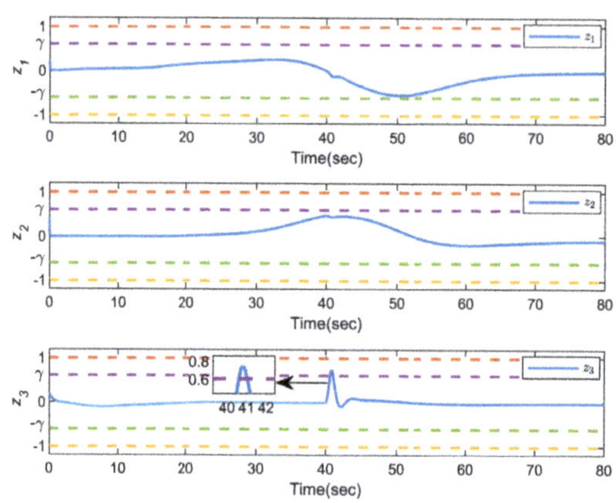

Figure 8. AFTC experiments: transformed errors and monitoring functions.

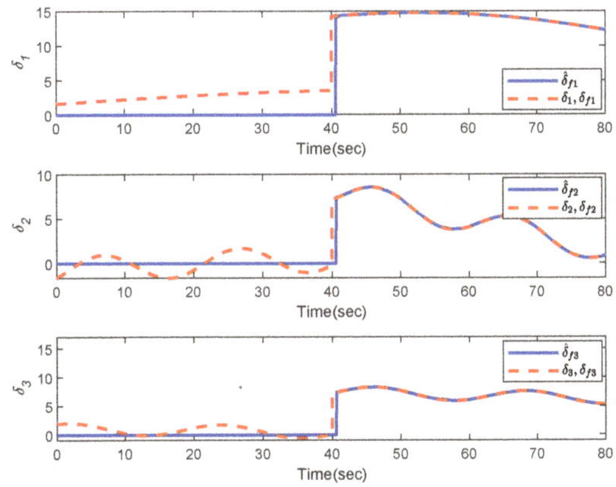

Figure 9. AFTC experiments: the lumped uncertainties and their estimations (after T_d).

In Figures 6–9, the detection results and the control performance under controller reconfiguration are depicted, which illustrates the system's behavior during the occurrence of component faults. Figure 6 depicts the curves of desired and actual trajectories of the ASV. The tracking errors and their constraints are illustrated in Figure 7, which reveals that the predefined performance is guaranteed at all times. Although there may be some variation in system performance when the fault occurs, it can be quickly recovered. Figure 8 focuses on the response of the transformed error. As seen from Figure 8, when the component fault occurs at $t = T_f = 40$ s, the transformed error z reacts to the fault faster than the tracking error e in Figure 7. This behavior indicates that the fault is detected and reflected in the transformed error before affecting the tracking performance. After the effect of component faults on ASV exceeds the detection threshold, the FE module provides an effective detection signal at $t = T_d = 40.8$ s. From Figure 9 we can see that the estimator is activated at $t = T_d$, indicating that the controller reconfiguration has been achieved, and the faults are well estimated. The control inputs are given in Figure 10. These figures demonstrate the system's response to component faults, the activation of the fault detection module, and the successful transition to the reconfigured controller.

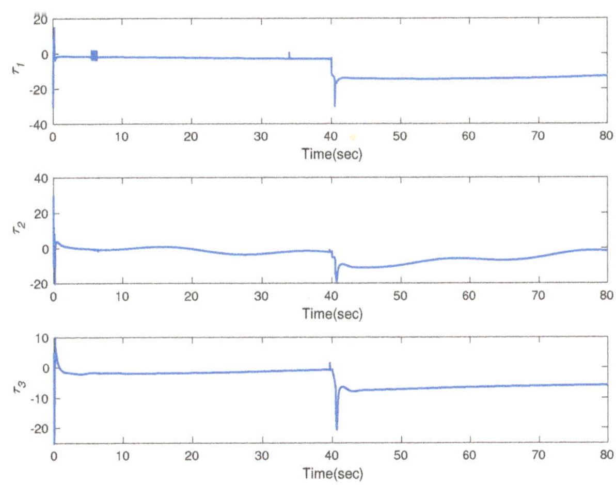

Figure 10. AFTC experiments: control inputs.

4.2. Robustness Verification

In this subsection, the lumped uncertainties are reset to three distinct frequencies in order to assess the robustness of the controller against various frequency perturbations. The frequencies used for testing are as follows:

$$(1) \qquad \delta = \begin{Bmatrix} 1.6 + 2\sin(0.3\pi t) + 0.5\cos(0.01\pi t) \\ -0.9 + 1.5\sin(0.3\pi t - \pi/6) + 1.5\sin(0.01\pi t) \\ \sin(0.3\pi t + \pi/3) + \cos(0.01\pi t) \end{Bmatrix}.$$

$$(2) \qquad \delta = \begin{Bmatrix} 1.6 + 2\sin(0.2\pi t) + 0.5\cos(0.01\pi t) \\ -0.9 + 1.5\sin(0.1\pi t - \pi/6) + 1.5\sin(0.01\pi t) \\ \sin(0.1\pi t + \pi/3) + \cos(0.01\pi t) \end{Bmatrix}.$$

$$(3) \qquad \delta = \begin{Bmatrix} 1.6 + 2\sin(0.01\pi t) \\ -0.9 + 1.5\sin(0.1\pi t - \pi/6) + 1.5\sin(0.01\pi t) \\ \sin(0.01\pi t + \pi/3) + \cos(0.01\pi t) \end{Bmatrix}.$$

The above formula is observed to provide a broader frequency range of disturbance. The simulation results are depicted in Figures 11 and 12. It can be inferred from Figure 11 that disturbances at different frequencies do not significantly impact the control performance, demonstrating the robustness of the method against interference. Figure 12 displays the estimation results of the estimator at various frequencies.

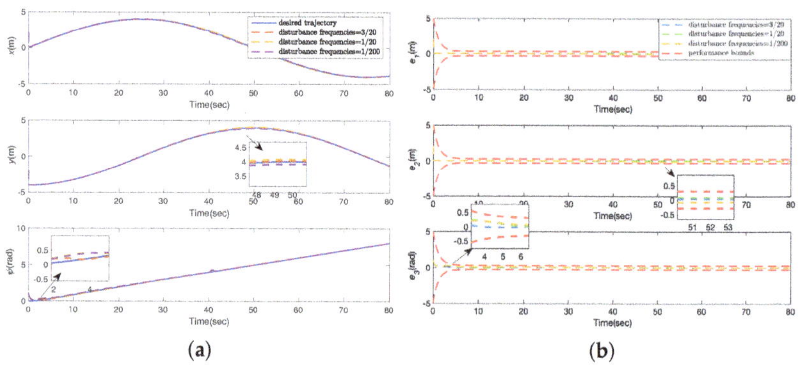

Figure 11. Robustness experiments: tracking performance under different frequency disturbance. (**a**) Trajectory tracking performance. (**b**) Tracking errors.

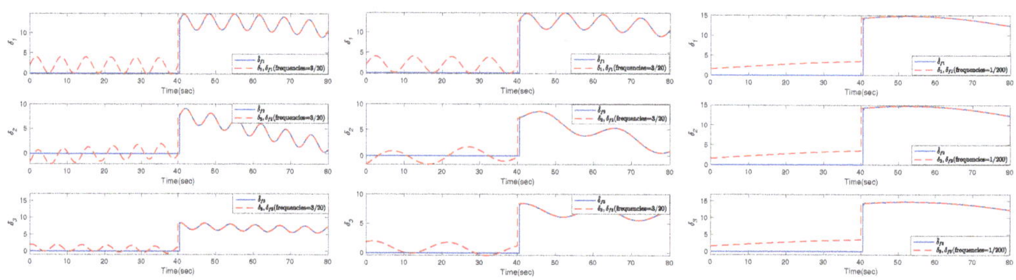

Figure 12. Robustness experiments: uncertainties estimation under different frequency disturbance. (**a**) Disturbance frequency = 3/20. (**b**) Disturbance frequency = 1/20. (**c**) Disturbance frequency = 1/200.

4.3. Advantages Highlight

In order to present a comprehensive assessment of the designed control strategy and demonstrate its superiority, additional numerical simulations were conducted in this subsection. Specifically, we performed a performance comparison between the designed control strategy and a classical backstepping control schemes used in [54]. For the backstepping method, the control input can be provide as follows:

$$\tau_b = MR^{-1}(-k_{b1}s - RM^{-1}g(\eta, \dot{\eta}) - \dot{\alpha} - \Phi) \tag{50}$$

where s is defined as the proposed controller, and $\alpha = -k_{b2}e + \dot{\eta}_d$. The remain equations are the same as the proposed controller. It is important to note that the proposed fault detection algorithm, being based on the FTPF, cannot be included in the comparison method. The simulation results of both controllers under fault conditions are presented in Figures 13 and 14. The simulation results for both controllers under fault conditions are depicted in Figures 13 and 14. As illustrated in Figure 13, the proposed controller exhibits a noticeably faster convergence rate compared to the backstepping controller. Furthermore, when component faults occur, the proposed ATFC scheme demonstrates superior fault tolerance, ensuring system stability. In Figure 14, the evolution of the tracking errors is presented, while the backstepping controller can partially guarantee the stability of the ASV, it is evident that without the FTPF, the tracking error cannot be maintained within a predefined range. Overall, these simulation results showcase the capability of the proposed controller to achieve satisfactory tracking performance while ensuring that all outputs remain within their specified ranges.

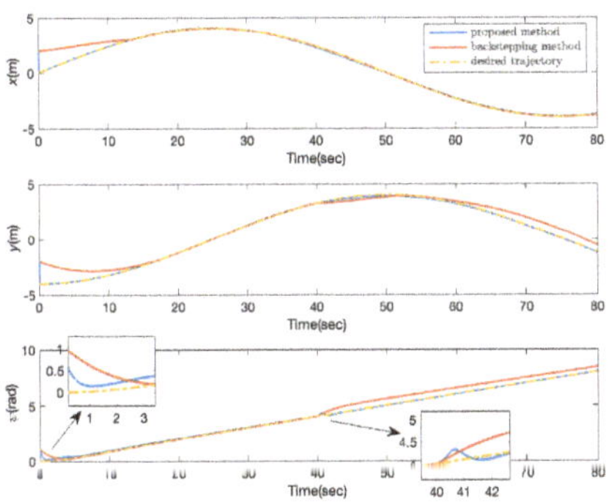

Figure 13. Comparison experiments: trajectories under the proposed controller and backstepping controller.

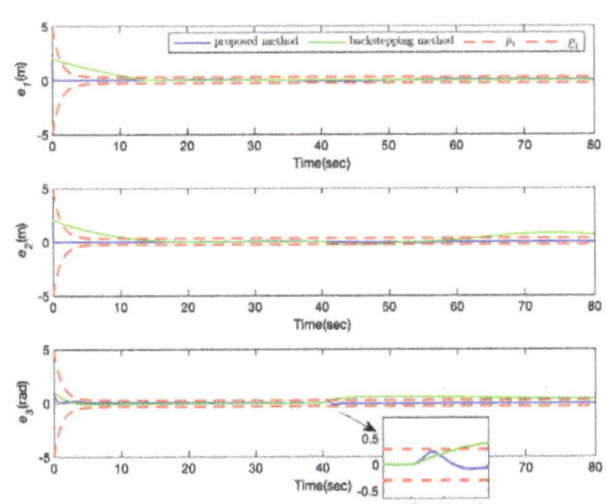

Figure 14. Comparison experiments: tracking errors under the proposed controller and backstepping controller.

5. Conclusions

In this paper, a novel AFTC scheme has been investigated to solve the predefined tracking performance problem of ASV with component faults. The integrated AFTC framework has been proposed to accomplish fault detection, fault estimation, and control reconfiguration autonomously. By introducing the error transformation and the Barrier Lyapunov function, a nominal controller was proposed to maintain the control performance under the normal fault-free condition. Within the guaranteed performance, a monitoring function has been designed to supervise the tracking behaviors and report the occurrence of faults. With the signal of the monitoring function, the reconfigured controller was activated, and the lumped uncertainty including fault was estimated precisely by a modified finite-time estimator. Finally, the effectiveness of the proposed AFTC controller has been verified by three aspects: fault-tolerant ability, robustness, and highlighted advantages. The simulation results demonstrate the system's fast response to faults, the activation of the fault detection module, and the successful transition to the reconfigured controller. The comparative simulations further show the superiority of the proposed method.

Author Contributions: Conceptualization, X.W. (Xuerao Wang) and Q.W.; Funding acquisition, X.W. (Xuerao Wang) and Y.O.; Methodology, Y.O. and X.W. (Xiao Wang); Project administration, X.W. (Xuerao Wang) and Q.W.; Software, X.W. (Xiao Wang); Supervision, Y.O. and Q.W.; Validation, Y.O. and X.W. (Xiao Wang); Writing—original draft, X.W. (Xuerao Wang); Writing—review and editing, X.W. (Xiao Wang). All authors have read and agreed to the published version of the manuscript.

Funding: This research was funded in part by the National Natural Science Foundation of China under Grants 62303012, 62303008.

Institutional Review Board Statement: Not applicable.

Informed Consent Statement: Not applicable.

Data Availability Statement: Data are contained within the article.

Conflicts of Interest: The authors declare no conflicts of interest.

Abbreviations

The following abbreviations are used in this manuscript:

ASV	autonomous surface vehicle
FTC	fault-tolerant control
AFTC	active fault-tolerant control
PFTC	passive fault-tolerant control
FD	fault detection
FE	fault estimation
LMI	linear matrix inequality
FTPF	finite-time performance function

References

1. Liu, Z.; Zhang, Y.; Yu, X.; Yuan, C. Unmanned surface vehicles: An overview of developments and challenges. *Annu. Rev. Control* **2016**, *41*, 71–93. [CrossRef]
2. Shi, Y.; Shen, C.; Fang, H.; Li, H. Advanced Control in Marine Mechatronic Systems: A Survey. *IEEE/ASME Trans. Mechatron.* **2017**, *22*, 1121–1131. [CrossRef]
3. Ren, Y.; Zhang, L.; Ying, Y.; Li, S.; Tang, Y. Model-Parameter-Free Prescribed Time Trajectory Tracking Control for Under-Actuated Unmanned Surface Vehicles with Saturation Constraints and External Disturbances. *J. Mar. Sci. Eng.* **2023**, *11*, 1717. [CrossRef]
4. Li, Z.; Xu, W.; Yu, J.; Wang, C.; Cui, G. Finite-Time Adaptive Heading Tracking Control for Surface Vehicles with Full State Constraints. *IEEE Trans. Circuits Syst. II Express Briefs* **2022**, *69*, 1134–1138. [CrossRef]
5. Zheng, Z.; Huang, Y.; Xie, L.; Zhu, B. Adaptive Trajectory Tracking Control of a Fully Actuated Surface Vessel with Asymmetrically Constrained Input and Output. *IEEE Trans. Control Syst. Technol.* **2018**, *26*, 1851–1859. [CrossRef]
6. Yan, X.; Jiang, D.; Miao, R.; Li, Y. Formation Control and Obstacle Avoidance Algorithm of a Multi-USV System Based on Virtual Structure and Artificial Potential Field. *J. Mar. Sci. Eng.* **2021**, *9*, 161. [CrossRef]

7. Gao, S.; Peng, Z.; Liu, L.; Wang, D.; Han, Q.L. Fixed-Time Resilient Edge-Triggered Estimation and Control of Surface Vehicles for Cooperative Target Tracking under Attacks. *IEEE Trans. Intell. Veh.* **2023**, *8*, 547–556. [CrossRef]
8. Zhou, Z.; Li, M.; Hao, Y. A Novel Region-Construction Method for Multi-USV Cooperative Target Allocation in Air-Ocean Integrated Environments. *J. Mar. Sci. Eng.* **2023**, *11*, 1369. [CrossRef]
9. Gu, N.; Wang, D.; Peng, Z.; Wang, J.; Han, Q.L. Disturbance observers and extended state observers for marine vehicles: A survey. *Control Eng. Pract.* **2022**, *123*, 105158. [CrossRef]
10. Zhang, G.; Chu, S.; Huang, J.; Zhang, W. Robust adaptive fault-tolerant control for unmanned surface vehicle via the multiplied event-triggered mechanism. *Ocean Eng.* **2022**, *249*, 110755. [CrossRef]
11. Liu, Z.; Ge, X.; Han, Q.; Wang, Y.; Zhang, X. Secure Cooperative Path Following of Autonomous Surface Vehicles under Cyber and Physical Attacks. *IEEE Trans. Intell. Veh.* **2023**, *8*, 3680–3691. [CrossRef]
12. Wu, W.; Tong, S. Fixed-time formation fault tolerant control for unmanned surface vehicle systems with intermittent actuator faults. *Ocean Eng.* **2023**, *281*, 114813. [CrossRef]
13. Liu, C.; Zhao, X.; Wang, X.; Ren, X. Adaptive fault identification and reconfigurable fault-tolerant control for unmanned surface vehicle with actuator magnitude and rate faults. *Int. J. Robust Nonlinear Control* **2023**, *33*, 5463–5483. [CrossRef]
14. Liu, L.; Wang, D.; Peng, Z. State recovery and disturbance estimation of unmanned surface vehicles based on nonlinear extended state observers. *Ocean Eng.* **2019**, *171*, 625–632. [CrossRef]
15. Jiang, J.; Yu, X. Fault-tolerant control systems: A comparative study between active and passive approaches. *Annu. Rev. Control* **2012**, *36*, 60–72. [CrossRef]
16. Wang, X.; Wang, Q.; Sun, C. Prescribed Performance Fault-Tolerant Control for Uncertain Nonlinear MIMO System Using Actor–Critic Learning Structure. *IEEE Trans. Neural Netw. Learn. Syst.* **2022**, *33*, 4479–4490. [CrossRef]
17. Shao, X.; Hu, Q.; Shi, Y.; Jiang, B. Fault-Tolerant Prescribed Performance Attitude Tracking Control for Spacecraft under Input Saturation. *IEEE Trans. Control Syst. Technol.* **2020**, *28*, 574–582. [CrossRef]
18. Yang, W.; Yu, W.; Zheng, W.X. Fault-Tolerant Adaptive Fuzzy Tracking Control for Nonaffine Fractional-Order Full-State-Constrained MISO Systems with Actuator Failures. *IEEE Trans. Cybern.* **2022**, *52*, 8439–8452. [CrossRef]
19. Saeed Nasrolahi, S.; Abdollahi, F.; Rezaee, H. Decentralized active sensor fault tolerance in attitude control of satellite formation flying. *Int. J. Robust Nonlinear Control* **2020**, *30*, 8340–8361. [CrossRef]
20. Cai, M.; He, X.; Zhou, D. An active fault tolerance framework for uncertain nonlinear high-order fully-actuated systems. *Automatica* **2023**, *152*, 110969. [CrossRef]
21. Hu, H.; Wang, B.; Cheng, Z.; Liu, L.; Wang, Y.; Luo, X. A novel active fault-tolerant control for spacecrafts with full state constraints and input saturation. *Aerosp. Sci. Technol.* **2021**, *108*, 106368. [CrossRef]
22. Hwang, I.; Kim, S.; Kim, Y.; Seah, C.E. A Survey of Fault Detection, Isolation, and Reconfiguration Methods. *IEEE Trans. Control Syst. Technol.* **2010**, *18*, 636–653. [CrossRef]
23. Qiu, A.; Al-Dabbagh, A.W.; Chen, T. A Tradeoff Approach for Optimal Event-Triggered Fault Detection. *IEEE Trans. Ind. Electron.* **2019**, *66*, 2111–2121. [CrossRef]
24. Shahriari-kahkeshi, M.; Sheikholeslam, F.; Askari, J. Adaptive fault detection and estimation scheme for a class of uncertain nonlinear systems. *Nonlinear Dyn.* **2015**, *79*, 2623–2637. [CrossRef]
25. Wang, N.; Lv, S.; Er, M.J.; Chen, W.H. Fast and Accurate Trajectory Tracking Control of an Autonomous Surface Vehicle with Unmodeled Dynamics and Disturbances. *IEEE Trans. Intell. Veh.* **2016**, *1*, 230–243. [CrossRef]
26. Walker, K.L.; Gabl, R.; Aracri, S.; Cao, Y.; Stokes, A.A.; Kiprakis, A.; Giorgio-Serchi, F. Experimental Validation of Wave Induced Disturbances for Predictive Station Keeping of a Remotely Operated Vehicle. *IEEE Robot. Autom. Lett.* **2021**, *6*, 5421–5428. [CrossRef]
27. Walker, K.L.; Giorgio-Serchi, F. Disturbance Preview for Non-Linear Model Predictive Trajectory Tracking of Underwater Vehicles in Wave Dominated Environments. In Proceedings of the 2023 IEEE/RSJ International Conference on Intelligent Robots and Systems (IROS), Detroit, MI, USA, 1–5 October 2023; pp. 6169–6176. [CrossRef]
28. Zhang, Z.H.; Yang, G.H. Distributed Fault Detection and Isolation for Multiagent Systems: An Interval Observer Approach. *IEEE Trans. Syst. Man Cybern. Syst.* **2020**, *50*, 2220–2230. [CrossRef]
29. Wang, X. Active Fault Tolerant Control for Unmanned Underwater Vehicle with Sensor Faults. *IEEE Trans. Instrum. Meas.* **2020**, *69*, 9485–9495. [CrossRef]
30. Rout, R.; Cui, R.; Han, Z. Modified Line-of-Sight Guidance Law with Adaptive Neural Network Control of Underactuated Marine Vehicles with State and Input Constraints. *IEEE Trans. Control Syst. Technol.* **2020**, *28*, 1902–1914. [CrossRef]
31. Wang, T.; Liu, Y.; Zhang, X. Extended state observer-based fixed-time trajectory tracking control of autonomous surface vessels with uncertainties and output constraints. *ISA Trans.* **2022**, *128*, 174–183. [CrossRef]
32. Zhang, J.; Yu, S.; Yan, Y. Fixed-time extended state observer-based trajectory tracking and point stabilization control for marine surface vessels with uncertainties and disturbances. *Ocean Eng.* **2019**, *186*, 106109. [CrossRef]
33. Fu, H.; Yao, W.; Cajo, R.; Zhao, S. Trajectory Tracking Predictive Control for Unmanned Surface Vehicles with Improved Nonlinear Disturbance Observer. *J. Mar. Sci. Eng.* **2023**, *11*, 1874. [CrossRef]
34. Liu, W.; Ye, H.; Yang, X. Model-Free Adaptive Sliding Mode Control Method for Unmanned Surface Vehicle Course Control. *J. Mar. Sci. Eng.* **2023**, *11*, 1904. [CrossRef]

35. Bechlioulis, C.P.; Rovithakis, G.A. Adaptive control with guaranteed transient and steady state tracking error bounds for strict feedback systems. *Automatica* **2009**, *45*, 532–538. [CrossRef]
36. Liu, Y.J.; Zeng, Q.; Tong, S.; Chen, C.P.; Liu, L. Actuator failure compensation-based adaptive control of active suspension systems with prescribed performance. *IEEE Trans. Ind. Electron.* **2019**, *67*, 7044–7053. [CrossRef]
37. Theodorakopoulos, A.; Rovithakis, G.A. Low-Complexity Prescribed Performance Control of Uncertain MIMO Feedback Linearizable Systems. *IEEE Trans. Autom. Control* **2016**, *61*, 1946–1952. [CrossRef]
38. Bikas, L.N.; Rovithakis, G.A. Combining Prescribed Tracking Performance and Controller Simplicity for a Class of Uncertain MIMO Nonlinear Systems with Input Quantization. *IEEE Trans. Autom. Control* **2019**, *64*, 1228–1235. [CrossRef]
39. Wang, H.; Li, M.; Zhang, C.; Shao, X. Event-Based Prescribed Performance Control for Dynamic Positioning Vessels. *IEEE Trans. Circuits Syst. II Express Briefs* **2021**, *68*, 2548–2552. [CrossRef]
40. Zhang, J.; Chai, T. Singularity-Free Continuous Adaptive Control of Uncertain Underactuated Surface Vessels with Prescribed Performance. *IEEE Trans. Syst. Man Cybern. Syst.* **2022**, *52*, 5646–5655. [CrossRef]
41. Liu, Y.; Liu, X.; Jing, Y.; Zhang, Z. A Novel Finite-Time Adaptive Fuzzy Tracking Control Scheme for Nonstrict Feedback Systems. *IEEE Trans. Fuzzy Syst.* **2019**, *27*, 646–658. [CrossRef]
42. Sui, S.; Tong, S. Finite-Time Fuzzy Adaptive PPC for Nonstrict-Feedback Nonlinear MIMO Systems. *IEEE Trans. Cybern.* **2023**, *53*, 732–742. [CrossRef]
43. Cheng, W.; Zhang, K.; Jiang, B. Fixed-Time Fault-Tolerant Formation Control for a Cooperative Heterogeneous Multiagent System with Prescribed Performance. *IEEE Trans. Syst. Man Cybern. Syst.* **2023**, *53*, 462–474. [CrossRef]
44. Wang, X. Active Fault Tolerant Control for Unmanned Underwater Vehicle with Actuator Fault and Guaranteed Transient Performance. *IEEE Trans. Intell. Veh.* **2021**, *6*, 470–479. [CrossRef]
45. Ouyang, H.; Lin, Y. Adaptive Fault-Tolerant Control and Performance Recovery against Actuator Failures with Deferred Actuator Replacement. *IEEE Trans. Autom. Control* **2021**, *66*, 3810–3817. [CrossRef]
46. Zhang, J.; Yang, G. Supervisory switching-based prescribed performance control of unknown nonlinear systems against actuator failures. *Int. J. Robust Nonlinear Control* **2020**, *30*, 2367–2385. [CrossRef]
47. Wang, W.; Wen, C. Adaptive actuator failure compensation control of uncertain nonlinear systems with guaranteed transient performance. *Automatica* **2010**, *46*, 2082–2091. [CrossRef]
48. Tee, K.P.; Ge, S.S.; Tay, E.H. Barrier Lyapunov functions for the control of output-constrained nonlinear systems. *Automatica* **2009**, *45*, 918–927. [CrossRef]
49. Yu, S.; Yu, X.; Shirinzadeh, B.; Man, Z. Continuous finite-time control for robotic manipulators with terminal sliding mode. *Automatica* **2005**, *41*, 1957–1964. [CrossRef]
50. Wei, H.; Shi, Y. Mpc-based motion planning and control enables smarter and safer autonomous marine vehicles: Perspectives and a tutorial survey. *IEEE/CAA J. Autom. Sin.* **2023**, *10*, 8–24. [CrossRef]
51. Guerreiro, B.J.; Silvestre, C.; Cunha, R.; Pascoal, A. Trajectory Tracking Nonlinear Model Predictive Control for Autonomous Surface Craft. *IEEE Trans. Control Syst. Technol.* **2014**, *22*, 2160–2175. [CrossRef]
52. Cui, Y.; Peng, L.; Li, H. Filtered Probabilistic Model Predictive Control-Based Reinforcement Learning for Unmanned Surface Vehicles. *IEEE Trans. Ind. Inform.* **2022**, *18*, 6950–6961. [CrossRef]
53. Skjetne, R.; Fossen, T.I.; Kokotović, P.V. Adaptive maneuvering, with experiments, for a model ship in a marine control laboratory. *Automatica* **2005**, *41*, 289–298. [CrossRef]
54. Ouyang, H.; Lin, Y. Adaptive fault-tolerant control for actuator failures: A switching strategy. *Automatica* **2017**, *81*, 87–95. [CrossRef]

Journal of
Marine Science and Engineering

MDPI

Article

Collision-Free Formation-Containment Tracking of Multi-USV Systems with Constrained Velocity and Driving Force

Jingchen Wang [1], Qihe Shan [1,*], Tieshan Li [2], Geyang Xiao [3] and Qi Xu [3]

1 School of Navigation, Dalian Maritime University, Dalian 116026, China; wjc1632021@dlmu.edu.cn
2 School of Automation Engineering, University of Electronic Science and Technology of China, Chengdu 611731, China; tieshanli@126.com
3 Research Institute of Intelligent Networks, Zhejiang Lab, Hangzhou 311121, China; xgyalan@outlook.com (G.X.); xuqi@zhejianglab.com (Q.X.)
* Correspondence: shanqihe@dlmu.edu.cn

Abstract: This paper studied the collision avoidance issue in the formation-containment tracking control of multi-USVs (unmanned surface vehicles) with constrained velocity and driving force. Specifically, based on a dual-layer control framework, it designed a multi-USV formation-containment tracking control strategy that accounts for constrained motion velocity and input driving force and validated the stability of this strategy using the Lyapunov method. Then, by utilizing zeroing control barrier function certificates, it considered collision avoidance among USVs with various roles as well as between each USV and static obstacles. A collision-free multi-USV formation-containment tracking control strategy considering constrained motion velocity and driving force was thus established, and its effectiveness was validated through the proposed simulation.

Keywords: multi-USV systems; formation-containment tracking control; quadratic programming; collision avoidance; control barrier function; dual-layer scheme

check for updates

Citation: Wang, J.; Shan, Q.; Li, T.; Xiao, G.; Xu, Q. Collision-Free Formation-Containment Tracking of Multi-USV Systems with Constrained Velocity and Driving Force. *J. Mar. Sci. Eng.* **2024**, *12*, 304. https://doi.org/10.3390/jmse12020304

Academic Editors: Carlos Guedes Soares, Xianbo Xiang, Lúcia Moreira and Haitong Xu

Received: 13 January 2024
Revised: 5 February 2024
Accepted: 7 February 2024
Published: 9 February 2024

1. Introduction

In recent years, with the continuous development of unmanned surface vehicle (USV) swarm coordination technology, the formation technology of multiple USVs has been increasingly applied in fields such as marine data collection [1], collaborative search and rescue [2,3], cooperative escorting [4], and collaborative transportation [5]. During the execution of various formation tasks by multiple USV clusters, collision avoidance among USVs, as well as avoidance of obstacles such as reefs, buoys, and ice floes on the sea surface, is a fundamental requirement [6–10]. Currently, substantial research has focused on collision avoidance within multi-USV formations [8,11–14]. Little attention has been paid to collision avoidance strategies for a specialized type of USV operation, known as formation-containment tracking. Consequently, devising effective strategies for these formation-containment tracking scenarios is a critical issue in ongoing multi-USV studies.

To achieve collision avoidance in a multi-USV formation, a variety of formation strategies could be considered. The leader–follower formation method based on consensus, due to its reliability and practicality, has been widely applied. In recent years, this approach has seen a wealth of developments [15–24]. Ren et al. [15] specifically devised a consensus-based formation control algorithm for second-order multi-vehicle systems, which means vehicle dynamics can be simplified to second-order integrator dynamics. Taking into account the practical engineering constraints on each agent's input driving force and motion velocity, Fu et al. [18] designed a leader–follower formation strategy with limited velocity and control inputs. Huang et al. [19] introduced a fixed-time USV leader–follower formation method representing a faster and more practically viable control strategy. Tang et al. [20] proposed a flexible serial formation protocol, based on the estimation of narrow waterways' curvature using an observer, to enable a USV fleet to navigate through narrow and winding waterways. Although the single leader–follower formation method is effective in some cases, it often falls short in accommodating large-scale USV formations or managing complex tasks. To enable systems to incorporate more USVs and undertake more complex tasks, a hierarchical concept emerged: formation containment. This approach enhances the conventional leader–follower structure by introducing three distinct roles: the highest-ranking virtual leader, the mid-ranking real leader, and the lowest-ranking follower. In this hierarchy, information flows unidirectionally from higher to lower ranks, with followers being specifically designed to converge within the convex polygonal regions formed by real leaders, thus enabling the handling of more complex operational tasks [25]. Hua et al. [26] developed a control protocol enabling linear multi-agent systems to achieve formation-containment tracking despite the leader's input being unknown. Wang et al. [27] introduced an innovative USV formation-containment strategy that employs a robust integral observer to estimate disturbances stemming from natural interferences and model uncertainties, complemented by an adaptive law specifically designed to offset actuator malfunctions. Hao et al. [28] has adopted a pioneering adaptive parameter fine-tuning strategy, which, even in the face of unknown global data and external disturbances within dynamically changing communication structures, can still precisely coordinate and maintain the stability of large-scale unmanned surface vessel (USV) formations. In actual marine settings, USVs are subject to various types of couplings, including dynamics and communication, complicating the control challenge. Liu et al. [29] introduced a sophisticated two-layer control framework, where the upper layer orchestrates formation containment using a fully actuated third-order integrator model, subsequently transmitting the generated trajectory to the lower layer in real time, and the lower layer leverages sliding mode control to enable under-actuated USVs to track the trajectory promptly, achieving effective formation containment. The hierarchical structure of formation containment offers new possibilities for addressing large-scale and complex tasks, while it also imposes novel requirements on collision avoidance strategies.

Collision avoidance is an indispensable aspect of any collaborative task involving USV swarms [30–32]. The artificial potential field method, a notable strategy for collision avoidance, effectively synergizes with the consensus formation of multi-vehicle systems [8,11–13,32–35]. Aranda-Bricaire et al. [11] developed an approach using repulsive vector fields (RVFs) grounded in the repulsive potential function (RPF) to facilitate collision-free formations in second-order multi-agent systems with input force and velocity constraints. Park et al. [13] advanced a multi-USV formation strategy that simultaneously addresses connectivity maintenance and collision avoidance, utilizing an innovative additional potential function to avert collisions. Ghommam et al. [12] proposed a practical approach for collision-free distributed formation control of under-actuated USVs, leveraging the repulsive potential function technique to enhance practical engineering applicability. Nevertheless, in collision avoidance tasks, distance plays a crucial role, especially the triggering distance for collision avoidance and the minimum safety distance. In these aspects, the control barrier function (CBF) method surpasses the artificial potential field (APF) approach, proving to be more aligned with practical collision avoidance applications. Firstly, regarding the criteria for triggering collision avoidance mechanisms, the APF method relies on a fixed triggering distance, initiating maneuvers based solely on proximity to obstacles. In contrast, the CBF method introduces a hazard coefficient as the trigger condition, closely tied to the relative distance and velocity between the USV and the obstacle. This meticulous consideration of dynamic factors ensures a more responsive and adaptable collision avoidance strategy. Secondly, there is a significant difference in how the two methods handle the minimum safety distance, typically set as the sum of the radius of two potentially colliding bodies. The APF method does not impose actual constraints on the obstacle's radius, potentially leading to scenarios where the USV intrudes into the obstacle boundary without proper parameter adjustment, posing a high risk in practical operations. On the other hand, the CBF method, by leveraging forward safety sets to explicitly define the obstacle radius, effectively eliminates the risk of USV intrusion into obstacle areas, thereby enhancing the system's safety and practicality. This makes the control barrier function (CBF) approach significantly more effective and friendly in collision avoidance in formation-containment tasks. Several notable accomplishments have been achieved using the control barrier function method for collision avoidance [36–39]. Gao et al. [36] implemented multi-target tracking for USVs by integrating the CBF with an extended state observer. Gong et al. [37] developed a technique employing a guiding vector field to steer the desired heading angle for reorganizing multi-USV formations and target tracking, concurrently utilizing a fixed-time CBF approach to evade both static and dynamic obstacles. Notably, Fu et al. [38] proposed a collision-free formation tracking method for second-order multi-agent systems that simultaneously considers connectivity maintenance and constraints on control inputs and velocity.

However, to our knowledge, the adoption of the CBF method for achieving collision-free multi-USV formation-containment tracking remains relatively unexplored. While the pioneering two-layer control framework in [16] presents a groundbreaking solution for such tasks, we believe there is still potential for refinement, especially within the upper distributed coordination layer. Here, the challenge of efficiently managing a fully actuated second-order point mass under velocity and input constraints for collision-free formation-containment tracking offers substantial scope for innovation. This paper concentrates on these areas, advancing the field with the following additional work and novel contributions:

1. **Expansion of the leader–follower controller**: Based on the practical second-order leader–follower formation tracking controller introduced in [18], this study expands the leader–follower controller to a formation-containment tracking controller that considers constraints on movement velocity and input driving force. Serving as the nominal controller in the distributed cooperative layer, it coordinates multi-USV collision-free formation-containment tracking tasks. Furthermore, the controller's asymptotic stability is demonstrated using the Lyapunov method. This hierarchical structure allows for the execution of more complex and flexible tasks, offering higher adaptability for complex maritime operations.
2. **Implementing collision avoidance with zeroing control barrier functions**: In this paper, zeroing control barrier functions are utilized within the distributed cooperative layer. When the collision risk coefficient of any USV falls below zero, the system triggers a quadratic programming solution that subtly alters the existing nominal controller, thereby efficiently and safely facilitating collision avoidance. This paper takes into account collision avoidance among vehicles with different roles in USV formation-containment tracking tasks, as well as between the USV fleet and static maritime obstacles. This enhances the adaptability and flexibility of USV fleets in avoiding collisions.

The remainder of this paper is organized as follows: Some lemmas, preliminaries, and the problem formulation are given in Section 2. Section 3 is split into two parts. In one part, the formation-containment tracking nominal controller is presented, whose velocity and input force are constrained. And the obstacle avoidance and collision avoidance strategies are given in the other part. In addition, a simulation example is given is Section 4. Finally, Section 5 makes some collations and conclusions.

2. Preliminaries and Problem Formulation

In this section, the relevant theories used in this paper are presented in four sub-modules: Section 2.1 introduces the two-tier control framework. Section 2.2 discusses the fundamentals of graph theory. In Section 2.3, the system studied in this paper, including some assumptions and lemmas, is described. Section 2.4 details the application of control barrier functions.

2.1. Dual-Layer USV Collaborative Motion Control Framework

Due to the inherent complexity of directly applying the three-degree-of-freedom (3-DOF) motion model for coordinated control of unmanned surface vehicles (USVs), especially when multiple USVs are performing in surface coordinated movements, tight dynamic coupling issues are encountered. Furthermore, the single USV three-degree-of-freedom (3-DOF) motion model examined in this paper is under-actuated, meaning that the model has fewer control inputs than degrees of freedom (or state variables). This under-actuation presents challenges that must be addressed during the control process. To overcome these difficulties, inspired by [29], this paper adopts a dual-layer USV fleet cooperative motion control framework, as shown in Figure 1. The upper layer is the distributed cooperative layer; each vehicle in this layer is abstracted as a mass point with limited input driving force and motion speed, described by a second-order integrator motion model. Without considering the influence of wind, water flow, and the movement of other vehicles, the formation-containment task is completed by the distributed consistency control strategy, and the motion reference trajectory generated by each vehicle is transmitted to the lower layer. The lower layer is the local dynamic control layer, which employs the nonlinear sliding mode method [29] to control the movement of an individual three-degree-of-freedom (3-DOF) unmanned surface vehicle (USV) motion model, performing real-time tracking based on the real-time reference trajectory generated by the corresponding vehicle in the upper layer. To achieve this objective, the three-degree-of-freedom dynamic model is employed. This model is described by the following equations, which delineate the motion of any unmanned vehicle i within the cluster [28]:

$$\begin{cases} \dot{x}_i = u_i \cos\varphi_i - v_i \sin\varphi_i \\ \dot{y}_i = u_i \sin\varphi_i - v_i \cos\varphi_i \\ \dot{\varphi}_i = r \\ m_{1i}\dot{u}_i = m_{2i}v_i r_i - d_{1i}u_i + \tau_{ui} \\ m_{1i}\dot{v}_i = m_{1i}u_i r_i - d_{2i}v_i \\ m_{3i}\dot{r}_i = (m_{1i} - m_{2i})u_i r_i - d_{3i}r_i + \tau_{ri} \end{cases} \tag{1}$$

where (x_i, y_i) represents the position of vehicle i's center of gravity; $\varphi_i \in [0, 2\pi]$ represents the heading angle of the vehicle; u_i, v_i, and r_i represent vehicle i's forward speed, sway speed, and yaw rate, respectively; "m_{1i}, m_{2i}, and m_{3i}" represent the inertia parameters of vehicle i in three coordinate directions, which can be calculated using a semi-empirical method; "d_{1i}, d_{2i}, and d_{3i}" represent the components of the fluid damping experienced by vehicle i on the water surface in three coordinate directions; τ_{ui} and τ_{ri} represent the control input forces of vehicle i, which are thrust for forward motion and torque for turning, respectively.

Although the precise dynamic modeling of the lower control layer is crucial for overall system performance, this study primarily focuses on the upper layer, the distributed cooperative layer. Within this layer, the work presented here is dedicated to generating real-time reference trajectories for each unmanned surface vehicle (USV).

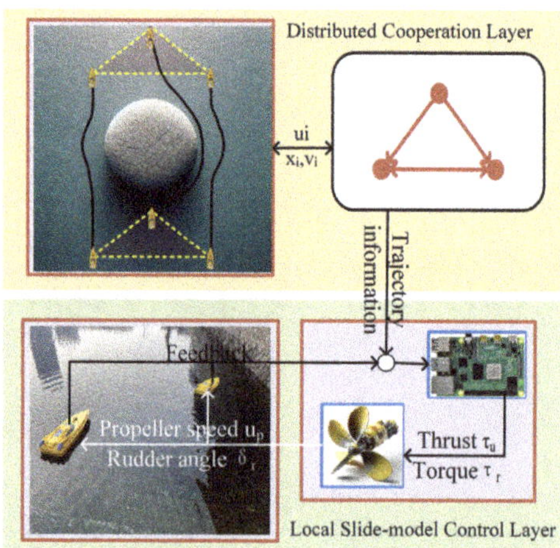

Figure 1. The structure of dual-layer USV collaborative motion control framework.

2.2. Graph Theory

The topology structure of M USVs (unmanned surface vehicles) is often expressed by a digraph $G = \{v_G, \varepsilon_G, a_G\}$, where $v_G = \{v_1, v_2, \ldots, v_M\}$ represents the set of nodes, $\varepsilon_G \subseteq \{(v_i, v_j) : v_i, v_j \in v; i \neq j\}$ is the set of edges, and $W = [a_{ij}] \in R^{M \times M}$ stands for the adjacency matrix. It is assumed that $e_{ij} = (v_i, v_j)$ is an edge of G, and there is a directed path from v_i to v_j if $a_{ij} = 1$. It is assumed that $a_{ij} = 1$ if and only if $e_{ij} \in \varepsilon_G$, and $a_{ij} = 0$ otherwise. The neighbors of i are represented by N_i, where $Ni = \{j \in v_G : (i,j) \in \varepsilon_G\}$. Let diagonal matrix $D = (d_{ij})_{M \times M}$, where $d_{ij} = \sum_{j \neq i} a_{ij}$. Laplacian matrix L is defined as $L = D - W$. If a root node has directional paths to all other nodes, the directional graph is said to contain a spanning tree. To illustrate the concepts mentioned above, the following Table 1 provides an example to aid understanding.

Table 1. Example topology and explanation of related concepts.

	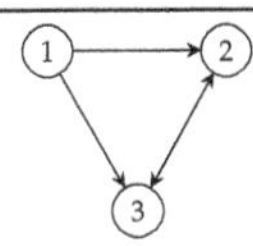
Vertex	$\{1,2,3\}$
Edge	$\{(1,2),(1,3),(2,3),(3,2)\}$
Neighbor	If two nodes are directly connected by an edge, they are considered neighbors. For example, the neighbors of node 1 are nodes 2 and 3, because there are edges (1, 2) and (1, 3) connecting them.
Adjacency matrix	$\begin{bmatrix} 0 & 1 & 1 \\ 0 & 0 & 1 \\ 0 & 1 & 0 \end{bmatrix}$
Degree(in) matrix	$\begin{bmatrix} 2 & 0 & 0 \\ 0 & 1 & 0 \\ 0 & 0 & 1 \end{bmatrix}$
Laplacian matrix	$\begin{bmatrix} 2 & -1 & -1 \\ 0 & 1 & -1 \\ 0 & -1 & 1 \end{bmatrix}$
Directed spanning tree	In a directed graph, a directed tree that originates from a root node and reaches all other nodes in the graph is known as a directed spanning tree. This means that starting from the root node, it is possible to reach every node in the graph via the edges of the tree. In the given graph, if node 1 is chosen as the root, then the structures formed by edges $\{(1,2),(1,3),(2,3)\}$ or $\{(1,2),(1,3),(3,2)\}$ can be considered directed spanning trees, as they allow all nodes to be reached from node 1.

2.3. Formation-Containment Tracking Task

The system composed by $M + N + 1$ USVs is considered. The USVs in the system can be divided into three categories, namely, the virtual leader or tracking leader, the real leaders or formation leaders, and the USV followers, as shown in Figure 2. The reference trajectory for the macroscopic motion of the entire multi-USV system is generated through the virtual leader, which is numbered 0. And the real leaders, which are numbered 1, ..., M, are required to track the trajectory while completing the specified formation shape. Note that the formation shape in this paper is not time-varying. The followers need to converge to the convex hull formed by the real leaders, which are numbered $M + 1$, ..., $N + M$. The information exchange among USVs is described by digraph G. The virtual leader has no neighbors, while a real leader can receive information from the virtual leader or the other real leaders, and a follower can receive information from the real leaders or the other followers.

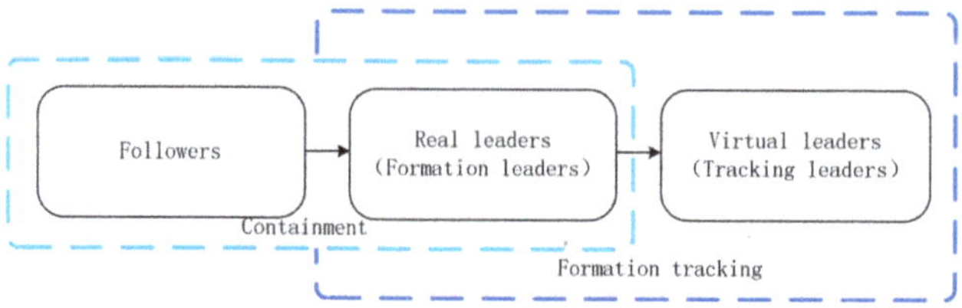

Figure 2. The structure of the multi-USV system.

The dynamics of the virtual leader in the distributed cooperative layer for the USVs can be expressed as follows, satisfying both the velocity and input constraints:

$$\begin{cases} \dot{x}_0 = v_0 \\ \dot{v}_0 = -k(v_0 - \bar{v}_0^r), \end{cases}$$

where $k = diag\{k_1, ..., k_n\}, k_i > 0, i = 1, 2, ..., n$; x_0 and v_0 are the position vector and the velocity vector of the vitual leader; $\bar{v}_0^r$ is the desired velocity signal of the vitual leader; and $v_{\text{lower}} \leq \bar{v}_0^r \leq v_{\text{upper}}$, where v_{lower} and v_{upper} are the minimum and the maximum values of the velocity of the virtual leader.

The dynamics of the ith USV can be described by

$$\begin{cases} \dot{x}_i = v_i \\ \dot{v}_i = u_i, \quad i = 1, \ldots, M+N \end{cases} \tag{2}$$

where $x_i(t) \in R^n$, $v_i(t) \in R^n$, and $u_i(t) \in R^n$ are the position vector, the velocity vector, and the control input vector for each agent, respectively. The formation of the real leaders can be described by $h = \left[h_1^T, h_2^T, \ldots, h_M^T\right]^T$, where $h_i = \left[h_{ix}^T, h_{iy}^T\right]^T \in R^{2n}, i = 1, ..., M$. h_i represents the relative position vector between formation leader i and the virtual leader.

Assumption 1. *For directed graph G of the communication among virtual leader and real leaders, there is a spanning tree, and the root node is the tracking leader.*

Assumption 2. *For each follower, there is at least one real leader who can reach it through a directed path.*

According to the topology characteristics of multi-USV systems, Laplacian matrix $L_G \in R^{(M+N+1)\times(M+N+1)}$ can be represented as

$$L_G = \begin{bmatrix} 0 & 0_{1\times M} & 0_{1\times N} \\ L_{01} & L_{11} & 0_{M\times N} \\ 0_{N\times 1} & L_{21} & L_{22} \end{bmatrix} \tag{3}$$

where $L_{01} \in R^{M\times 1}$, $L_{11} \in R^{M\times M}$, $L_{21} \in R^{N\times M}$, and $L_{22} \in R^{N\times N}$.

Let $L_{11} = L'_{11} + A_0$ and $A_0 = diag\{a_{10}, ..., a_{M0}\}$, where $a_{i0} > 0$ if the virtual leader directly interacts with real leader i, or $a_{i0} = 0$ otherwise, and L'_{11} represents the Laplacian only among the real leaders, which ignores the virtual leader and the communication links it emits.

Let $L_{21} = [-A_{L1F}, -A_{L2F}, \cdots, -A_{LMF}]$, where each A_{LiF} represents the communication connection relationship between the ith real leader and all the followers. Similarly, let $L_{22} = L'_{22} + \sum_{i=1}^{M}\{A_i\}$, where L'_{22} represents the Laplacian only among the followers, which ignores the real leaders and the communication links they emit, and each $A_i = diag\{A_{LiF}\}$.

To facilitate the proof in the next section, the following lemmas and definitions are given.

Lemma 1 ([40])**.** *If assumption 1 holds, then L_{11} is of full rank. And there is a diagonal matrix $Q = \mathrm{diag}\{q_1, q_2, ..., q_M\}$, such that matrixes Q and L_{11} satisfy $\varphi = QL_{11} + L_{11}^T Q > 0$, where $[q_1, q_2, \ldots, q_M]^T = (L_{11}^T)^{-1}\mathbf{1}_M$.*

Lemma 2 ([41]). *If assumption 2 holds, then L_{22} is of full rank. There is a diagonal matrix $P = \text{diag}\{p_1, p_2, ..., p_N\}$, such that matrixes P and L_{22} satisfy $\phi = PL_{22} + L_{22}^T P > 0$, where $[p_1, p_2, ..., p_N]^T = (L_{22}^T)^{-1}\mathbf{1}_N$.*

Lemma 3 ([42]). *If assumption 1 holds, then all eigenvalues of L_{11} have positive real parts.*

Lemma 4 ([43]). *If assumption 2 holds, then all eigenvalues of L_{22} have positive real parts, and the sum of each row of $-L_{22}^{-1}L_{21}$ equals 1. Additionally, every element of $-L_{22}^{-1}L_{21}$ is non-negative.*

Definition 1. *For the real leaders, if*

$$\lim_{t\to+\infty}(x_i - h_i - x_0) = 0, i = 1, \ldots, M \tag{4}$$

then the real leaders are considered to have completed the formation tracking task.

Definition 2. *For the followers, if there is a set of non-negative constants λ_{ij} satisfying $\sum_{j=1}^{M}\lambda_{ij} = 1$, and if the equation*

$$\lim_{t\to+\infty}\left(x_i - \sum_{j=1}^{M}\lambda_{ij}x_j\right) = 0, i = M+1, \ldots, M+N \tag{5}$$

holds, then the followers are considered to have completed the containment task.

Definition 3. *If conditions (1) and (2) hold, then system (3) is considered to have completed the formation-containment tracking task.*

Remark 1. *In addition to the formation-containment tracking target, obstacle avoidance and collision avoidance are also the key concerns of the system in the process of completing the tasks. To ensure the normal execution of system tasks, firstly, a nominal controller which can perform the task of formation-containment tracking while constraining the input as well as the velocity of each USV is designed; secondly, various possible collisions will be presented as constraints to keep the system safe.*

2.4. Collision Avoidance Using Control Barrier Function (CBF) Method

To ensure collision-free states in a dynamical system, we focus on systems of the form $\dot{x} = f(x) + g(x)u$, where $x \in R^n$ represents the system states and $u \in U \subset R^n$ denotes the control inputs. Safety set S is defined using a zeroing control barrier function (ZCBF), which guarantees that the system states always remain within this set. Lemma 1 outlines the approach to achieve this goal.

Lemma 5 ([39]). *For a given ZCBF candidate h that meets specific conditions, any Lipschitz continuous controller $u{:}d \to R$ ensuring $u \in S(x)$ will maintain system states $x(t)$ within safety set S and ensure asymptotic stability in d.*

Furthermore, to guarantee the forward invariance of safety set S, the following inequality must be satisfied:

$$L_f h(x) + L_g h(x)u + \alpha h^3(x) \geq 0 \tag{6}$$

where $\alpha > 0$ and $L_f h(x)$ and $L_g h(x)$ represent the Lie derivatives of $h(x)$.

3. Main Results

In this section, a formation-containment tracking strategy based on neighbor information exchange is presented in part A. This controller is then constrained to avoid obstacles when necessary based on a control obstacle function in part B.

3.1. Formation-Containment Tracking Strategy

Lemma 6. *We consider the following dynamic system:*

$$\dot{v}_i = -k(v_i + v_c) \tag{7}$$

where $\|v_c(t)\| \leq v_m$. If the system satisfies the initial condition $\|v_i(t_0)\|_\infty \leq v_m$, then the constraint $\|v_i(t)\| \leq v_m$ will hold all the time. At the same time, the acceleration of the system is also constrained.

Proof. We suppose that the lemma is not true, which means that $v_i(t) > v_m$ for a period of time (t_1, t_2). We assume that $v_i(t_1) = v_m$ and $v_i(t_2) > v_m$. According to the Lagrange mean value theorem, there exists a time $t_1 < t_3 < t_2$ such that $\dot{v}_i(t_3) = \frac{v_i(t_2) - v_i(t_1)}{t_2 - t_1} > 0$. However, since $\|v_c(t)\| \leq v_m$, we have $\dot{v}_i(t_3) = -k(v_i(t_3) + v_c) < 0$. Therefore, the hypothesis is not valid. And $\|\dot{v}_i\|_\infty \leq 2v_m\|k\|_\infty$; by choosing $\|k\|_\infty \leq \frac{u_{\max}}{2v_m}$, the acceleration of the system will be constrained. □

We consider the following nominal controller:

$$u_i = \begin{cases} -k\Big(v_i - v_f + \sigma C \operatorname{sgn}\Big(\sum_{j=0}^{M} a_{ij}(v_i - v_j) + \\ k\sum_{j=0}^{M} a_{ij}\big((x_i - h_i) - (x_j - h_j)\big)\Big)\Big) i = 1, 2, \ldots, M, \\ -k\Big(v_i - v_f + C \operatorname{sgn}\Big(k\sum_{j=1}^{M+N} a_{ij}(x_i - x_j) + \\ \sum_{j=1}^{M+N} a_{ij}(v_i - v_j)\Big)\Big) i = M+1, M+2, \ldots, M+N, \end{cases} \tag{8}$$

where $k = \operatorname{diag}\{k_1, \ldots, k_n\}, 0 < k_i \leq \frac{u_l}{\|v_{\text{upper}} - v_{\text{lower}}\|}, i = 1, 2, ..., n$; u_l is the upper bound of the driving force; v_{upper} and v_{lower} represent the upper bound and the lower bound of the velocity of every single USV, respectively; $C = \frac{1}{2}\operatorname{diag}\{v_{\text{upper}} - v_{\text{lower}}\}, 0 < \sigma < 1$; and v_f represents the ultimate mean velocity.

Theorem 1. *Based on Lemma 6, multi-USV system (2) can solve the formation-containment tracking problem while the system is constrained in velocity and input under protocol (8), where $0 < k_i \leq \frac{u_l}{\|v_{upper} - v_{lower}\|}, 0 < \sigma < 1, 0 < \delta < 1$.*

Proof. First, the demonstration of the consistency between virtual leaders and real leaders is as follows.

The dynamic equation of Virtual Leader 0 is $\dot{v}_0 = -k(v_0 - \bar{v}_0^r)$, where $\bar{v}_0^r = v_f - Cv_b$ and $\|v_b\|_\infty \leq \sigma\delta$; it requires that $0 < \sigma < 1$ and $0 < \delta < 1$.

For the real leaders, we let

$$\begin{aligned} e_{1i}^x &= x_i - d_i - x_0 \\ e_{1i}^v &= v_i - v_0 \end{aligned} \tag{9}$$

denote the error in the tracking formation.

Differentiating these two error quantities with respect to time t yields

$$\begin{aligned} \dot{e}_{1i}^x &= e_{1i}^v \\ \dot{e}_{1i}^v &= -ke_{1i}^v + kCv_b - \sigma kC \operatorname{sgn}\Bigg[\sum_{j=0}^{M} a_{ij}(v_i - v_j) \\ &\quad + k\sum_{j=0}^{M} a_{ij}\big((x_i - d_i) - (x_j - d_j)\big)\Bigg] \end{aligned} \tag{10}$$

We let $e_1^x = \left[e_{11}^{xT}, e_{12}^{xT}, \ldots, e_{1M}^{xT}\right]^T$ and $e_1^v = \left[e_{11}^{vT}, e_{12}^{vT}, \ldots, e_{1M}^{vT}\right]^T$ and rewrite (10) in vector form as

$$\begin{aligned}
\dot{e}_1^x &= e_1^v \\
\dot{e}_1^v &= -(I_M \otimes k)e_1^v + (I_M \otimes kC)\bar{v}_b \\
&\quad - (I_M \otimes kC)\sigma \operatorname{sgn}[(L_{11} \otimes I_n)(I_M \otimes k)e_1^x + (L_{11} \otimes I_n)e_1^v]
\end{aligned} \tag{11}$$

where $\bar{v}_b(t) = \left[v_b^T(t), v_b^T(t), \cdots, v_b^T(t)\right]^T$. We let $\varsigma = (I_M \otimes k)e_1^x + e_1^v$; then, it follows that

$$\begin{aligned}
\dot{e}_1^v = &-(I_M \otimes k)e_1^v + (I_M \otimes kC)\bar{v}_b \\
&-(I_M \otimes kC)\sigma \operatorname{sgn}[(L_{11} \otimes I_n)\varsigma]
\end{aligned} \tag{12}$$

By moving the $(I_M \otimes k)e_1^v$ term to the left side and letting $(I_M \otimes k)e_1^x + e_1^v = \varsigma$, it follows that

$$\dot{\varsigma} = (I_M \otimes kC)\bar{v}_b - (I_M \otimes kC)\sigma \operatorname{sgn}[(L_{11} \otimes I_n)\varsigma] \tag{13}$$

Then, we multiply both sides of (13) by the $(L_{11} \otimes I_n)$ term and let $(L_{11} \otimes I_n)\varsigma = \xi$.

$$\dot{\xi} = (A_0 \otimes kC)\bar{v}_b - (L_{11} \otimes kC)\sigma \operatorname{sgn} \xi \tag{14}$$

Matrix L_{11} can be represented as $L_{11} = \bar{L}_{11} + \delta A_0$, where $\bar{L}_{11} = L'_{11} + (1-\delta)A_0$. Due to $0 < \delta < 1$, $\bar{L}_{11}$ expresses the same virtual leader–real leaders communication connection as L_{11}. By using Lemma 2, there are $[\bar{q}_1, \bar{q}_2, \cdots, \bar{q}_M]^T$, which satisfy that $\bar{Q} = \operatorname{diag}\{\bar{q}_1, \bar{q}_2, \cdots, \bar{q}_N\} > 0$ and $\bar{\varphi} = \bar{Q}\bar{L}_{11} + \bar{L}_{11}^T\bar{Q} > 0$.

We consider the Lyapunov candidate function $V_1 = \sum_{i=0}^{M} \bar{q}_i \|\xi_i\|_1$. We differentiate the function along the trajectory of (14) to obtain

$$\begin{aligned}
\dot{V}_1 &= \sum_{i=0}^{M} \bar{q}_i \operatorname{sgn}(\xi_i)\dot{\xi}_i \\
&= \operatorname{sgn}(\xi)^T(\bar{Q} \otimes I_n)[(A_0 \otimes kC)\bar{v}_b(t) \\
&\quad -(L_{11} \otimes kC)\sigma \operatorname{sgn}(\xi)] \\
&= \operatorname{sgn}(\xi)^T(\bar{Q}A_0 \otimes kC)\bar{v}_b(t) \\
&\quad - \operatorname{sgn}(\xi)^T(\bar{Q}\bar{L}_{11} \otimes kC)\sigma \operatorname{sgn}(\xi) \\
&\quad + \operatorname{sgn}(\xi)^T(\bar{Q}_A \otimes kC)\sigma\delta \operatorname{sgn}(\xi) \\
&= -\operatorname{sgn}(\xi)^T(\bar{Q}\bar{L}_{11} \otimes kC)\sigma \operatorname{sgn}(\xi) \\
&\quad + \operatorname{sgn}(\xi)^T(\bar{Q}A_0 \otimes kC)[\bar{v}_b(t) - \sigma\delta \operatorname{sgn}(\xi)] \\
&\leq -\operatorname{sgn}(\xi)^T(\bar{\varphi} \otimes kC)\sigma \operatorname{sgn}(\xi) \\
&\leq 0
\end{aligned} \tag{15}$$

where we use the fact that $\|v_b\|_\infty \leq \sigma\delta$. If $\dot{V}_1$ is equal to 0, then ξ is equal to 0. If not, then $\dot{V}_1 < 0$, which means that ξ is equal to 0 in a finite time. Since $L_{11} > 0$, it leads to $e_1^x \to 0$ and $e_1^v \to 0$ as $t \to \infty$. And the system has completed the formation tracking task in a finite time. Then, it will be demonstrated that containment control can be achieved. For the real leaders, it follows that $\dot{v}_i = -k\left(v_i - v_f + Cv_{bi}\right)$, where

$v_{bi} = \sigma \operatorname{sgn}\left[k\sum_{j=0}^{M} a_{ij}((x_i - d_i) - (x_j - d_j)) + \sum_{j=0}^{M} a_{ij}(v_i - v_j)\right]$, which requires that $\|v_{bi}\|_\infty \leq \sigma$. For the followers, we let

$$
\begin{aligned}
e_{2i}^{x} &= \sum_{j=1}^{M+N} a_{ij}(x_i - d_i - x_j + d_j) \\
e_{2i}^{v} &= \sum_{j=1}^{M+N} a_{ij}(v_i - v_j)
\end{aligned} \tag{16}
$$

denote the error in containment.

We let $e_2^x = \left[e_{2(M+1)}^{xT}, e_{2(M+2)}^{xT}, \cdots, e_{2(M+N)}^{xT}\right]^T$ and $e_2^v = \left[e_{2(M+1)}^{vT}, e_{2(M+2)}^{vT}, \cdots, e_{2(M+N)}^{vT}\right]^T$. We rewrite (18) in vector form as

$$
\begin{aligned}
\dot{e}_2^x &= e_2^v \\
e_2^v &= (L_{21} \otimes I_n)\dot{v}_{RL} + (L_{22} \otimes I_n)\dot{v}_F
\end{aligned} \tag{17}
$$

where $\dot{v}_{RL} = \left[\dot{v}_1^T, \dot{v}_2^T, \cdots, \dot{v}_M^T\right]^T$, $\dot{v}_F = \left[\dot{v}_{M+1}^T, \dot{v}_{M+2}^T, \cdots, \dot{v}_{M+N}^T\right]^T$.

By substituting $\dot{v}_i = -k\left(v_i - v_f + Cv_{bi}\right)$ into (19), it follows that

$$
\begin{aligned}
\dot{e}_2^v = &-(I_N \otimes k)(L_{21} \otimes I_n)v_{RL} - (I_N \otimes k)(L_{22} \otimes I_n)v_F \\
&- (L_{21} \otimes kC)\bar{v}_B(t) - (L_{22} \otimes kC)\operatorname{sgn}[e_2^v + (I_N \otimes k)e_2^x]
\end{aligned} \tag{18}
$$

where $\bar{v}_B(t) = \left[v_{b1}^T(t), v_{b2}^T(t), \cdots, v_{bM}^T(t)\right]^T$.

By rearranging the terms, we can derive

$$
\begin{aligned}
\dot{e}_2^v + (I_N \otimes k)e_2^v = &-(L_{21} \otimes kC)\bar{v}_B(t) \\
&- (L_{22} \otimes kC)\operatorname{sgn}[e_2^v + (I_N \otimes k)e_2^x]
\end{aligned}
$$

By letting $e_2^v + (I_N \otimes k)e_2^x = \xi$, it follows that

$$
\dot{\xi} = -(L_{21} \otimes kC)\bar{v}_B(t) - (L_{22} \otimes kC)\operatorname{sgn}[\xi] \tag{19}
$$

Matrix $L_{22} = \bar{L}_{22} + \sigma\sum_{i=1}^{M} A_i$, where $\bar{L}_{22} = L_{22}' + (1-\sigma)\sum_{i=1}^{M} A_i$. Similarly, because of $0 < \sigma < 1$, $\bar{L}_{22}$ expresses the same real leader–follower communication connection as L_{22}. By using Lemma 3, there are $[\bar{p}_1, \bar{p}_2, \cdots, \bar{p}_N]^T$, which satisfy that $\bar{P} = \operatorname{diag}\{\bar{p}_1, \bar{p}_2, \cdots, \bar{p}_N\} > 0$ and $\bar{\phi} = \bar{P}\bar{L}_{22} + \bar{L}_{22}^T\bar{P} > 0$.

Now, we consider the Lyapunov candidate equation $V_2 = \sum_{i=M+1}^{M+N} \bar{p}_i\|\xi_i\|_1$ and differentiate the function along the trajectory of (18) to obtain

$$
\begin{aligned}
\dot{V}_2 &= \sum_{i=M+1}^{M+N} \bar{p}_i \operatorname{sgn}(\xi_i)\dot{\xi}_i \\
&= \operatorname{sgn}(\xi)^T(\bar{P} \otimes I_n)[-(L_{21} \otimes kC)\bar{v}_B(t) \\
&\quad -(L_{22} \otimes kC)\operatorname{sgn}(\xi)]
\end{aligned}
$$

We utilize the property of $L_{21}\bar{v}_B(t) = -\sum_{i=1}^{M} A_i\bar{v}_{bi}(t)$, where each $\bar{v}_{bi}(t) = \left[v_{bi}^T(t), v_{bi}^T(t), \cdots, v_{bi}^T(t)\right]^T \in R^{M\times n}$. It derives

$$\begin{aligned}
\dot{V}_2 = &\operatorname{sgn}(\xi)^T(\bar{P}\otimes I_n)\left[-\sum_{i=1}^{M}(-A_i\otimes kC)\bar{v}_{bi}(t)\right.\\
&\left.-(L_{22}\otimes kC)\operatorname{sgn}(\xi)\right]\\
&=\operatorname{sgn}(\xi)^T(\bar{P}\otimes I_n)\left[\sum_{i=1}^{M}(A_i\otimes kC)\bar{v}_{bi}(t)\right.\\
&\left.-(\bar{L}_{22}\otimes kC)\operatorname{sgn}(\xi)-\sigma\sum_{i=1}^{M}(A_i\otimes kC)\operatorname{sgn}(\xi)\right]\\
&=-\operatorname{sgn}(\xi)^T(\bar{P}\bar{L}_{22}\otimes kC)\operatorname{sgn}(\xi)\\
&+\sum_{i=1}^{M}\operatorname{sgn}(\xi)^T(\bar{P}A_i\otimes kC)[\bar{v}_{bi}(t)-\sigma\operatorname{sgn}(\xi)]\\
&\leq -\frac{1}{2}\operatorname{sgn}(\xi)^T(\bar{\phi}\otimes kC)\operatorname{sgn}(\xi)\\
&\leq 0
\end{aligned} \tag{20}$$

where we use the fact that $\|v_{bi}\|_\infty \leq \sigma$. If $\dot{V}_2$ is equal to 0, then $\bar{\xi}$ is equal to 0. If not, then $\dot{V}_2 < 0$, which means that $\bar{\xi}$ is equal to 0 in a finite time. That means that $e_2^x \to 0$ and $e_2^v \to 0$ as $t \to \infty$. And the system has completed the containment task. Thus, the formation-containment tracking problem is solved. □

Remark 2. *The nominal controller of the system is designed by Lemma 6 to conform to the basic form where the velocities and control inputs are constrained. Real-time information exchange among the USVs is then used to achieve distributed control. To better alleviate to system jitters caused by nonlinear terms and to better achieve consistency, we may replace the sign function with (21) in practical applications.*

$$\operatorname{sat}_\omega(x) = \begin{cases} -1, x < -\frac{1}{\omega} \\ \omega x, |x| < \frac{1}{\omega} \\ 1, x > \frac{1}{\omega} \end{cases} \tag{21}$$

3.2. Barrier Function-Based Collision Avoidance Controller

In the actual multi-USV system, the acceleration and velocity of USVs are limited. This should also considered in the distributed cooperation layer. Considering the difference between leaders and followers, it is assumed that the real leaders' velocity (v_L) and acceleration (u_L) are limited by μ_{ml} and η_{ml}, which means that $\|v_L\|_\infty \leq \mu_{ml}$ and $\|u_L\|_\infty \leq \eta_{ml}$. Similarly, the followers' velocity (v_F) and acceleration (u_F) satisfy $\|v_F\|_\infty \leq \mu_{mf}$ and $\|u_F\|_\infty \leq \eta_{mf}$. The controller of each agent is designed as $u_i = \bar{u}_i, i = 0, \ldots, N+M$, where $\bar{u}_i$ satisfies the control input of the safety set. Next, the security zone needs to be divided.

For convenience, the USVs and obstacles in the distributed cooperative layer are regarded as circles when dealing with obstacle avoidance and collision avoidance. Taking into account the difference between the USVs and the obstacles, it is supposed that the radius of the USVs is r_a and that the radius of the ith obstacle is r_i.

It is assumed that the safe distance among the USVs is d_a, where $d_a = 2r_a$. Considering the collision avoidance requirements, the distance between any two USVs must be satisfy $\|\Delta x_{ij}(t)\| \geq d_a$ all the time, where $d_a = 2r_a$. For a pair of USVs i and j, where $i, j = 1, 2, \ldots, N+M$, the velocity difference can be expressed as $\Delta v_{ij} = v_i - v_j$.

The collision avoidance maintenance condition is converted into the forward invariance of some sets. A comprehensive consideration of possible collisions throughout the system has led to the classification of collisions into four scenarios.

Assumption 3. *For each USV, the convergence position satisfies the safety distance requirement under a fixed topology. That means that* $\min_{t\to\infty}\|x_i(t)-x_j(t)\| \geq d_a$*, where* $i = 1,\ldots, M+N, j = 1,\ldots,M+N,$ *and* $i \neq j$.

1. Case 1: First, we consider avoidance among the followers. According to the derivation similar to [39], collision avoidance set of USV i can be expressed as

$$S_{ij} = \left\{(x_i, v_i) \in R^{2n} \mid h_{ij} \geq 0\right\} \tag{22}$$

where $i, j \in \{M+1,\ldots,M+N\}$ and

$$h_{ij} = \sqrt{4\eta_{mf}(\|\Delta x_{ij}\| - d_a)} + \frac{\Delta x_{ij}^T}{\|\Delta x_{ij}\|}\Delta v_{ij} \tag{23}$$

where h_{ij} is the level set function of set S_{ij}. This means that when a collision between two USVs is about to occur, each of the two USVs will prevent the collision with its maximum acceleration. Additionally, the forward invariance of S_{ij} will be ensured if the ZCBF constraint in (6) is satisfied. By combining (6) and (23), the following inequality can be obtained:

$$\begin{aligned} -\Delta x_{ij}^T\Delta\bar{u}_{ij} \leq \alpha h_{ij}^3\|\Delta x_{ij}\| - \frac{\left(\Delta v_{ij}^T\Delta x_{ij}\right)^2}{\|\Delta x_{ij}\|^2} \\ + \frac{2\eta_{mf}\Delta v_{ij}^T\Delta x_{ij}}{\sqrt{4\eta_{mf}(\|\Delta x_{ij}\| - D_a)}} + \|\Delta v_{ij}\|^2 \end{aligned} \tag{24}$$

where $\Delta\bar{u}_{ij} = \bar{u}_i - \bar{u}_j$. It can also be written in the following form:

$$-\Delta x_{ij}^T(\bar{u}_i - \bar{u}_j) \leq b_{ij}$$

where $b_{ij} = \alpha h_{ij}^3\|\Delta x_{ij}\| - \frac{\left(\Delta v_{ij}^T\Delta x_{ij}\right)^2}{\|\Delta x_{ij}\|^2} + \frac{2\eta_{mf}\Delta v_{ij}^T\Delta x_{ij}}{\sqrt{4\eta_{mf}(\|\Delta x_{ij}\|-d_a)}} + \|\Delta v_{ij}\|^2$. Thus, we can constrain the safety barrier of USV i and j distribution as

$$-\Delta x_{ij}^T\bar{u}_i \leq \frac{1}{2}b_{ij}, \Delta x_{ij}^T\bar{u}_j \leq \frac{1}{2}b_{ij} \tag{25}$$

Constraint (25) can be written in the linear form $X_{ij}\bar{u}_i \leq \hat{b}_{ij}$ for each follower, where $X_{ij} = -\Delta x_{ij}^T$ and $\hat{b}_{ij} = \frac{1}{2}b_{ij}$.

2. Case 2: Similarly, taking into account collision avoidance among real leaders, the following safety set can be obtained:

$$\bar{S}_{ij} = \left\{(x_i, v_i) \in R^{2n} \mid \bar{h}_{ij} \geq 0\right\}$$

where $i, j \in \{1,\ldots,M\}$ and

$$\bar{h}_{ij} = \sqrt{4\eta_{ml}(\|\Delta x_{ij}\| - d_a)} + \frac{\Delta x_{ij}^T}{\|\Delta x_{ij}\|}\Delta v_{ij} \tag{26}$$

Then, the linear constraint can be obtained as $\bar{X}_{ij}\bar{u}_i \leq \frac{1}{2}\bar{b}_{ij}$ for each real leader, where $\bar{b}_{ij} = \alpha h_{ij}^3\|\Delta x_{ij}\| - \frac{\left(\Delta v_{ij}^T\Delta x_{ij}\right)^2}{\|\Delta x_{ij}\|^2} + \frac{2\eta_{ml}\Delta v_{ij}^T\Delta x_{ij}}{\sqrt{4\eta_{ml}(\|\Delta x_{ij}\|-D_a)}} + \|\Delta v_{ij}\|^2$ and $X_{ij} = -\Delta x_{ij}^T$. Since the virtual leader is not real, there is no need to consider its collision avoidance.

3. Case 3: For a follower i and a real leader j, the collision avoidance set is

$$\tilde{S}_{ij} = \left\{(x_i, v_i) \in R^{2n} \mid \tilde{h}_{ij} \geq 0\right\}$$

where $i \in \{1,\ldots,M\}, j \in \{M+1,\ldots,M+N\}$, and

$$\tilde{h}_{ij} = \sqrt{2\left(\eta_{mf} + \eta_{ml}\right)(\|\Delta x_{ij}\| - d_a)} + \frac{\Delta x_{ij}^T}{\|\Delta x_{ij}\|}\Delta v_{ij} \tag{27}$$

The linear constraint can be obtained as $\tilde{X}_{ij}\bar{u}_i \leq \frac{1}{2}\tilde{b}_{ij}$, where $\tilde{b}_{ij} = \alpha h_{ij}^3\|\Delta x_{ij}\| - \frac{\left(\Delta v_{ij}^T \Delta x_{ij}\right)^2}{\|\Delta x_{ij}\|^2} + \frac{(\eta_{ml}+\eta_{mf})\Delta v_{ij}^T \Delta x_{ij}}{\sqrt{2(\eta_{ml}+\eta_{mf})(\|\Delta x_{ij}\| - D_a)}} + \|\Delta v_{ij}\|^2$ and $\tilde{X}_{ij} = -\Delta x_{ij}^T$.

4. Case 4: In addition to collision avoidance among USVs, collision avoidance between obstacles and USVs should also be taken into consideration. It is assumed that the safe distance between an USV i and an obstacle O_j is d_s^j, where $d_s^j = r_a + r_j$. Considering the collision avoidance requirements, the distance between any USV and any obstacle must satisfy $\left\|\Delta x_{iO_j}(t)\right\| \geq d_s^j$ all the time.
First, considering collision avoidance between obstacles and followers, the safety set between a follower USV i and an obstacle O_j is represented by the following expression:

$$S_{iO_j} = \left\{(x_i, v_i) \in R^{2n} \mid h_{iO_j} \geq 0\right\}$$

where $i \in \{M+1,\ldots,M+N\}$ and

$$h_{iO_j} = \sqrt{2\left(\eta_{mf} + \eta_{mO}\right)\left(\left\|\Delta x_{iO_j}\right\| - d_s^j\right)} + \frac{\Delta x_{iO_j}^T}{\left\|\Delta x_{iO_j}\right\|}\Delta v_{iO_j} \tag{28}$$

where h_{iO_j} is the level set function of set S_{iO_j}. This means that when a collision between an USV and an object is about to occur, the USV will prevent the collision with its maximum acceleration. The velocity (v_O) and acceleration (η_O) of the obstacles can be detected by sensors mounted on the USV. Furthermore, the static obstacles can be regarded as special cases of the dynamic obstacles with zero velocity and acceleration. Additionally, the forward invariance will be ensured if the ZCBF constraint in (6) is satisfied. By combining (6) and (28), the following inequality can be obtained:

$$-\Delta x_{iO_j}^T \bar{u}_i \leq \alpha h_{iO_j}^3\left\|\Delta x_{iO_j}\right\| - \frac{\left(\Delta v_{iO_j}^T \Delta x_{iO_j}\right)^2}{\left\|\Delta x_{iO_j}\right\|^2} + \left\|\Delta v_{iO_j}\right\|^2 + \frac{\left(\eta_{mf} + \eta_{mO}\right)\Delta v_{iO_j}^T \Delta x_{iO_j}}{\sqrt{2\left(\eta_{mf} + \eta_{mO}\right)\left(\left\|\Delta x_{iO_j}\right\| - d_s^j\right)}} \tag{29}$$

Inequality (29) can also be written in the following form:

$$-\Delta x_{iO_j}^T \bar{u}_i \leq b_{iO_j} \tag{30}$$

where $b_{iO_j} = \alpha h^3_{iO_j}\left\|\Delta x_{iO_j}\right\| - \frac{\left(\Delta v^T_{iO_j}\Delta x_{iO_j}\right)^2}{\left\|\Delta x_{iO_j}\right\|^2} + \left\|\Delta v_{iO_j}\right\|^2 + \frac{(\eta_{mf}+\eta_{mO})\Delta v^T_{iO_j}\Delta x_{iO_j}}{\sqrt{2(\eta_{mf}+\eta_{mO})\left(\left\|\Delta x_{iO_j}\right\|-d^j_s\right)}}$. We let X_{iO_j} stand for $-\Delta x^T_{iO_j}$; then, constraint (30) can be written in the linear form $X_{iO_j}\bar{u}_i \leq b_{iO_j}$ for each follower.

Then, considering the avoidance of collisions between a real leader j and obstacles, the following safety set can be obtained:

$$\hat{S}_{iO_j} = \left\{(x_i, v_i) \in R^{2n} \mid \hat{h}_{iO_j} \geq 0\right\}$$

where $i \in \{1, \ldots, M\}$ and

$$\hat{h}_{iO_j} = \sqrt{2(\eta_{ml} + \eta_{mO})\left(\left\|\Delta x_{iO_j}\right\| - d^j_s\right)} + \frac{\Delta x^T_{iO_j}}{\left\|\Delta x_{iO_j}\right\|}\Delta v_{iO_j} \tag{31}$$

Then, we can obtain the linear constraint $\hat{X}_{iO_j}\bar{u}_j \leq \hat{b}_{iO_j}$ for each real leader, where $\hat{b}_{iO_j} = \alpha h^3_{iO_j}\left\|\Delta x_{iO_j}\right\| - \frac{\left(\Delta v^T_{iO_j}\Delta x_{iO_j}\right)^2}{\left\|\Delta x_{iO_j}\right\|^2} + \left\|\Delta v_{iO_j}\right\|^2 + \frac{(\eta_{ml}+\eta_{mO})\Delta v^T_{iO_j}\Delta x_{iO_j}}{\sqrt{2(\eta_{ml}+\eta_{mO})\left(\left\|\Delta x_{iO_j}\right\|-d^j_s\right)}}$.

In order to avoid collisions, the system needs to satisfy the constraints of the above four cases at all times. To ensure the effectiveness of the nominal controller, collision avoidance strategies should be triggered as little as possible, and the nominal controller should be modified as little as possible. Since the constraints are linear, the following quadratic programming (QP) problem can be formulated in a least squares sense:

$$\begin{aligned} \bar{u}^*_i &= \underset{\bar{u}_i \in R^n}{\arg\min} J(\bar{u}_i) \\ \text{s.t. } & X_{ij}\bar{u}_i \leq \hat{b}_{ij}, i \in L, j \in L \\ & \bar{X}_{ij}\bar{u}_i \leq \bar{b}_{ij}, i \in F, j \in F \\ & \tilde{X}_{ij}\bar{u}_i \leq \tilde{b}_{ij}, i \in F, j \in L \\ & X_{iO_j}\bar{u}_i \leq b_{iO_j}, i \in F, j \in OB \\ & \hat{X}_{iO_j}\bar{u}_i \leq \hat{b}_{iO_j}, i \in L, j \in OB \end{aligned} \tag{32}$$

where $J(\bar{u}_i) = \|u_i - \bar{u}_i\|^2$, u_i is the nominal controller, $\bar{u}_i$ is the controller that satisfies the security strategy constraints, L is the set of real leaders, F is the set of followers, and OB is the set of obstacles. The actual controller is shown in the following expression:

$$u_i = \bar{u}^*_i \tag{33}$$

where $i = 0, 1, \ldots, N + M$.

Theorem 2. *Given the formation-containment tracking system, if for all USV $i = 1, \ldots, M + N$, the control input is given by (33), where $\bar{u}^*_i$ can be solved by QP problem (32), then the system is able to avoid collisions under assumption 3.*

Proof. If all controllers of the USVs satisfy the decentralized safety certificates, then security set S is forward-invariant. If triggered, they are confined to the safety set [39]. As a result, the system is able to achieve obstacle and collision avoidance. □

Remark 3. *The priority of collision avoidance and obstacle avoidance control for USVs is higher than that of formation-containment control based on consensus for multiple USVs. When consid-*

ering containment control problems, if the topology of the system is fixed, the final convergence position of the followers should be calculated in advance. If the final convergence position of the followers does not satisfy the safety distance requirement, the CBF (control barrier function) will take precedence in maintaining a safe distance between follower and follower, consequently disrupting the preset consensus formation among the followers. However, the pre-formed position of the real leader USV should satisfy the requirement of a safe position; otherwise, it may deviate from the preset trajectory due to the inability to achieve consistent velocity.

Remark 4. *At this point, the conditions that the real leaders and followers need to meet to avoid obstacles and collisions have all been given. However, the proposed constraint takes into account all pairs of USVs, which can be a very large number. From a topology structure perspective, this requires all the information interactions of the USVs. With the increase in the number of USVs, the related computing and sensing requirements will increase significantly. We consider a fact that if the USVs are far away enough, they will not collide in a period of time. Thus, each USV i only needs to consider other USVs and obstacles in a certain range.*

4. Simulation

In this section, an example is used to illustrate the effectiveness of the control strategy in the distributed cooperation layer. The multi-USV system in this simulation example is composed of one virtual leader, four real leaders, and three followers. The communication topology of this system is shown in Figure 3, where USV 0 is the virtual leader, USVs 1–4 are real leaders, and USVs 5–7 are followers. The initial position vector and initial velocity vector of USVs 1–7 are $x(0) = [0,0,10,0,6,10,0,10,20,0,5,9,1,20]^T$ and $v(0) = [1,2,2.5,1.5,2.25,1,1.5,2,3,2.75,1,3,1.75,2.5]^T$, respectively. The formation shape given by the real leaders is a rectangle, and the specific value vector is $h = [0,0,20,0,20,20,0,20]^T$. The velocity limits of each real leader and follower are both $6\,\mathrm{m/s}$, and the input limits are both $10\,\mathrm{m/s^2}$. The dynamic equation of the virtual leader is $\dot{v}_0 = -k(v_0 - 1)$. The initial position vector and initial velocity vector of the virtual leader are $[1,0]^T$ and $[2,1]^T$, respectively. And the velocity limits and acceleration limits given by the virtual leader are $3\,\mathrm{m/s}$ and $6\,\mathrm{m/s^2}$, with $\sigma = 0.75$. According to Theorem 1, this leads to $k = 1$. There is only one obstacle in the motion environment of the system. The radius of the USV in the distributed cooperative layer is 1 m, and the radius of the obstacle is 10 m. This means that the safe distance among USVs is 2 m and the safe distance between any USV and any obstacle is 11 m.

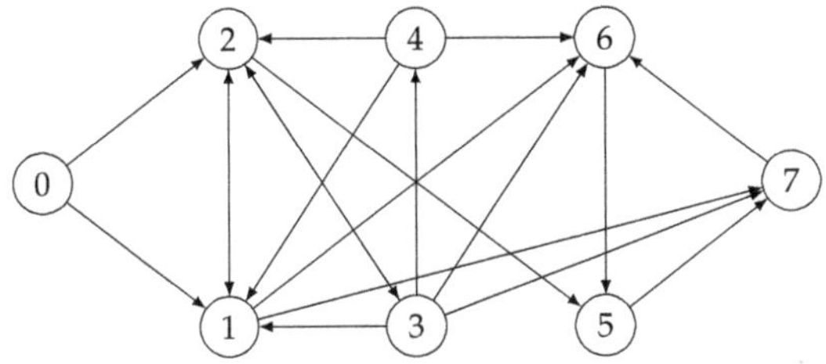

Figure 3. The topology of the multi-USV system.

The trajectories of USVs in the distributed cooperative layer obtained by the multi-USV system driven by controller (33) are shown in Figure 4. The system can be driven by the control strategy to perform the formation-containment tracking control task: real leaders can form the expected formations; followers can converge into the convex hull; the motion of the whole system is controlled by the virtual leader. At the same time, all USVs can avoid obstacles. The smooth trajectories of the agents represent no significant abrupt changes in the state of the USVs.

More details on three aspects are presented in the following simulation: the implementation of formation-containment tracking, the velocities and input constraints of the USVs, and the effectiveness of the obstacle and collision avoidance strategy.

The errors in formation-tracking control and containment control can be seen in Figures 5–8. Figures 5 and 6 show the formation-tracking position errors and velocity errors of the real leaders. The errors arise from the need for real leaders to track virtual leader states and form time-invariant formations, as shown in Equation (9). Figures 7 and 8 show the containment position errors and velocity errors of the followers. The errors arise from the need for followers to track the states of the real leaders, as shown in Equation (16). The asymptotic convergence of all errors to zero indicates that the system is finally stable and completes the formation-containment tracking control task. The oscillations in these figures are produced by avoiding obstacles and other USVs. It is worth noticing that the convergence of the followers occurs after the convergence of the real leaders, but the time interval can be ignored in the application of the algorithm.

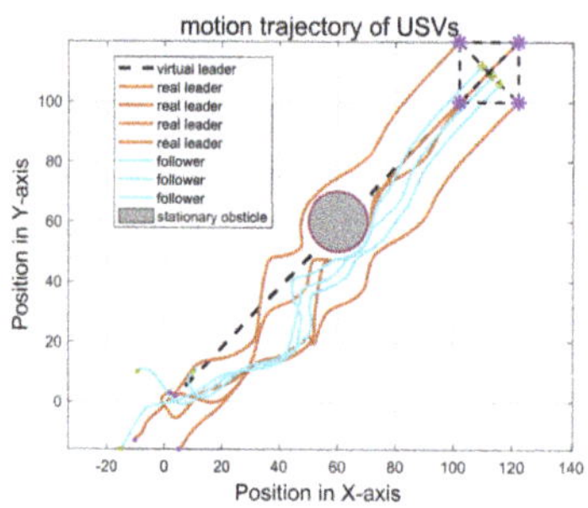

Figure 4. The trajectories of USVs in the distributed cooperative layer. The purple asterisk represents the position of the formation leader at $t = 100$ s.

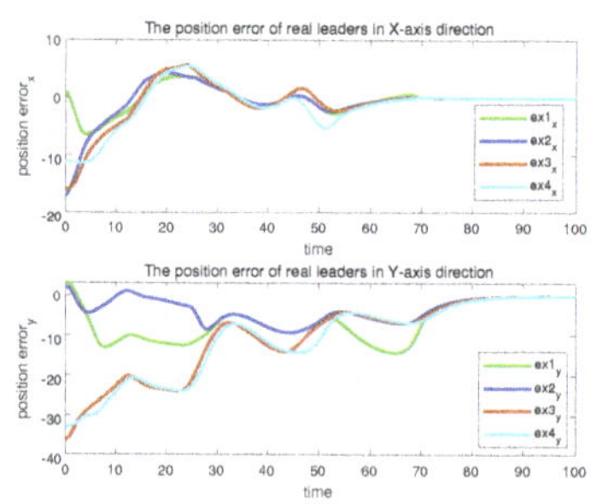

Figure 5. The position errors of the real leaders.

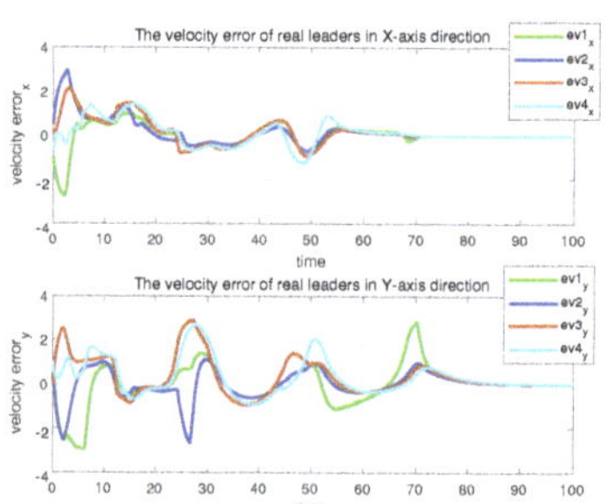

Figure 6. The velocity errors of the real leaders.

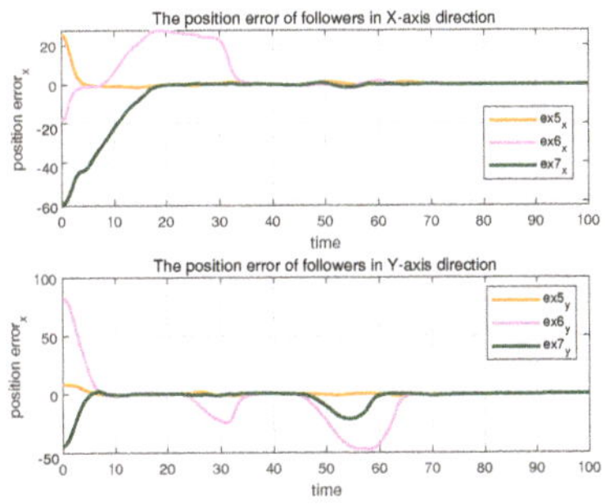

Figure 7. The position errors of the followers.

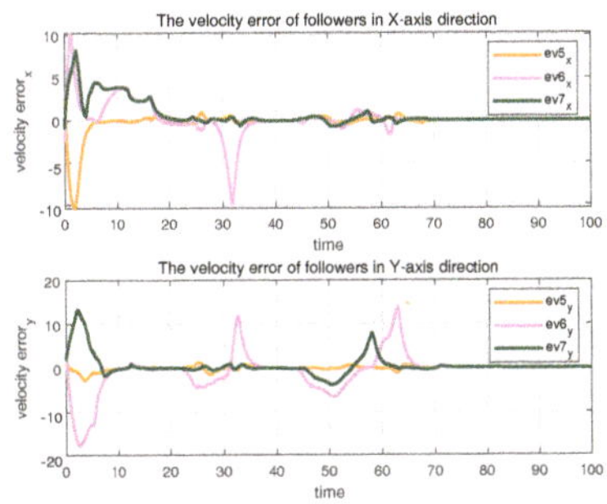

Figure 8. The velocity errors of the followers.

The accelerations and velocities of each USV are limited to the required range, as shown in Figures 8 and 9, where dashed lines indicate acceleration and velocity limits. Although the controller input oscillates during obstacle avoidance, the velocities and accelerations of all USVs eventually converge to the same level. The oscillations are caused by the controller solving a quadratic programming problem during obstacle avoidance, the solution of which leads to abrupt changes in the controller inputs. Even during the obstacle avoidance process, the velocities and accelerations of all the USVs do not exceed the required limits.

To illustrate more clearly that in this example, the USVs can avoid obstacles, the distance between each USV and the center of the obstacle is plotted, as shown in Figure 10, where the dashed lines represent the safe distance. We further illustrate the effectiveness of obstacle avoidance by adding a time axis, zooming in on the local trajectory figure, and increasing the safety distance incrementally, as shown in Figure 11. The figure shows that the trajectory of each USV is smooth when avoiding obstacles and confirms that the velocities and inputs of each USV are limited. With the increase in safety distance, the implementation of the algorithm is still effective. Therefore, in practice, the setting of safety distance can be adjusted appropriately for better security.

Due to the different limited accelerations and velocities of the followers and leaders, the distance between the USVs is plotted to illustrate the collision avoidance effectiveness of the USVs, as shown in Figures 12 and 13, with the dotted line representing the safe distance. In this example, the initial position of the real leader is set farther away from the follower, and the likelihood of collisions occurring is reduced. However, the whole system still realizes collision avoidance with the USVs avoiding obstacles.

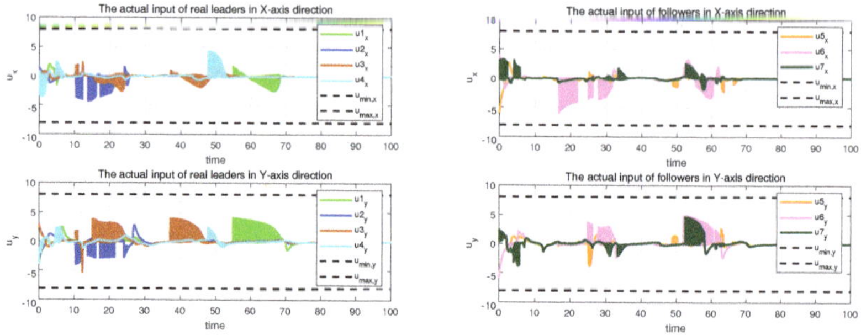

Figure 9. The inputs of the USVs with the controller.

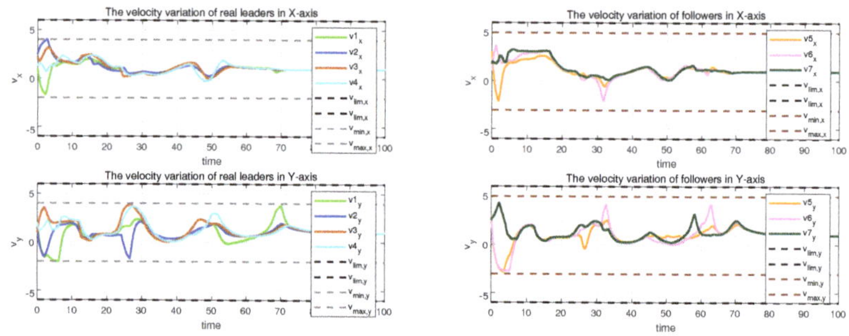

Figure 10. The velocities of the USVs with the controller.

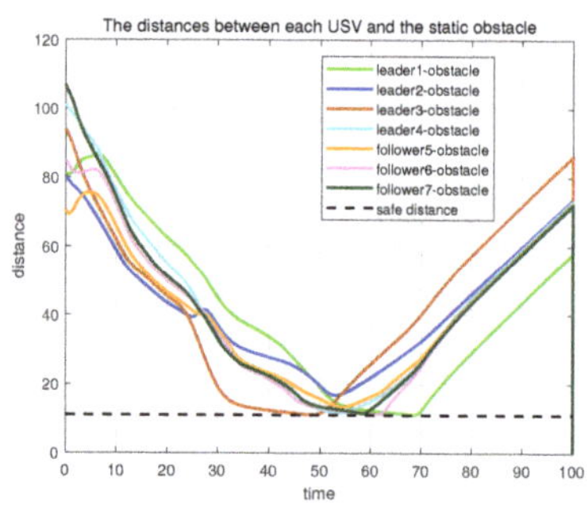

Figure 11. The distance of each USV from the center of the obstacle.

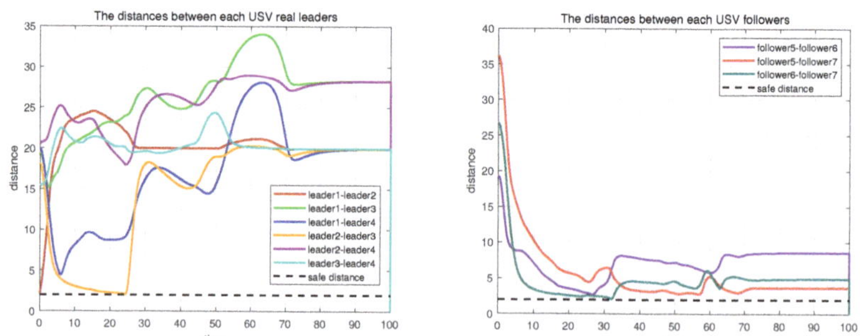

Figure 12. The distance between the followers and the leaders.

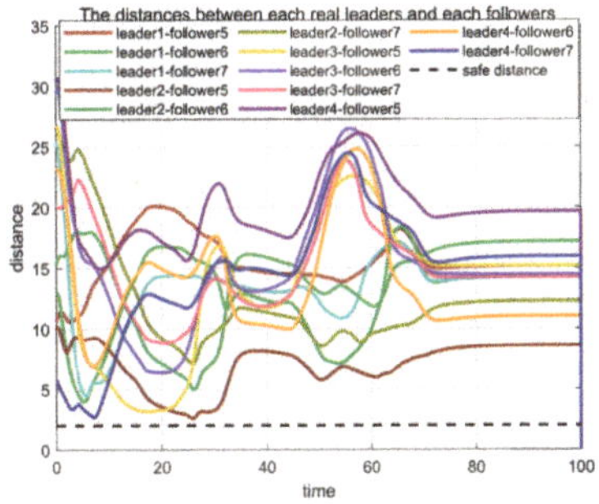

Figure 13. The distance between the followers and leaders.

5. Conclusions

This paper investigates the distributed formation-containment tracking task for collision-free multi-USV (unmanned surface vehicle) systems under constraints on velocity and driving force. Initially, a standard controller based on consensus theory is proposed, designed to guide the entire multi-USV system, subject to velocity and input constraints, to complete the formation-containment tracking task. Additionally, a collision avoidance approach based on control barrier functions is introduced. This method accounts for avoiding collisions both among USVs with different roles within the formation-containment mission and against static obstacles. The effectiveness of the proposed strategies is validated through simulation experiments. Future work will focus on generalizing the results to agents with nonlinear dynamics and taking into account interference and uncertainty.

Author Contributions: Conceptualization, J.W.; Methodology, J.W.; Software, Q.S.; Validation, J.W. and Q.X.; Formal analysis, J.W.; Investigation, J.W.; Resources, J.W., Q.S. and G.X.; Data curation, Q.S.; Writing—original draft, J.W.; Writing—review & editing, Q.S.; Visualization, Q.S.; Supervision, Q.S., T.L., G.X. and Q.X.; Project administration, Q.S., T.L., G.X. and Q.X.; Funding acquisition, Q.S. and T.L. All authors have read and agreed to the published version of the manuscript.

Funding: This work was supported in part by National Natural Science Foundation of China under grant 52371360, grant 52201407, and grant 51939001; in part by the High Level Talents Innovation Support Plan of Dalian (Young Science and Technology Star Project) under grant 2021RQ058; in part by Funds for the Central Universities under grant 3132023147; in part by Fundamental Research Funds for the Central Universities under grant 3132023103; and in part by the Zhejiang Lab Open Research Project under grant K2022QA0AB03.

Data Availability Statement: Data are contained within the article.

Conflicts of Interest: The authors declare no conflict of interest.

References

1. Xie, J.J.; Luo, J.; Peng, Y.; Xie, S.R.; Pu, H.Y.; Li, X.M.; Su, Z.; Liu, Y.; Zhou, R. Data driven hybrid edge computing-based hierarchical task guidance for efficient maritime escorting with multiple unmanned surface vehicles. *J. Abbr.* **2020**, *13*, 1788–1798. [CrossRef]
2. Jia, Q.; Li, R.H.; Li, J.W.; Li, Z.; Liu, J.Q. Vessel traffic scheduling optimization for passenger RoRo terminals with restricted harbor basin. *Ocean. Coast Manag.* **2023**, *246*, 106904. [CrossRef]
3. Dewan, C.; Sibilia, E.A. Global containments and local leakages: Structural violence and the toxic flows of shipbreaking. *Environ. Plan. C-Politics Space* **2023**, *231*, 208202 . [CrossRef]
4. Patterson, R.G.; Lawson, E.; Udyawer, V.; Brassington, G.B.; Groom, R.A.; Campbell, H.A. Uncrewed Surface Vessel Technological Diffusion Depends on Cross-Sectoral Investment in Open-Ocean Archetypes: A Systematic Review of USV Applications and Drivers. *Front. Mar. Sci.* **2022**, *8*, 736984. [CrossRef]
5. Fu, S.S.; Gu, S.Y.; Zhang, Y.; Zhang, M.Y.; Weng, J.X. Towards system-theoretic risk management for maritime transportation systems: A case study of the yangtze river estuary. *Ocean Eng.* **2023**, *286*, 115637. [CrossRef]
6. Chen, Y.J.; Wang, T.F.; Xiao, Y.A.; Wang, Y.; Yan, G.A. An Approach of Consensus-Based Double-Layer Blockchain System for Multi-Ship Collision Risk Mitigation Considering COLREGs. *Appl. Sci.* **2023**, *13*, 11178. [CrossRef]

7. Zhang, Z.Q.; Wang, J.F.; Guo, L.L.; Yang, W.J.; Wang, F.; Xu, P. Research on Minimum Non-Collision Distance and Protection Strategy for Normal Train to Avoid Rear-End Accidents With Braking-Fault Train. *IEEE Trans. Intell. Transp. Syst.* **2023**, *24*, 4306–4319. [CrossRef]
8. Wang, Y.Y.; Wang, X.; Zhou, W.X.; Yan, H.C.; Xie, S.R. TThreat potential field based Pursuit-Evasion Games for underactuated Unmanned Surface Vehicles? *Ocean Eng.* **2023**, *285*, 115381. [CrossRef]
9. Niu, Y.H.; Zhu, F.X.; Wei, M.X.; Du, Y.F.; Zhai, P.Y. A Multi-Ship Collision Avoidance Algorithm Using Data-Driven Multi-Agent Deep Reinforcement Learning. *J. Mar. Sci. Eng.* **2023**, *11*, 2101. [CrossRef]
10. Zhou, M.X.; Bachmayer, R.; DeYoung, B. Surveying a Floating Iceberg with the USV SEADRAGON. *Front. Mar. Sci.* **2021**, *8*, 549566. [CrossRef]
11. Aranda-Bricaire, E.; González-Sierra, J. Formation with Non-Collision Control Strategies for Second-Order Multi-Agent Systems. *Entropy* **2023**, *25*, 904. [CrossRef]
12. Ghommam, J.; Saad, M.; Mnif, F.; Zhu, Q.M. Guaranteed Performance Design for Formation Tracking and Collision Avoidance of Multiple USVs with Disturbances and Unmodeled Dynamics. *IEEE Syst. J.* **2021**, *15*, 4346–4357. [CrossRef]
13. Park, B.S.; Yoo, S.J. An Error Transformation Approach for Connectivity-Preserving and Collision-Avoiding Formation Tracking of Networked Uncertain Underactuated Surface Vessels. *IEEE Trans. Cybern.* **2023**, *53*, 5454–5465. [CrossRef]
14. Wu, W.; Tong, S.C. Collision-Free Adaptive Fuzzy Formation Control for Stochastic Nonlinear Multiagent Systems. *IEEE Trans. Syst. Man Cybern. Syst.* **2023**, *53*, 5454–5465. [CrossRef]
15. Ren, W. Consensus strategies for cooperative control of vehicle formations. *IET Control Theory Appl.* **2007**, *1*, 505–512. [CrossRef]
16. Fang, X.; Wen, G.H.; Yu, X.H.; Chen, G.R. Formation control for unmanned surface vessels: A game-theoretic approach. *Asian J. Control* **2022**, *24*, 498–509. [CrossRef]
17. Hu, B.B.; Zhang, H.T.; Liu, B.; Meng, H.F.; Chen, G.R. Distributed Surrounding Control of Multiple Unmanned Surface Vessels with Varying Interconnection Topologies. *IEEE Trans. Control Syst. Technol.* **2022**, *30*, 400–407.. [CrossRef]
18. Fu, J.; Wen, G.; Yu, W.; Huang, T.; Yu, X. Consensus of Second-Order Multiagent Systems with both Velocity and Input Constraints. *IEEE Trans. Ind. Electron.* **2019**, *66*, 7946–7955. [CrossRef]
19. Huang, C.; Zhang, X.; Zhang, G.; Deng, Y. Robust practical fixed-time leader–follower formation control for underactuated autonomous surface vessels using event-triggered mechanism. *Ocean Eng.* **2021**, *233*, 109026.. [CrossRef]
20. Tang, C.; Zhang, H.T.; Wang, J. Flexible Formation Tracking Control of Multiple Unmanned Surface Vessels for Navigating through Narrow Channels with Unknown Curvatures. *IEEE Trans. Ind. Electron.* **2023**, *70*, 2927–2938.. [CrossRef]
21. Wang, J.C.; Shan, Q.H.; Zhu, J.; Cheng, X.F.; Wei, B.Z. Algebraic-Connectivity-Based Multi-USV Distributed Formation Method via Adding a Reverse Edge. *Mathematics* **2023**, *11*, 2942. [CrossRef]
22. Shao, J.; Shi, L.; Cheng, Y.; Li, T. Asynchronous Tracking Control of Leader-Follower Multiagent Systems with Input Uncertainties Over Switching Signed Digraphs. *IEEE Trans. Cybern.* **2021**, *52*, 6379–6390. [CrossRef] [PubMed]
23. Yang, Y.; Xiao, Y.; Li, T. Attacks on Formation Control for Multiagent Systems. *IEEE Trans. Cybern.* **2021**, *52*, 12805–12817. [CrossRef] [PubMed]
24. Li, Z.; Li, T.; Feng, G.; Zhao, R.; Shan, Q. Neural network-based adaptive control for pure-feedback stochastic nonlinear systems with time-varying delays and dead-zone input. *IEEE Trans. Syst. Man Cybern. Syst.* **2020**, *50*, 5317–5329. [CrossRef]
25. Li, T.; Bai, W.; Liu, Q.; Long, Y.; Chen, C.L.P. Distributed Fault-Tolerant Containment Control Protocols for the Discrete-Time Multi-Agent Systems via Reinforcement Learning Method. *IEEE Trans. Neural Netw. Learn. Syst.* **2023**, *8*, 3979–3991. [CrossRef]
26. Hua, Y.; Dong, X.; Han, L.; Li, Q.; Ren, Z. Formation-containment tracking for general linear multi-agent systems with a tracking-leader of unknown control input. *Syst. Control Lett.* **2018**, *122*, 67–76. [CrossRef]
27. Wang, Y.; Liu, C. Distributed finite-time adaptive fault-tolerant formation–containment control for USVs with dynamic event-triggered mechanism. *Ocean Eng.* **2023**, *280*, 114524. [CrossRef]
28. Hao, Y.; Lin, Z.; Hu, K.; Liu, L. Layered fully distributed formation-containment tracking control for multiple unmanned surface vehicles. *Ocean Eng.* **2023**, *270*, 113658. [CrossRef]
29. Liu, Z.-W.; Hou, H.; Wang, Y.-W. Formation-containment control of multiple underactuated surface vessels with sampling communication via hierarchical sliding mode approach. *ISA Trans.* **2022**, *124*, 458–467. [CrossRef]
30. Liu, Y.Y.; Lin, X.G.; Zhang, C. Affine Formation Maneuver Control for Multi-Heterogeneous Unmanned Surface Vessels in Narrow Channel Environments. *J. Mar. Sci. Eng.* **2023**, *11*, 1811. [CrossRef]
31. Wu, W.; Tong, S.C. Fixed-time formation fault tolerant control for unmanned surface vehicle systems with intermittent actuator faults. *Ocean Eng.* **2023**, *281*, 114813. [CrossRef]
32. Tan, G.G.; Zhuang, J.Y.; Zou, J.; Wan, L. Adaptive adjustable fast marching square method based path planning for the swarm of heterogeneous unmanned surface vehicles (USVs). *Ocean Eng.* **2023**, *268*, 113432. [CrossRef]
33. Wei, X.W.; Wang, H.; Tang, Y.X. Deep hierarchical reinforcement learning based formation planning for multiple unmanned surface vehicles with experimental results. *Ocean Eng.* **2023**, *268*, 115577. [CrossRef]
34. Yan, X.; Jiang, D.P.; Miao, R.L.; Li, Y.L. Formation Control and Obstacle Avoidance Algorithm of a Multi-USV System Based on Virtual Structure and Artificial Potential Field. *J. Mar. Sci. Eng.* **2021**, *9*, 161. [CrossRef]
35. Zhang, R.B.; Wen, N.F.; Yu, H.Y.; Wang, L.Y.; Wu, J.W.; Liu, G.Q. USVs Cooperative Collision Avoidance Based on Man-Machine Interaction and the Artificial Potential Field Method. In Proceedings of the 2019 Chinese Automation Congress (Cac2019), Hangzhou, China, 22–24 November 2019; pp. 1552–1557.

36. Gao, S.; Peng, Z.; Wang, H.; Liu, L.; Wang, D. Safety-Critical Model-Free Control for Multi-Target Tracking of USVs with Collision Avoidance. *IEEE/CAA J. Autom. Sin.* **2022**, *9*, 1323–1326. [CrossRef]
37. Gong, X.; Liu, L.; Peng, Z. Safe-critical formation reconfiguration of multiple unmanned surface vehicles subject to static and dynamic obstacles based on guiding vector fields and fixed-time control barrier functions. *Ocean Eng.* **2022**, *250*, 110821. [CrossRef]
38. Fu, J.; Wen, G.; Yu, X.; Wu, Z.G. Distributed Formation Navigation of Constrained Second-Order Multiagent Systems with Collision Avoidance and Connectivity Maintenance. *IEEE Trans. Cybern.* **2022**, *52*, 2149–2162..2020.3000264. [CrossRef] [PubMed]
39. Wang, L.; Ames, A.D.; Egerstedt, M. Safety Barrier Certificates for Collisions-Free Multirobot Systems. *IEEE Trans. Robot.* **2017**, *33*, 661–674. [CrossRef]
40. Zhang, H.; Lewis, F.L. Adaptive cooperative tracking control of higher-order nonlinear systems with unknown dynamics. *Automatica* **2012**, *48*, 1432–1439. [CrossRef]
41. Li, Z.; Wen, G.; Duan, Z.; Ren, W. Designing Fully Distributed Consensus Protocols for Linear Multi-Agent Systems with Directed Graphs. *IEEE Trans. Autom. Control* **2015**, *60*, 1152–1157. [CrossRef]
42. Li, Z.; Duan, Z.; Chen, G.; Huang, L. Consensus of Multiagent Systems and Synchronization of Complex Networks: A Unified Viewpoint. *IEEE Trans. Circuits Syst. I Regul. Pap.* **2010**, *57*, 213–224. [CrossRef]
43. Meng, Z.; Ren, W.; You, Z. Distributed finite-time attitude containment control for multiple rigid bodies. *Automatica* **2010**, *46*, 2092–2099. [CrossRef]

Journal of
Marine Science and Engineering

MDPI

Article

Formation Control for UAV-USVs Heterogeneous System with Collision Avoidance Performance

Yuyang Huang [1], Wei Li [1], Jun Ning [1,*] and Zhihui Li [2]

1 College of Navigation, Dalian Maritime University, Dalian 116026, China; huangyuyang@dlmu.edu.cn (Y.H.); li_wei@dlmu.edu.cn (W.L.)
2 College of Information Technology, Jilin Normal University, Siping 136000, China; lizhihui1650@163.com
* Correspondence: junning@dlmu.edu.cn

Abstract: This paper investigates the cooperative formation trajectory tracking problem for heterogeneous unmanned aerial vehicle (UAV) and multiple unmanned surface vessel (USV) systems with collision avoidance performance. Firstly, a formation control protocol based on extended state observer (ESO) is proposed to ensure that the UAV and the USVs track the target trajectory simultaneously in the XY plane. Then, the collision avoidance control strategy of USV formation based on artificial potential field (APF) theory is designed. Specifically, the APF method is improved by reconstructing the repulsive potential field to make the collision avoidance action of USVs more in line with the requirements of International Regulations for Preventing Collisions at Sea (COLREGs). Following that, an altitude controller for the UAV is proposed to maintain the cooperative formation of the heterogeneous systems. Based on the input-to-state stability, the stability of the proposed control structure is proven, and all the signals in the closed-loop system are ultimately bounded. Finally, a simulation study is provided to show the efficacy of the proposed strategy.

Keywords: heterogeneous formation control system; UAV-USVs; extended state observer; collision avoidance; artificial potential field method

check for updates

Citation: Huang, Y.; Li, W.; Ning, J.; Li, Z. Formation Control for UAV-USVs Heterogeneous System with Collision Avoidance Performance. *J. Mar. Sci. Eng.* **2023**, *11*, 2332. https://doi.org/10.3390/jmse11122332

Academic Editor: Weicheng Cui

Received: 9 October 2023
Revised: 24 November 2023
Accepted: 7 December 2023
Published: 10 December 2023

1. Introduction

In the past few years, unmanned systems have gained significant prominence in the realm of industrial advancement, attributed to their diverse usage in areas such as reconnaissance [1,2], marine pollution tracking [3,4], and advanced traffic management [5,6]. These systems principally encompass unmanned aerial vehicles (UAVs), unmanned ground vehicles (UGVs), unmanned surface vehicles (USVs), and autonomous underwater vehicles (AUVs) [7,8]. Although each unmanned system can handle tasks, its capacity limits its ability to deal with more complex tasks. Therefore, a homogeneous or heterogeneous system is more effective than a single unmanned system in completing intricate tasks [9–11].

The development of USV has been widely employed in ocean engineering due to its benefits, which include increased loading capacity, more convenience, and cheaper mission costs [12,13]. However, finding the rescue target in rescue missions and maritime searches is difficult owing to the restricted observation range of USVs [14,15]. Fortunately, incorporating UAVs into USV systems to construct heterogeneous multi-agent systems can compensate for this shortcoming [16,17]. A USV usually boosts its capacity to operate at sea by using the UAV's flexibility and utilizing the powerful target search ability while simultaneously widening the communication range of the sea between USVs, therefore, the heterogeneous systems extend the effective working area [18,19]. As a result, it is critical to investigate the formation challenges of a heterogeneous system, which has primarily engineering importance.

For the cooperative tracking control problem of UAVs and USVs, many research results have been addressed, among which typical control methods include the leader following method [20], virtual structure method [21], behavior-based approach [22], and model prediction method [23]. In [24], a series of coordinate transformations have been developed to convert the tracking error dynamics of a marine aerial-surface heterogeneous system into translation–rotation cascade forms. This eliminates variability between the UAV and the USV in terms of both translation and rotation dynamics. In [25], the research proposes an adaptive fault-tolerant time-varying formation control scheme. This scheme is particularly tailored to address challenges posed by actuator failures, parameter uncertainties, and external disturbances within the framework of a directed communication topology.

In the area of cooperative control of heterogeneous formations of UAVs and USVs, a globally applicable fixed-duration adaptive neural network is designed in [18] that employs a nonsingular rapid terminal sliding approach for formation control. This protocol is engineered to accurately follow a specified trajectory and achieve a predefined formation arrangement within a set timeframe, effectively addressing a range of uncertainties. This is accomplished by harnessing the combined strengths of adaptive methodologies and Radial Basis Function (RBF) neural networks. In [26], the research proposes a velocity estimation-based control strategy that comprises a distributed observer for estimating each vehicle's reference velocity. In [23], a distributed model predictive control algorithm tailored for heterogeneous systems is presented, characterized by a directed topological structure.

It is worth noting that the above research results tend to focus on the establishment of air–sea cooperation and the realization of trajectory tracking without considering the problem of collision avoidance [27,28]. During the mission, a minimum safe distance has to be maintained between any two unmanned systems to prevent collisions and to avoid damage to the overall performance of the search and rescue cooperative mission. Therefore, another key issue for the heterogeneous formation system is collision avoidance. Collision avoidance includes not only avoiding collisions between unmanned systems or between unmanned systems and other individuals but also avoiding collisions between unmanned systems and obstacles. Collision avoidance can be further divided into static collision avoidance based on sensor information and dynamic collision avoidance based on an unknown environment. Collision avoidance between aerial vehicles and obstacles can be divided into two ways: overall collision avoidance and changing formation collision avoidance.

There are relatively few research results on the cooperative formation control of heterogeneous systems with collision avoidance and obstacle avoidance. To avoid collision between ships, [29,30] propose an unmanned ship formation control method based on guaranteed performance, but the collision avoidance of obstacles is not addressed. For the collision avoidance problem between UAVs with static and dynamic obstacles, a cooperative controller for multiple unmanned ships based on artificial potential function is proposed in [12,31], while in the area of cooperative collision avoidance and obstacle avoidance for heterogeneous systems, research results are more limited. In [32], a distributed heterogeneous cooperative tracking control approach is proposed based on the leader-following method, and the artificial potential field (APF) algorithm is used to construct a control strategy with a collision avoidance mechanism. However, the research results between vehicles ignore the nautical engineering practice and poorly consider the COLREGs.

Based on the preceding analysis, this paper investigates the cooperative formation trajectory tracking problem for UAV-USV heterogeneous systems with model uncertainty and external disturbances. To solve this problem, a formation control protocol based on extended state observer (ESO) is proposed to ensure that UAV and USV track the target trajectory simultaneously. The collision avoidance control strategy of USV formation based on improved APF theory is designed. The collision avoidance problem between multiple USV formations formed under UAV coordination is solved by innovatively introducing the ship encounter situation and danger evaluation index into the artificial potential field. The key contributions can be summarized as follows:

(1) Compared with the existing results in [12,29–31], which only study the cooperative trajectory tracking problem of UAV and USV heterogeneous systems, this paper

explores the collision avoidance protocol for USV formation under UAV cooperation with navigation practice.

(2) Compared with the existing results in [18,25,26,32], which estimate that the system's indeterminate terms rely on RBF neural networks and fuzzy logic, etc., this paper employs ESO to realize the accurate compensation of uncertainties and external disturbances in heterogeneous systems.

(3) Compared with the existing results in [31–37], this paper innovatively introduces the ship encounter situation and danger evaluation index into the APF approach, and the improved APF method for heterogeneous cooperative control collision avoidance decision is more in line with the navigation practice.

The organization of this paper is structured as follows. Section 2 formulates the problem. Section 3 expounds on the controller design and the closed-loop system stability analysis. Section 4 demonstrates the simulation illustrations. Section 5 concludes this paper.

2. Preliminaries and Problem Statement

2.1. Problem Formulation

Consider the heterogeneous systems consisting of one UAV and N USVs. First, the dynamic models of the UAV and N USVs are presented. They are used to illustrate a unified dynamic model for the heterogeneous system. For ease of use, let $\Pi = \{1, 2, \ldots, N\}$. According to the results in [38], the dynamic model of the quadrotor UAV can be described as

$$\begin{cases} \ddot{p}_{ax} = (\sin\phi_a \sin\psi_a + \cos\phi_a \sin\theta_a \cos\psi_a)\frac{u_{ap}}{m_a} - \frac{d_{ax}\dot{p}_{ax}}{m_a} + \Delta_{ax}, \\ \ddot{p}_{ay} = (\sin\phi_a \cos\psi_a + \cos\phi_a \sin\theta_a \sin\psi_a)\frac{u_{ap}}{m_a} - \frac{d_{ay}\dot{p}_{ay}}{m_a} + \Delta_{ay}, \\ \ddot{p}_{az} = (\cos\theta_a \cos\phi_a)\frac{u_{ap}}{m_a} - \frac{d_{az}\dot{p}_{az}}{m_a} - g_a + \Delta_{az}, \end{cases} \tag{1}$$

$$\begin{cases} \ddot{\phi}_a = \dot{\theta}_a\dot{\psi}_a\frac{J_{ay}-J_{az}}{J_{ax}} - \frac{J_{ar}}{J_{ax}}\dot{\theta}_a\bar{d}_a + \frac{\tau_{a\phi}}{J_{ax}} - \frac{d_{a\phi}\dot{\phi}_a}{J_{ax}}, \\ \ddot{\theta}_a = \dot{\phi}_a\dot{\psi}_a\frac{J_{az}-J_{ax}}{J_{ay}} - \frac{J_{ar}}{J_{ay}}\dot{\phi}_a\bar{d}_a + \frac{\tau_{a\theta}}{J_{ay}} - \frac{d_{a\theta}\dot{\theta}_a}{J_{ay}}, \\ \ddot{\psi}_a = \dot{\phi}_a\dot{\theta}_a\frac{J_{ax}-J_{ay}}{J_{az}} + \frac{\tau_{a\psi}}{J_{az}} - \frac{d_{a\psi}\dot{\psi}_a}{J_{az}}, \end{cases} \tag{2}$$

where $[\phi_a, \theta_a, \psi_a]^T$ is the attitude state, $[p_{ax}, p_{ay}, p_{az}]^T$ is the position state, $\tau_{a\phi}, \tau_{a\theta}, \tau_{a\psi}$ are the three control torques, u_{ap} is the control thrust, g_a is the gravitational acceleration, m_a denotes the mass, $\bar{d}_a$ is the overall residual rotor angle, J_{ax}, J_{ay}, J_{az} represent the moments of the inertia, $d_{ax}, d_{ay}, d_{az}, d_{a\phi}, d_{a\theta}, d_{a\psi}$ represent the translational drag coefficients, $\Delta_{ax}, \Delta_{ay}, \Delta_{az}$ are the external disturbances encountered, and J_{ar} denotes the moment of rotor's inertia.

Inspired by the results in [25], the UAV's rotational dynamics can be stabilized separately. Therefore, in light of external disturbances and parametric uncertainties, the UAV model (1) is redefined in the following manner

$$\ddot{p}_a = g_a u_a + f_a + \Delta_a \tag{3}$$

where $p_a = [p_{ax}, p_{ay}, p_{az}]^T$ is the position , $f_a = [-d_{ax}\dot{p}_{ax}/m_a, -d_{ay}\dot{p}_{ay}/m_a, -d_{az}\dot{p}_{az}/m_a - g_a]^T$, $g_a = \text{diag}\{1/m_a, 1/m_a, 1/m_a\}$, $\Delta_a = [\Delta_{ax}, \Delta_{ay}, \Delta_{az}]^T$, $u_a = [u_{ax}, u_{ay}, u_{az}]^T$ denotes the new control signal which is given as [39]

$$\begin{cases} u_{ax} = (\sin\phi_a \sin\psi_a + \cos\phi_a \sin\theta_a \cos\psi_a)u_p \\ u_{\text{ay}} = (\sin\phi_a \cos\psi_a + \cos\phi_a \sin\theta_a \sin\psi_a)u_p \\ u_{az} = (\cos\theta_a \cos\phi_a)u_p \end{cases}$$

In the horizontal plane, the i-th ($i \in \Pi$) USV's kinematic and dynamic equations are described as [40]

$$\begin{cases} \dot{x}_{bi} = \mu_{bi}\cos\psi_{bi} - v_{bi}\sin\psi_{bi}, \\ \dot{y}_{bi} = \mu_{bi}\sin\psi_{bi} + v_{bi}\cos\psi_{bi}, \\ \dot{\psi}_{bi} = r_{bi} \end{cases} \tag{4}$$

$$\begin{cases} \dot{\mu}_{bi} = f_{\mu bi}(\alpha_i) + \frac{1}{m_{\mu bi}}\left(\tau^f_{\mu bi} + w_{\mu bi}\right), \\ \dot{v}_{bi} = f_{vbi}(\alpha_i) + \frac{1}{m_{vbi}} w_{vbi} \\ \dot{r}_{bi} = f_{rbi}(\alpha_i) + \frac{1}{m_{rbi}}\left(\tau^f_{rbi} + w_{rbi}\right) \end{cases} \tag{5}$$

and

$$\begin{cases} f_{\mu bi}(\alpha_i) = \frac{1}{m_{\mu bi}}\left(m_{vbi}v_{bi}r_{bi} - d_{\mu bi}\mu_{bi} - d_{\mu bi1}|\mu_{bi}|\mu_{bi}\right) \\ f_{vbi}(\alpha_i) = \frac{1}{m_{vbi}}\left(-m_{\mu bi}\mu_{bi}r_{bi} - d_{vbi}v_{bi} - d_{vbi1}|v_{bi}|v_{bi}\right), \\ f_{rbi}(\alpha_i) = \frac{1}{m_{rbi}}\left(\left(m_{\mu bi} - m_{vbi}\right)\mu_{bi}v_{bi} - d_{rbi}r_{bi} - d_{rbi1}|r_{bi}|r_{bi}\right), \end{cases} \tag{6}$$

where ψ_{bi} denotes the yaw angle; (x_{bi}, y_{bi}) denotes the position; $\alpha_i = [\mu_{bi}, v_{bi}, r_{bi}]^T$ denote the surge, sway, and yaw velocity, respectively; $m_{\mu bi}, m_{vbi}, m_{rbi}$ represent the inertial mass; $f_{\mu bi}(\alpha_i), f_{vbi}(\alpha_i), f_{rbi}(\alpha_i)$ denote the nonlinear dynamics consisting of Coriolis forces and the unmodeled hydrodynamics; $w_{\mu bi}, w_{vbi}, w_{rbi}$ represent the bounded disturbances; and $\tau^f_{\mu bi}$ and τ^f_{rbi} represent the surge force and the yaw moment.

Because the USVs' motion model given in (4) and (5) is underactuated, a hand position technique is employed to compensate. The USVs' front point $\left(p_{bix}, p_{biy}\right)$ is defined as the hand point, which can be expressed as

$$\begin{cases} p_{bix} = x_{bi} + L_{bi}\cos\psi_{bi} \\ p_{biy} = y_{bi} + L_{bi}\sin\psi_{bi} \end{cases} \tag{7}$$

where L_{bi} denotes the distance between the new defined hand point $\left(p_{bix}, p_{biy}\right)$ and the actual position (x_{bi}, y_{bi}), which is depicted in Figure 1.

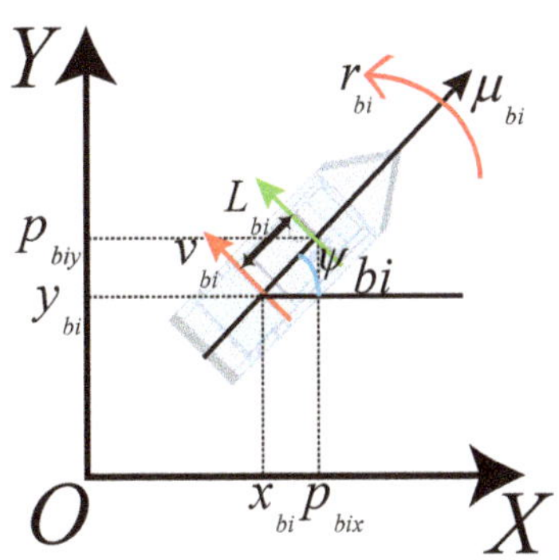

Figure 1. The kinematic model of USV.

The second derivative of (7) yields the following result

$$\begin{cases} \ddot{p}_{bix} = \dot{\mu}_{bi}\cos\psi_{bi} - (\dot{v}_{bi} + L_{bi}\dot{r}_{bi})\sin\psi_{bi} - \mu_{si}r_{bi}\sin\psi_{bi} \\ \qquad - \left(v_{bi}r_{bi} + L_{bi}r_{bi}^2\right)\cos\psi_{bi} \\ \ddot{p}_{biy} = \dot{\mu}_{bi}\sin\psi_{bi} - (\dot{v}_{bi} + L_{bi}\dot{r}_{bi})\cos\psi_{bi} + \mu_{si}r_{bi}\cos\psi_{bi} \\ \qquad - \left(v_{bi}r_{bi} + L_{bi}r_{bi}^2\right)\sin\psi_{bi} \end{cases} \tag{8}$$

Substituting (6) into (8) gives that

$$\begin{cases} \ddot{p}_{bix} = f_{bix}(\beta) + \frac{\cos\psi_{bi}}{m_{\mu bi}}\tau_\mu^f - \frac{L_{bi}\sin\psi_{bi}}{m_{rbi}}\tau_r^f + w_{bix} \\ \ddot{p}_{biy} = f_{biy}(\beta) + \frac{\sin\psi_{bi}}{m_{\mu bi}}\tau_\mu^f + \frac{L_{bi}\cos\psi_{bi}}{m_{rbi}}\tau_r^f + w_{biy} \end{cases} \tag{9}$$

where

$$\begin{cases} f_{bix}(\beta) = f_\mu(\alpha)\cos\psi_{bi} - (f_v(\alpha) + L_{bi}f_r(\alpha))\sin\psi_{bi} \\ \qquad - \mu_{bi}r_{bi}\sin\psi_{bi} - \left(v_{bi}r_{bi} + L_{bi}r_{bi}^2\right)\cos\psi_{bi} \\ f_{biy}(\beta) = f_\mu(\alpha)\sin\psi_{bi} + (f_v(\alpha) + L_{bi}f_r(\alpha))\cos\psi_{bi} \\ \qquad + \mu_{bi}r_{bi}\cos\psi_{bi} - \left(v_{bi}r_{bi} + L_{bi}r_{bi}^2\right)\sin\psi_{bi} \end{cases}$$

$$\begin{cases} w_{bix}(\beta) = \frac{w_{\mu bi}}{m_{\mu bi}}\cos\psi_{bi} - \left(\frac{w_{vbi}}{m_{\mu bi}} + \frac{L_{bi}w_{rbi}}{m_{rbi}}\right)\sin\psi_{bi} \\ w_{biy}(\beta) = \frac{w_{\mu bi}}{m_{\mu bi}}\sin\psi_{bi} + \left(\frac{w_{vbi}}{m_{\mu bi}} + \frac{L_{bi}w_{rbi}}{m_{rbi}}\right)\cos\psi_{bi} \end{cases}$$

with $\beta = [\mu_{bi}, v_{bi}, r_{bi}, \psi_{bi}]^T$.

Based on (9), the i-th USV position dynamics can be described as

$$\ddot{p}_{bi} = f_{bixy} + \Omega_{bi}(\psi_{bi})\omega_{bi}u_{bi} + w_{bixy} \tag{10}$$

where $p_{bi} = \left[p_{bix}, p_{biy}\right]^T$ is the i-th USV position, $f_{bixy} = \left[f_{bix}, f_{biy}\right]^T$, $\Omega_{bi}(\psi_{bi}) = [\cos\psi_{bi}, -\sin\psi_{bi}; \sin\psi_{bi}, \cos\psi_{bi}]$, $u_{bi} = [\tau_\mu, \tau_r]^T$, $\omega_{bi} = \mathrm{diag}\left\{1/m_{\mu bi}, L_{bi}/m_{rbi}\right\}$, $w_{bixy} = \left[w_{bix}, w_{biy}\right]^T$.

Combining (3) and (10), the unified model of the heterogeneous systems can be described as

$$\begin{cases} \dot{x}_{i1} = x_{i2} \\ \dot{x}_{i2} = F_{xi} + G_{xi}u_{xi} + \Delta_{xi} \end{cases} \tag{11}$$

When (11) represents the UAV model. Where $x_{i1} = \left[p_{ax}, p_{ay}\right]^T = x_{a1} \in R^2$, $x_{i2} = [\dot{p}_{ax}, \dot{p}_{ay}]^T = x_{a2} \in R^2$, $F_{xi} = F_{axy} = [-d_x\dot{p}_{ax}/m_a, -d_y\dot{p}_{ay}/m_a]^T$, $G_{xi} = G_{axy} = \mathrm{diag}\{1/m_a, 1/m_a\}$, $\Delta_{xi} = \Delta_{axy} = [\Delta_{ax}, \Delta_{ay}]^T$, $u_{xi} = u_{axy} = \left[u_{ax}, u_{ay}\right]^T$.

When (11) represents the USV model. Where $x_{i1} = \left[p_{bix}, p_{biy}\right]^T = x_{bi1} \in R^2$, $x_{i2} = \left[\dot{p}_{bix}, \dot{p}_{biy}\right]^T = x_{bi2} \in R^2$, $F_{xi} = f_{bixy} = \left[f_{bix}, f_{biy}\right]^T$, $G_{xi} = \Omega_{bi}(\psi_{bi})\omega_{bi}$, $\Delta_{xi} = w_{bixy} = \left[w_{bix}, w_{biy}\right]^T$, $u_{xi} = u_{bi} = \left[\tau_\mu, \tau_r\right]^T$.

The virtual leader is defined and its motion model is described as follows

$$\begin{cases} \dot{x}_{l1} = x_{l2} \\ \dot{x}_{l2} = F_l \end{cases} \tag{12}$$

where $x_{l1} \in R^2, x_{l2} \in R^2$ denote the position and velocity state vectors, respectively. $F_l \in R^2$ is a smooth unknown nonlinear function.

Define the system's error variable as

$$\begin{cases} \bar{x}_{i1} = x_{i1} - x_{l1} - \delta_i \\ \bar{x}_{i2} = x_{i2} - x_{l2} \end{cases} \tag{13}$$

where $\delta_i \in R^2$ represents the desired relative position vector between i-th of the agent of the heterogeneous systems and the virtual leader.

Assumption 1. *The aerodynamic drag coefficients d_{ix}, d_{iy}, and d_{iz} are bounded and unknown. The unknown nonlinear function F_l is bounded. Then, there exists a positive constant α_l such that $F_l < \alpha_l$.*

Assumption 2. *The quadrotor UAV experiences external perturbations $\Delta_{ax}, \Delta_{ay}, \Delta_{az}$, which are confined within certain bounds, fulfilling the conditions $\|\Delta_{ax}\| \leq \bar{\Delta}_{ax}$, $\|\Delta_{ay}\| \leq \bar{\Delta}_{ay}$, $\|\Delta_{az}\| \leq \bar{\Delta}_{az}$. Here, $\bar{\Delta}_{ax}$, $\bar{\Delta}_{ay}$, and $\bar{\Delta}_{az}$ represent unknown positive constants. The USV experiences external disturbances $w_{\mu bi}, w_{vbi}, w_{rbi}$ which are confined within certain limits, fulfilling the conditions $\|w_{\mu bi}\| \leq \bar{\varpi}_{\mu bi}$, $\|w_{vbi}\| \leq \bar{\varpi}_{vbi}$, $\|w_{rbi}\| \leq \bar{\varpi}_{rbi}$. In this context, $\bar{\varpi}_{\mu bi}, \bar{\varpi}_{vbi}$, and $\bar{\varpi}_{rbi}$ signify unknown positive constants.*

Assumption 3. *The velocity and acceleration of the agent and the virtual leader are bounded, and they satisfy $v_l \neq 0, v_i \neq 0$.*

Assumption 4. *G_{xi} is a symmetric matrix and its eigenvalues $\lambda_1, \lambda_2, \ldots, \lambda_m$ are satisfied such that $0 < \vartheta_i < \lambda_1 < \lambda_2 \cdots < \lambda_m < \infty$, where ϑ_i is a positive constant.*

Theorem 1 ([41])**.** *For any point in time, the function $V(t)$ is continuous and positive, with its initial state being within limits. Given that the inequality $\dot{V}(t) \leq -\gamma V(t) + \mu$ is satisfied, where $\gamma > 0$ and $\mu > 0$, it can be deduced that*

$$V(t) \leq \frac{\mu}{\gamma}(1 - e^{-\gamma t}) + V(0)e^{-\gamma t}.$$

Theorem 2 ([42])**.** *At all moments, the function $S(t)$ remains positive and continuous, with its initial condition $S(0)$ being constrained. Should the condition $\dot{S}(t) > qS(t)$ be valid for $t - t_0 \geq 0$ with $q > 0$, it leads to the inference that*

$$S(t) > e^{q(t-t_0)}S(t_0).$$

2.2. Algebraic Graph Theory

$G = (V_g, \varepsilon_g, A_g)$ is an undirected graph in this paper, with $V_g = \{v_1, v_2, \ldots, v_n\}$ being the node-set, $\varepsilon_g \subseteq V_g \times V_g$ being the edge set, and $A_g = [a_{ij}]$ being the adjacency matrix, of which all the elements are non-negative. The adjacency matrix A_g depicts the effectiveness of communication from agent j to agent i, where a_{ij} denotes the communication weight corresponding to the edge, and diagonal element $a_{ii} = 0$, $a_{ij} > 0$. As an undirected graph, the adjacency element $a_{ij} = a_{ji}$ must be guaranteed to be satisfied. If any pair of unique nodes can be connected by an undirected path, then the graph is connected. The Laplacian matrix $L_g = [l_{ij}] \subset R^{n \times n}$ is defined as follows

$$L_g = D_g - A_g$$

where $D_g = \text{diag}\{d_1, d_2, \ldots, d_n\}, d_i = \sum_{j=1}^{n} a_{ij}$ and $i = 1, 2, \ldots, n$. Defining the leader adjacency weight matrix as $B_g = \text{diag}\{b_1, b_2, \ldots, b_n\}$, where $b_i > 0$, indicates an exchange of information between agent i and the leader, otherwise, $b_i = 0, i = 1, 2, \ldots, n$. It is presupposed that there is always at least one agent linked to the leader and $b_1 + b_2 + \cdots + b_n > 0$.

Theorem 3 ([43]). *G must be irreducible for the graph to be linked.*

Theorem 4 ([43]). $\tilde{L_g} = L_g + B_g = \begin{bmatrix} l_{11} + b_1 & \cdots & l_{1n} \\ \vdots & \ddots & \vdots \\ l_{n1} & \cdots & l_{nn} + b_n \end{bmatrix}$ *are positive, where* B_g *is the leader adjacency weight matrix if the Laplacian matrix* L_g *of the undirected graph G is irreducible.*

2.3. Improved Artificial Potential Field and Virtual Repulsion

The essential concept of the APF technique is that each agent is viewed as a high-potential field for the control objective of collision avoidance. Any agent that is close to the others will be repelled from their potential fields by the repulsive force. Specifically, the APF method has fewer design parameters and a significantly simpler controller architecture. Some results regard the agent as a particle, which makes collision avoidance issues impractical. In actuality, the domain of agents should be considered during an encounter situation. As illustrated in Figure 2, the domain of USV can be defined as three concentric circles with unequal radii.

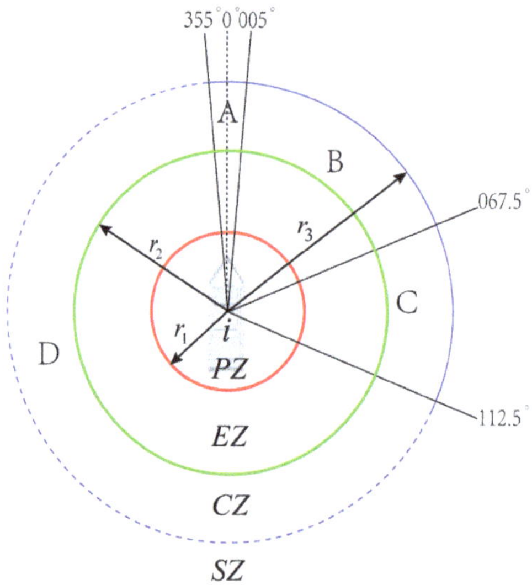

Figure 2. Repulsive potential field partition.

As is shown in Figure 2, the repulsive force field is separated into the following four areas. The area outside the first circular area is defined as the Safe Zone, where d_{ij} is greater than the detection distance r_3 of the potential field, there is no repulsive force in this area. The area inside the first circular area and outside the second circular is defined as the Negotiation Collision Avoidance Zone, where the distance d_{ij} is larger than r_2 and less than or equal to r_3. If the incoming ship is in zones A, B, or C in the area, the i-th USV is a giving-way vessel or has the same avoidance responsibility, and the i-th USV should take collision avoidance actions to avoid the collision. If the incoming ship is in zone D, the i-th USV is a stand-on vessel. The area inside the second circular area and outside the third circular area is defined as the Emergency Collision Avoidance Zone, where the distance d_{ij} is larger than r_1 and less than or equal to r_2. In the area where the distance between the other ship and the i-th USV is larger than and less than r_2, the i-th USV should perform emergency collision avoidance actions that can momentarily violate the COLREGs.

The area within the third circular area is defined as the Prohibited Zone, which is shown by the red circular limit with radius r_1 in Figure 2. All other USVs are prohibited from entering this area.

Where $d_{ij} = x_i - x_j$ represents the relative position variable between the i-th USV and the incoming j-th USV, $\|d_{ij}\|$ represents the distance variable, and r_3 is the collision hazard detection distance, we conclude that the j-th USV is defined as a collision avoidance neighbor Π^c of the i-th USV. Once its collision avoidance neighbor Π^c holds $\|d_{ij}\| = \|x_i - x_j\| \leq 2r_1$, it can be defined that the systems have been collided.

When the judgment condition $355^\circ \leq T_r < 360^\circ$ or $0^\circ \leq T_r < 67.5^\circ$, $r_2 < d_{ij} \leq r_3$ are met, the incoming j-th USV is in Zone A and Zone B. Meanwhile, when the judgment condition $67.5^\circ < T_r \leq 112.5^\circ$, $r_2 < d_{ij} \leq r_3$ are met, the incoming j-th USV is in Zone C. The i-th USV is a give-way ship. According to the COLREGs, it should turn right to give way. The repulsive force field is shown as follows [34]:

$$\varphi_{ij}(p,v) = \eta_d R_j \left(e^{\theta_m - \theta} - 1\right)\left(\frac{1}{d_{ij} - r_2} - \frac{1}{\rho_0}\right) \tag{14}$$

when the judgment condition $r_1 < d_{ij} \leq r_2$ is met, the i-th USV needs emergency collision avoidance. The repulsive force field is shown as follows:

$$\varphi_{ij}(p,v) = \eta_d R_j \left[\left(\frac{1}{d_{ij} - r_1} - \frac{1}{\rho_0}\right)^2 + (\|v_{ot}\| \cos\theta)^2\right] \tag{15}$$

When $d_{ij} > r_2$ or the incoming j-th USV is in Zone D, the repulsion field is not defined, that is, the repulsion is zero.

Where p and v are the position and speed, T_r denotes the relative position of the incoming j-th USV and the i-th USV, η_d is the positive scaling factor for the USVs, R_j is the radius of the puffing circle of the other ship, ρ_0 denotes the repulsive potential field's influence radius of other ships, θ_m is the angle between the maximum relative position line, θ is the angle between p_{ot}, v_{ot}, p_{ot} denotes the relative position line of the i-th USV and other ships, and v_{ot} denotes the relative velocity.

The virtual repulsive force f_{ij}^{ca} is defined as the negative gradient of the repulsive potential function $\varphi_{ij}(p,v)$ as follows:

$$f_{ij}^{ca} = -\nabla_{(p,v)} \varphi_{ij}(p,v) \tag{16}$$

where $-\nabla_{(p,v)}$ represents a negative gradient along p and v.

The total virtual repulsive force term of the whole systems is induced from (14)–(16) as follows:

$$u_i^{ca} = -\frac{\omega_i}{\vartheta_i} \sum_{j \in \Pi^c} \nabla_{(p,v)} \varphi_{ij}(p,v) \tag{17}$$

where ω_i denotes the positive gain parameter and ϑ_i is a positive constant.

3. Main Results

This section introduces a formation control protocol utilizing an extended state observer, aimed at guiding heterogeneous systems to follow a predefined trajectory in the XY plane, accounting for model uncertainties and external disturbances. Subsequently, a decentralized formation controller is developed to manage the height-tracking control of the UAV along the Z axis.

3.1. Controller Design Based on ESO

Rewrite (11) as

$$\begin{cases} \dot{x}_{i1} = x_{i2} \\ \dot{x}_{i2} = F_i + G_{xi}u_{xi} \end{cases} \tag{18}$$

where $F_i = [F_{Ti}, F_{Li}]^T = F_{xi} + \Delta_{xi}$, F_{Ti} and F_{Li} denote the transverse and longitudinal orientation of F_i and F_1 denotes the sum of the nonlinear unknown term and the external disturbance encountered of the UAV. F_2, F_3, F_4, F_5 denote the sum of the nonlinear unknown term and the external disturbance encountered by different USVs, respectively.

For the purpose of approximating F_i, the design of an extended state observer has been formulated as follows

$$\begin{cases} \dot{\hat{x}}_{i1} = \hat{x}_{i2} + \frac{\alpha_{i1}}{\varepsilon_i} k_{xi} \\ \dot{\hat{x}}_{i2} = G_{xi}u_{xi} + \hat{F}_i + \frac{\alpha_{i2}}{\varepsilon_i^2} k_{xi} \\ \dot{\hat{F}}_i = \frac{\alpha_{i3}}{\varepsilon_i^3} k_{xi} \end{cases} \tag{19}$$

where $\hat{x}_{i1}, \hat{x}_{i2}, \hat{F}_i$ represent the observer state, $\varepsilon_i > 0$, $\alpha_{i1} \in \mathbb{R}, \alpha_{i2} \in \mathbb{R}, \alpha_{i3} \in \mathbb{R}$ are positive constants, $k_{xi} = x_{i1} - \hat{x}_{i1}$. According to (19), we have

$$\begin{cases} \dot{\tilde{x}}_{i1} = \tilde{x}_{i2} + \frac{\alpha_{i1}}{\varepsilon_i} k_{xi} \\ \dot{\tilde{x}}_{i2} = \tilde{F}_i + \frac{\alpha_{i2}}{\varepsilon_i^2} k_{xi} \\ \dot{\tilde{F}}_i = \frac{\alpha_{i3}}{\varepsilon_i^3} k_{xi} - \dot{F}_i \end{cases} \tag{20}$$

where $\tilde{x}_{i1} = \hat{x}_{i1} - x_{i1}$ is the ESO estimation error.

Define η_i as

$$\eta_i = \begin{bmatrix} \eta_{i1} & \eta_{i2} & \eta_{i3} \end{bmatrix}^T$$

where

$$\eta_{i1} = \frac{x_{i1} - \hat{x}_{i1}}{\varepsilon_i^2}, \quad \eta_{i2} = \frac{x_{i2} - \hat{x}_{i2}}{\varepsilon_i}, \quad \eta_{i3} = F_i - \hat{F}_i$$

The observation error equation of state can be written as

$$\varepsilon_i \dot{\eta}_i = \overline{A}_i \eta_i + \varepsilon_i \overline{B}_i \dot{F}_i \tag{21}$$

where

$$\overline{A}_i = \begin{bmatrix} -\alpha_{i1} & 1 & 0 \\ -\alpha_{i2} & 0 & 1 \\ -\alpha_{i3} & 0 & 0 \end{bmatrix}, \quad \overline{B}_i = \begin{bmatrix} 0 \\ 0 \\ 1 \end{bmatrix}$$

The error in position and velocity for heterogeneous systems within the XY plane is defined as follows

$$\begin{aligned} e_i^{x_1} = & \sum_{j \in \Pi} a_{ij}\left((x_{1i} - x_{1j}) - (\delta_i - \delta_j)\right) \\ & + b_i(x_{1i} - x_{1l} - \delta_i) \end{aligned} \tag{22}$$

$$e_i^{x_2} = \sum_{j \in \Pi} a_{ij}(x_{2i} - x_{2j}) + b_i(x_{2i} - x_{2l}) \tag{23}$$

The formation control term is designed as follows

$$u_i^f = \frac{1}{\vartheta_i}[-\hat{F}_i - k_i(e_i^{x_1} + e_i^{x_2})] \tag{24}$$

Furthermore, the distributed formation controller with collision avoidance performance can be defined as follows

$$u_i = u_i^f + u_i^{ca} \tag{25}$$

Remark 1. *In the formation controller (25), the distributed formation control term u_i^f is used by tracking the time-varying velocity variable of the leader. The virtual repulsive force term u_i^{ca} is used to achieve the collision avoidance objective among the agents. By designing the action functions (14) and (15) of the APF, they are non-conflicting items.*

Using (17) and (24), distributed formation controller (25) can be given as

$$u_i = \frac{1}{\vartheta_i}[-\hat{F}_i - k_i(e_i^{x_1} + e_i^{x_2})] - \frac{\omega_i}{\vartheta_i}\sum_{j\in\Pi^c} \nabla_{(p,v)}\varphi_{ij}(p,v) \tag{26}$$

Take the derivative of (13) and substitute (11) and (12) into it. Define the derivatives of the error variables and write them in vector form as follows

$$\dot{\bar{Z}} = -[M \otimes I_m]\bar{Z} + \begin{bmatrix} 0_{nm} \\ F_{xi} \end{bmatrix} + \begin{bmatrix} 0_{nm} \\ U \end{bmatrix} + \begin{bmatrix} 0_{nm} \\ \Delta_{xi} \end{bmatrix} - \begin{bmatrix} 0_{nm} \\ F_l \end{bmatrix} \tag{27}$$

where $\dot{\bar{Z}} = \left[\dot{\bar{x}}_{11}^T, \ldots, \dot{\bar{x}}_{n1}^T, \dot{\bar{x}}_{12}^T, \ldots, \dot{\bar{x}}_{n2}^T\right]^T$, $U = G_{xi}u_{xi}$, $M = \begin{bmatrix} 0_n & -I_n \\ 0_n & 0_n \end{bmatrix}$, $\otimes$ denotes the Kronecker product.

3.2. Altitude Controller Design for UAV

The error system of altitude is defined as

$$e_{zp} = p_{az} - c_p \tag{28}$$

$$e_{zv} = v_{az} - c_v \tag{29}$$

where the desired position signal is c_p and the desired velocity signal is c_v.

The adaptive laws and control input of formation tracking are constructed as

$$u_{az} = \hat{H}_{az}^{-1}\left(-\sigma e_\zeta - \frac{e_\zeta \hat{\kappa}_{11}\omega_1^T\omega_1}{2\varsigma_1^2} - \frac{e_\zeta\hat{\kappa}_{12}}{2\varsigma_2^2} + \dot{c}_v - k_\zeta e_{zv}\right) \tag{30}$$

$$\dot{\hat{\kappa}}_{11} = l_{11}\left(-k_{11}\hat{\kappa}_{11} + \frac{e_\zeta^T e_\zeta \omega_1^T\omega_1}{2\varsigma_1^2}\right) \tag{31}$$

$$\dot{\hat{\kappa}}_{12} = l_{12}\left(-k_{12}\hat{\kappa}_{12} + \frac{e_\zeta^T e_\zeta}{2\varsigma_2^2}\right) \tag{32}$$

$$\dot{\hat{H}}_{az} = \text{Proj}_{[\underline{H}_{az},\bar{H}_{az}]}\{\mathcal{F}\} = \begin{cases} 0, & \text{if } \hat{H}_{az} = \bar{H}_{az} \\ & \text{and } \mathcal{F} \geq 0 \\ & \text{or } \hat{H}_{az} = \underline{H}_{az} \\ & \text{and } \mathcal{F} \leq 0 \\ \mathcal{F}, & \text{otherwise} \end{cases} \tag{33}$$

where $\mathcal{F} = l_{13}(-k_{13}\hat{H}_{az} + e_{\zeta}u_{az}), \underline{H}_{az}$, and $\bar{H}_{az}$ are the parameter H_{az}'s lower bound and upper bound of, respectively, where $\sigma, k_{11}, k_{12}, k_{13}, k_{\zeta}, l_{11}, l_{12}, l_{13}, \varsigma_1, \varsigma_2$ denote positive parameters.

3.3. Stability Analysis

- **Part A. Proof of the stability of the extended state observer**

Consider the following Lyapunov function

$$V_0 = \sum_{i=1}^{N+1} V_{0i} \tag{34}$$

where V_{0i} is given as

$$V_{0i} = \varepsilon_i {\eta_i}^T P_i \eta_i \tag{35}$$

$$\begin{aligned} \dot{V}_{0i} &= \varepsilon_i \dot{\eta}_i{}^T P_i \eta_i + \varepsilon_i \eta_i^T P_i \dot{\eta}_i \\ &= (\overline{A}_i \eta_i + \varepsilon_i \overline{B}_i \dot{F}_i)^T P_i \eta_i + {\eta_i}^T P_i (\overline{A}_i \eta_i + \varepsilon_i \overline{B}_i \dot{F}_{ni}) \\ &= {\eta_i}^T \overline{A}_i^T P_i \eta_i + \varepsilon_i (\overline{B}_i \dot{F}_{ni})^T P \eta_i \\ &+ {\eta_i}^T P_i \overline{A}_i \eta_i + \varepsilon {\eta_i}^T P_i \overline{B}_i \dot{F}_{ni} \\ &= {\eta_i}^T \left(\overline{A}_i^T P_i + P_i \overline{A}_i \right) \eta_i + 2\varepsilon_i {\eta_i}^T P_i \overline{B}_i \dot{F}_{ni} \\ &\leqslant -{\eta_i}^T Q_i \eta_i + 2\varepsilon_i \| P_i \overline{B}_i \| \cdot \| \eta_i \| \cdot | \dot{F}_i | \end{aligned} \tag{36}$$

and

$$\dot{V}_{0i} \leqslant -\lambda_{\min}(Q_i) \|\eta_i\|^2 + 2\varepsilon_i |\dot{F}_i| \| P_i \overline{B}_i \| \|\eta_i\| \tag{37}$$

Therefore, the convergence condition of the observer satisfies

$$\|\eta_i\| \leqslant \frac{2\varepsilon_i |\dot{F}_i| \| P \overline{B}_i \|}{\lambda_{\min}(Q_i)} \tag{38}$$

where Q_i denotes any given symmetric positive definite matrix, $\lambda_{\min}(Q_i)$ is the minimum eigenvalue of Q_i, and there is a three-by-three symmetric positive definite matrix P_i that satisfies the following equation

$$\overline{A}_i^T P_i + P_i \overline{A}_i + Q_i = 0 \tag{39}$$

Remark 2. *Based on Assumptions 2 and 3, both the velocity and acceleration of the agent, as well as the external disturbances, are bounded. From $F_i = F_{xi} + \Delta_{xi}$, it is evident that F_i encompasses the agent's velocity and acceleration, as well as external disturbances. So, $|\dot{F}_i|$ is bounded.*

- **Part B. Proof of the stability of the system**

Define the Lyapunov quadratic scalar function as follows

$$V_1 = \frac{1}{2}\bar{Z}^T(P_g \otimes I_m)\bar{Z} \tag{40}$$

where $P = \begin{bmatrix} 2\tilde{L}_g & \tilde{L}_g \\ \tilde{L}_g & \tilde{L}_g \end{bmatrix}$, $\tilde{L}_g = L_g + B_g$.

According to Theorem 4, the Lyapunov quadratic scalar function V_1 is a positive definite function.

Substitute (27) into $\dot{V}_1$ and expand the partial matrix as follows

$$\begin{aligned} \dot{V}_1 =& \bar{Z}^T(P_g \otimes I_m)\dot{\bar{Z}} \\ =& -\frac{1}{2}\bar{Z}^T\left(\left(M^TP_g + P_gM\right) \otimes I_m\right)\bar{Z} + \bar{Z}^T(P_g \otimes I_m)\begin{bmatrix} 0_{nm} \\ F_{xi} + U + \Delta_{xi} - F_l \end{bmatrix} \\ =& -\bar{Z}^T\left(\begin{bmatrix} 0_n & -\tilde{L}_g \\ -\tilde{L}_g & -\tilde{L}_g \end{bmatrix} \otimes I_m\right)\bar{Z} + \sum_{i=1}^{n}(e_i^{x_1} + e_i^{x_2})^T(F_{xi} + G_{xi}u_{xi} + \Delta_{xi} - F_l) \end{aligned} \tag{41}$$

Substituting the distributed formation controller (24) into (41) , we can obtain

$$\begin{aligned} \dot{V}_1 =& -\bar{Z}^T\left(\begin{bmatrix} 0_n & -\tilde{L}_g \\ -\tilde{L}_g & -\tilde{L}_g \end{bmatrix} \otimes I_m\right)\bar{Z} \\ & + \sum_{i=1}^{n}(e_i^{x_1} + e_i^{x_2})^T F_{xi} - \sum_{i=1}^{n}(e_i^{x_1} + e_i^{x_2})^T F_l + \sum_{i=1}^{n}(e_i^{x_1} + e_i^{x_2})^T \\ & \times \left(-\hat{F}_i - k_i(e_i^{x_1} + e_i^{x_2}) - \omega_i \sum_{j \in \Pi^c} \nabla_{(p,v)}\varphi_{ij}(p,v)\right) \\ =& -\bar{Z}^T\left(\begin{bmatrix} 0_n & -\tilde{L}_g \\ -\tilde{L}_g & -\tilde{L}_g \end{bmatrix} \otimes I_m\right)\bar{Z} - \sum_{i=1}^{n}(e_i^{x_1} + e_i^{x_2})^T F_l \\ & - \sum_{i=1}^{n} k_i\|(e_i^{x_1} + e_i^{x_2})\|^2 - \sum_{i=1}^{n}(e_i^{x_1} + e_i^{x_2})^T \times \omega_i \sum_{j \in \Pi^c} \nabla_{(p,v)}\varphi_{ij}(p,v) \end{aligned} \tag{42}$$

In the case of an undirected graph, the Laplacian matrix exhibits symmetry, and the virtual repulsive force term u_i^{ca} becomes negligible once collision avoidance is achieved through the repulsive potential function $\varphi_{ij}(p,v)$

Therefore, the aforementioned term $V' = -\sum_{i=1}^{n}(e_i^{x_1} + e_i^{x_2})^T \omega_i \sum_{j \in \Pi^c} \nabla_{(p,v)} \varphi_{ij}(p,v) = 0$, based on Assumption 1 and Young's inequality as follows

$$-(e_i^{x_1} + e_i^{x_2})^T F_l \leq \xi_i\|e_i^{x_1} + e_i^{x_2}\|^2 + \frac{\alpha_l^2}{4\xi_i}$$

where ξ_i is a positive constant.

And then the results (42) can be derived as follows

$$\begin{aligned} \dot{V}_1 \leq& -\bar{Z}^T\left(\begin{bmatrix} 0_n & -\tilde{L}_g \\ -\tilde{L}_g & -\tilde{L}_g \end{bmatrix} \otimes I_m\right)\bar{Z} + \sum_{i=1}^{n}\xi_i\|e_i^{x_1} + e_i^{x_2}\|^2 + \sum_{i=1}^{n}\frac{\alpha_l}{4\xi_i} - \sum_{i=1}^{n}k_i\|(e_i^{x_1} + e_i^{x_2})\|^2 \\ =& -\bar{Z}^T\left(\begin{bmatrix} 0_n & -\tilde{L}_g \\ -\tilde{L}_g & -\tilde{L}_g \end{bmatrix} \otimes I_m\right)\bar{Z} + \sum_{i=1}^{n}(\xi_i - k_i)\|e_i^{x_1} + e_i^{x_2}\|^2 + \sum_{i=1}^{n}\frac{\alpha_l^2}{4\xi_i} \end{aligned} \tag{43}$$

Let ρ_i satisfy the limit of $\rho_i \leq k_i - \xi_i$ and $\rho_i > 0$, the inequality is transformed into

$$\begin{aligned}\dot{V}_1 &\leq -\bar{Z}^T\left(\begin{bmatrix} 0_n & -\tilde{L}_g \\ -\tilde{L}_g & -\tilde{L}_g \end{bmatrix} \otimes I_m\right)\bar{Z} - \sum_{i=1}^{n} \rho_i \|e_i^{x_1} + e_i^{x_2}\|^2 + \sum_{i=1}^{n} \frac{\alpha_l^2}{4\xi_i} \\ &= -\bar{Z}^T\left(\left(\rho_i\begin{bmatrix} \tilde{L}_g^2 & \tilde{L}_g^2 \\ \tilde{L}_g^2 & \tilde{L}_g^2 \end{bmatrix} - \begin{bmatrix} 0_n & \tilde{L}_g \\ \tilde{L}_g & \tilde{L}_g \end{bmatrix}\right) \otimes I_m\right)\bar{Z} + \sum_{i=1}^{n} \frac{\alpha_l^2}{4\xi_i} \\ &= -\bar{Z}^T\left(\begin{bmatrix} \rho_i\tilde{L}_g^2 & \rho_i\tilde{L}_g^2 - \tilde{L}_g \\ \rho_i\tilde{L}_g^2 - \tilde{L}_g & \rho_i\tilde{L}_g^2 - \tilde{L}_g \end{bmatrix} \otimes I_m\right)\bar{Z} + \sum_{i=1}^{n} \frac{\alpha_l^2}{4\xi_i}\end{aligned} \tag{44}$$

According to the Lemma of linear matrix inequality in [41], $\rho_i\tilde{L}_g^2 - \left(\rho_i\tilde{L}_g^2 - \tilde{L}_g\right) = \tilde{L}_g^2 > 0$, $\rho_i\tilde{L}_g^2 - \tilde{L}_g > 0$, so the matrix $\begin{bmatrix} \rho_i\tilde{L}_g^2 & \rho_i\tilde{L}_g^2 - \tilde{L}_g \\ \rho_i\tilde{L}_g^2 - \tilde{L}_g & \rho_i\tilde{L}_g^2 - \tilde{L}_g \end{bmatrix} > 0$. Therefore, the inequality (44) can be rewritten as follows

$$\begin{aligned}\dot{V}_1 \leq &- \bar{Z}^T\left(\left(\rho_i\begin{bmatrix} \tilde{L}_g^2 & \tilde{L}_g^2 \\ \tilde{L}_g^2 & \tilde{L}_g^2 \end{bmatrix} - \begin{bmatrix} 0_n & \tilde{L}_g \\ \tilde{L}_g & \tilde{L}_g \end{bmatrix}\right) \otimes I_m\right)\bar{Z} + \sum_{i=1}^{n} \frac{\alpha_l^2}{4\xi_i} \\ = &- \bar{Z}^T((\rho_i\Theta - \Xi) \otimes I_m)\bar{Z} + \Delta\end{aligned} \tag{45}$$

where

$$\Theta = \begin{bmatrix} \tilde{L}_g^2 & \tilde{L}_g^2 \\ \tilde{L}_g^2 & \tilde{L}_g^2 \end{bmatrix}, \Xi = \begin{bmatrix} 0_n & \tilde{L}_g \\ \tilde{L}_g & \tilde{L}_g \end{bmatrix}, \Delta = \sum_{i=1}^{n} \frac{\alpha_l^2}{4\xi_i}.$$

Set $\rho_i > \frac{1}{\lambda_{\min}^{\Theta}}\left(\lambda_{\max}^{\Xi} + \frac{\Im}{2}\lambda_{\max}^{P}\right)$, where $\lambda_{\min}^{\Theta}, \lambda_{\max}^{\Xi}, \lambda_{\max}^{P}$ denote the smallest eigenvalue of the matrix Θ, the largest eigenvalue of the matrix Ξ, and the largest eigenvalue of the matrix P, respectively, and $\Im = \min\{\sigma_1\Upsilon_1, \ldots, \sigma_n\Upsilon_n\}$.

Then, we can obtain

$$\begin{aligned}\dot{V}_1 &\leq -\Im\bar{Z}^T(P \otimes I_m)\bar{Z} + \Delta \\ &= -\Im V_1 + \Delta\end{aligned} \tag{46}$$

In accordance with Theorem 1, the subsequent inequality can be presented in the following manner

$$V_1(t) \leq V_1(0)e^{-\Im t} + \frac{\Delta}{\Im}\left(1 - e^{-\Im t}\right) \tag{47}$$

Consequently, by choosing suitable parameters as indicated in [44], one can achieve the performance of adaptive leader-following formation.

- **Part C. Proof of collision avoidance**

The collision avoidance performance is analyzed only for the i-th ($i \in \Pi$) USV and its collision avoidance neighbor the j-th ($j \in \Pi^c$) USV, and the others can be analyzed by the same way in [42].

Define the energy function holds quadratic form as follows

$$S(t) = \frac{1}{2}v_i^T(t)v_i(t) + \frac{1}{2}d_{ik}^T(t)d_{ik}(t) \tag{48}$$

Taking the time derivative of (48) , we can get

$$\dot{S}(t) = \omega_i v_i^T(t)\nabla_{(p,v)}\varphi_{ij}(p,v) + d_{ik}^T(t)(v_i(t) - v_k(t)) - v_i^T(t)\hat{F}_i - k_i v_i^T(t)\left(e_i^{x_1}(t) + e_i^{x_2}(t)\right) \tag{49}$$

where $d_{ik}(t)$ is the relative position variable.

Because of the designing of the repulsive potential function $\varphi_{ij}(p,v)$, $\varphi_i(d_{ik}(t))$ will approach infinity by designing the parameter $\bar{\omega}_i > 0$ if the i-th USV is closing to the j-th USV. Therefore, if the i-th USV and the j-th USV are close to each other, and design the appropriate parameter, the inequality can be met in the following manner:

$$\begin{aligned} v_i^T(t)\nabla_{(p,v)}\varphi_{ij}(p,v) > &\frac{1}{2}d_{ik}^T(t)d_{ik}(t) + \frac{1}{2}v_i^T(t)v_i(t) \\ &- \frac{1}{\omega_i}d_{ik}^T(t)(v_i(t) - v_k(t)) + \frac{1}{\omega_i}v_i^T(t)\hat{F}_i \\ &+ \frac{k_i}{\omega_i}v_i^T(t)(e_i^{x_1}(t) + e_i^{x_2}(t)) \end{aligned} \tag{50}$$

Substituting the aforementioned inequality (50) into (49), the subsequent inequality can be derived

$$\dot{S}(t) > \omega_i S(t)$$

Based on the principles outlined in Theorem 2, the subsequent inequality can be derived

$$d_{ik}^T(t)d_{ik}(t) > 2e^{\omega_i(t-t_0)}S(t) - v_i^T(t)v_i(t)$$

Since the previously indicated term $v_i^T(t)v_i(t)$ is continuous and bounded at this dwell time. By designing the gain parameter $\omega_i > 0$ large enough, we get can the inequality $2e^{\omega_i(t-t_0)}S(t) - v_i^T(t)v_i(t) > (2r_1)^2$ and the result can be obtained as $\|d_{ik}(t)\| > 2r_1$. As a result, the collision avoidance performance can be guaranteed by the proposed formation control with collision avoidance strategy.

- **Part D. Proof of the decentralized formation controller**

If the feasibility condition $c_v - \dot{c}_p = 0$ is satisfied, it then allows the time derivative of Equations (28) and (29) to be expressed as follows

$$\dot{e}_{zp} = e_{zv} \tag{51}$$

$$\dot{e}_{zv} = F_{az} + G_{az}u_{az} + \Delta_{az} - \dot{c}_v. \tag{52}$$

where e_{zv} may be regarded as the virtual control input in the system (51). The stability of the system (51) can be assured by constructing the virtual control input $\zeta = -k_\zeta e_{zp}$.

A positive Lyapunov function is considered as

$$V_{zp} = \frac{1}{2}e_{zp}^T e_{zp} \tag{53}$$

A new error is defined as

$$e_\zeta = e_{zv} - \zeta \tag{54}$$

By computing the time derivative of Equation (53) and incorporating the control input from Equation (30) into this calculation, one can obtain

$$\dot{e}_\zeta = -\sigma e_\zeta + F_{az} + \Delta_{az} + \tilde{H}_{az}u_{az} - \frac{e_\zeta \hat{\kappa}_{11}\omega_1^T\omega_1}{2\varsigma_1^2} - \frac{e_\zeta \hat{\kappa}_{12}}{2\varsigma_2^2}. \tag{55}$$

V_z is given as

$$V_z = \frac{1}{2}e_{zp}^T e_{zp} + \frac{1}{2}e_\zeta^T e_\zeta + \frac{\tilde{\kappa}_{11}^2}{2l_{11}} + \frac{\tilde{\kappa}_{12}^2}{2l_{12}} + \frac{\mathrm{Tr}(\tilde{H}_{az}^T\tilde{H}_{az})}{2l_{13}} \tag{56}$$

with $\tilde{\kappa}_{11} = \kappa_{11} - \hat{\kappa}_{11}, \tilde{\kappa}_{12} = \kappa_{12} - \hat{\kappa}_{12}$ and $\tilde{H}_{az} = H_{az} - \hat{H}_{az}$.

$$\dot{V}_z \leq -k_\zeta e_{zp}^T e_{zp} - \sigma e_\zeta^T e_\zeta + e_\zeta^T F_{az} + e_\zeta^T \Delta_{az} + e_\zeta^T \tilde{H}_{az} u_{az} - \frac{e_\zeta^T e_\zeta \hat{\kappa}_{11} \omega_1^T \omega_1}{2\varsigma_1^2} - \frac{e_\zeta^T e_\zeta \hat{\kappa}_{12}}{2\varsigma_2^2} + \frac{\tilde{\kappa}_{11}\dot{\hat{\kappa}}_{11}}{l_{11}} + \frac{\tilde{\kappa}_{12}\dot{\hat{\kappa}}_{12}}{l_{12}} + \frac{\mathrm{Tr}\left(\tilde{H}_{az}^T \dot{\hat{H}}_{az}\right)}{l_{13}} \tag{57}$$

Given that F_{az} is an undefined function, as per the universal approximation theorem cited in [45], for any arbitrarily small constant ϵ_1, there exists a fuzzy logic system $\theta_1^{*T}\omega_1$ that can be represented as $F_{az} = \theta_1^{*T}\omega_1 + \epsilon_1$. In this context, θ_1^{*T} represents the optimal weight, ω_1 is the fuzzy basis vector, ϵ_1 denotes the fuzzy system's approximation error, and it is established that $|\epsilon_1| \leq \bar{\epsilon}_1$.

Regarding Young's inequality, it can be deduced that

$$e_\zeta^T \theta_1^{*T} \omega_1 \leq \frac{e_\zeta^T e_\zeta \kappa_{11} \omega_1^T \omega_1}{2\varsigma_1^2} + \frac{\varsigma_1^2}{2} \tag{58}$$

$$e_\zeta^T (\epsilon_1 + \Delta_{az}) \leq \frac{e_\zeta^T e_\zeta \kappa_{12}}{2\varsigma_2^2} + \frac{\varsigma_2^2}{2} \tag{59}$$

where $\kappa_{11} = \theta_1^{*T}\theta_1^*$ and $\kappa_{12} = (\bar{\epsilon}_1 + \bar{\Delta}_{az})^T(\bar{\epsilon}_1 + \bar{\Delta}_{az})$.

Substituting (58) and (59) into (57), we can obtain

$$\begin{aligned}
\dot{V}_{iz} \leq & -k_\zeta e_{zp}^T e_{zp} - \sigma e_\zeta^T e_\zeta + \frac{e_\zeta^T e_\zeta \kappa_{11} \omega_1^T \omega_1}{2\varsigma_1^2} + \frac{e_\zeta^T e_\zeta \kappa_{12}}{2\varsigma_2^2} \\
& - \frac{e_\zeta^T e_\zeta \hat{\kappa}_{11} \omega_1^T \omega_1}{2\varsigma_1^2} - \frac{e_\zeta^T e_\zeta \hat{\kappa}_{12}}{2\varsigma_2^2} - \tilde{\kappa}_{12}\left(-k_{12}\hat{\kappa}_{12} + \frac{e_\zeta^T e_\zeta}{2\varsigma_2^2}\right) \\
& - \tilde{\kappa}_{11}\left(-k_{11}\hat{\kappa}_{11} + \frac{e_\zeta^T e_\zeta \omega_1^T \omega_1}{2\varsigma_1^2}\right) + e_\zeta^T \tilde{H}_{az} u_{az} \\
& - \mathrm{Tr}\left(\tilde{H}_{az}^T(-k_{13}\hat{H}_{az} + e_\zeta u_{az})\right) + \frac{\varsigma_1^2}{2} + \frac{\varsigma_2^2}{2} \\
\leq & -k_\zeta e_{zp}^T e_{zp} - \sigma e_\zeta^T e_\zeta - \frac{k_{11}}{2}\tilde{\kappa}_{11}^2 - \frac{k_{12}}{2}\tilde{\kappa}_{12}^2 - \frac{k_{13}}{2}\|\tilde{H}_{az}\|_F^2 \\
& + \frac{k_{11}}{2}\kappa_{11}^2 + \frac{k_{12}}{2}\kappa_{12}^2 + \frac{k_{13}}{2}\|H_{az}\|_F^2 + \frac{\varsigma_1^2}{2} + \frac{\varsigma_2^2}{2} \\
\leq & -\vartheta_z V_z + v_z
\end{aligned} \tag{60}$$

where $\theta_z = \min\{2k_\zeta, 2\sigma, l_{11}k_{11}, l_{12}k_{12}, l_{13}k_{13}\} > 0, v_z = \frac{k_{11}}{2}\kappa_{11}^2 + \frac{k_{12}}{2}\kappa_{22}^2 + \frac{k_{13}}{2}\|H_{az}\|_F^2 + \frac{\varsigma_1^2}{2} + \frac{\varsigma_2^2}{2}$.

Based on Equation (60), $\dot{V}_z$ is derived as follows

$$\dot{V}_z \leq -\vartheta_z V_z + v_z \tag{61}$$

According to the boundedness theorem, the closed-loop system solution is uniformly eventually bounded. Using e_{zp} and e_ζ, it is possible to deduce that the UAV's velocity error and altitude tracking error are uniformly eventually bounded.

4. Simulation Result

In this section, a simulation analysis for the heterogeneous systems with collision avoidance performance under parameter uncertainty and external disturbance is provided to illustrate the effect of the proposed approach. One UAV and four USVs make up the heterogeneous system. The simulation experiments described in this section were executed on the MATLAB R2020a simulation platform.

The communication topology is defined in Figure 3, where agent l represents the virtual leader, agent 0 represents the UAV, and agents 1–4 represent the four USVs. If $B_g = \text{diag}\{0,0,0,0,1\}$ is the adjacency weight matrix between agents and the virtual leader, then the adjacency matrix A_g and the Laplacian matrix L_g are as follows:

$$A_g = \begin{bmatrix} 0 & 1 & 1 & 0 & 1 \\ 1 & 0 & 1 & 1 & 0 \\ 1 & 1 & 0 & 0 & 0 \\ 0 & 1 & 0 & 0 & 0 \\ 1 & 0 & 0 & 0 & 0 \end{bmatrix} \quad L_g = \begin{bmatrix} 3 & -1 & -1 & 0 & -1 \\ -1 & 3 & -1 & -1 & 0 \\ -1 & -1 & 2 & 0 & 0 \\ 0 & -1 & 0 & 1 & 0 \\ -1 & 0 & 0 & 0 & 1 \end{bmatrix}$$

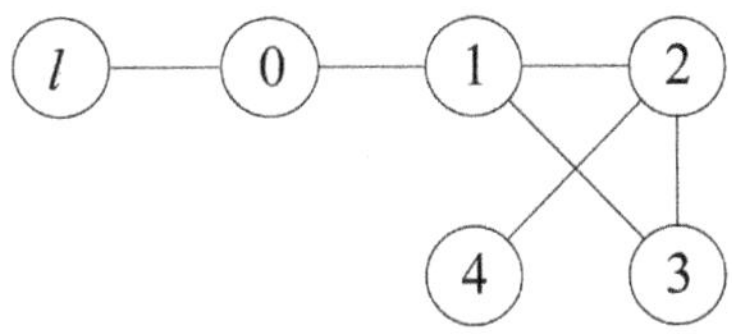

Figure 3. Communication topology graph.

The system parameters of a UAV and USVs are given in Tables 1 and 2, respectively. APF design parameters $r_{in} = 0.5, r_{\text{out}} = 2, \bar{\omega}_i = 0.3, \omega_i = 1$, where $i = 1, \ldots, 4$. The external disturbances are given as $\Delta_{axy} = [0.2\cos(0.5t), 0.8\cos(t)]^T$, $\Delta_{az} = 0.3\cos(0.2t)$, $w_{bixy} = [1.1\cos(0.5t), -0.2\sin(2t)]^T$. The total duration of the simulation runs $T_z = 80$ s and the sampling time $T_s = 0.6$ s.

Table 1. The model parameters of UAV.

Parameter	Value	Unit
m_a	2	kg
g_a	9.8	$\text{m} \cdot \text{s}^{-2}$
J_{ax}, J_{ay}, J_{az}	1.5	$\text{N} \cdot \text{s}^2 \cdot \text{rad}^{-1}$
d_{ax}, d_{ay}, d_{az}	0.012	$\text{N} \cdot \text{s}^2 \cdot \text{rad}^{-1}$

Table 2. The model parameters of USV.

Parameter	Value	Unit
$m_{\mu bi}$	25.8	kg
m_{vbi}	33.8	kg
m_{rbi}	2.76	kg
$d_{\mu bi}$	0.725	$\text{kg} \cdot \text{s}^{-1}$
d_{vbi}	0.89	$\text{kg} \cdot \text{s}^{-1}$
d_{rbi}	−1.9	$\text{kg} \cdot \text{m}^{-2} \cdot \text{s}^{-1}$
$d_{\mu bi1}$	−1.33	$\text{kg} \cdot \text{s}^{-1}$
d_{vbi1}	−36.47	$\text{kg} \cdot \text{s}^{-1}$
d_{rbi1}	−0.75	$\text{kg} \cdot \text{m}^{-2} \cdot \text{s}^{-1}$

The initial position state vector of the UAV and USVs are defined as $x_0(0) = [6,2,10]^T$, $x_1(0) = [8,0,0]^T, x_2(0) = [4,0,0]^T$, $x_3(0) = [4,4,0]^T, x_4(0) = [8,4,0]^T, \delta_0 = [0,0]^T, \delta_1 = [0,3\aleph]^T, \delta_2 = [-\sqrt{3}\aleph, 0]^T, \delta_3 = [\sqrt{3}\aleph, 0]^T, \delta_4 = [0,\aleph]^T, \aleph = \frac{3}{2}$.

The simulation results by using the proposed control method for heterogeneous systems with collision avoidance strategy are shown in Figures 4–11. The trajectories of the four USVs and the UAV in the 3D environment are depicted in Figure 4. The black line represents the UAV's trajectory, while the other color lines represent the trajectories of the four USVs. From the given starting point, the USVs maintain a safe distance and establish a designated configuration between each other. Figure 5 shows the trajectories of heterogeneous systems without collision avoidance. At the sampling time $T_s = 0.6$ s, Figure 5 shows that three USVs have been collided. Simulation trajectory plots demonstrate the importance of collision avoidance for heterogeneous systems.

Figure 6 and 7 represent the position tracking errors and velocity errors in x-label and y-label, respectively. Under the influence of external disturbances and parameter uncertainties, the formation trajectory tracking errors are capable of converging to a minimal residual set. Figures 8 and 9 depict the surge force and yaw torque. The outcomes of the simulations indicate that the USVs are able to continue tracking the virtual leader even after activating the improved APF for collision avoidance.

Figures 10 and 11 further display the distances between USVs without and with considering the collision avoidance. Figure 10 shows that the USVs have collided, because of the distance $\|d_{ij}\|_{\min} < 2r_1 = 1$. Figure 11 shows that the distances of USVs can always maintain $\|d_{ij}\|_{\min} > 2r_1 = 1$ at any time. The simulation diagrams further confirm that the proposed formation control with collision avoidance strategy has an excellent performance. The performance of ESO in estimating F_i is depicted in Figure 12, which clearly demonstrates the effectiveness of the proposed solution for the accurate compensation of uncertainties and external disturbances within heterogeneous systems.

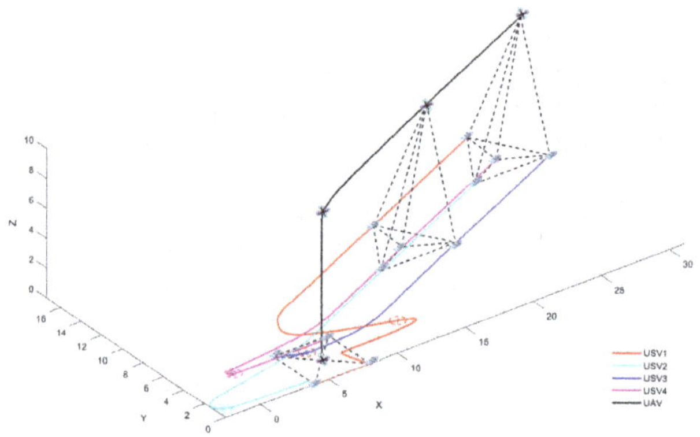

Figure 4. Formation evolution of heterogeneous systems with collision avoidance.

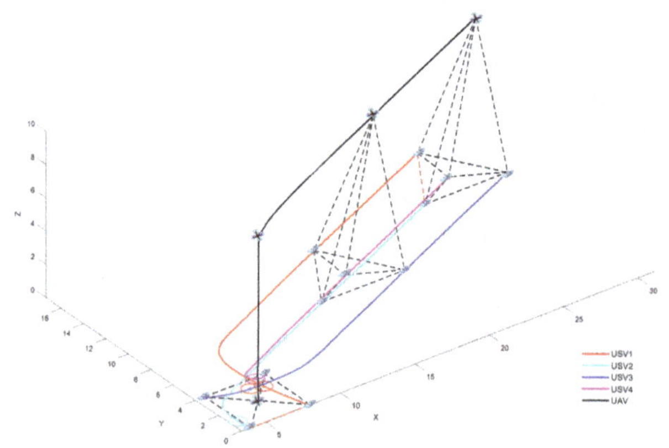

Figure 5. Formation evolution of heterogeneous systems without collision avoidance.

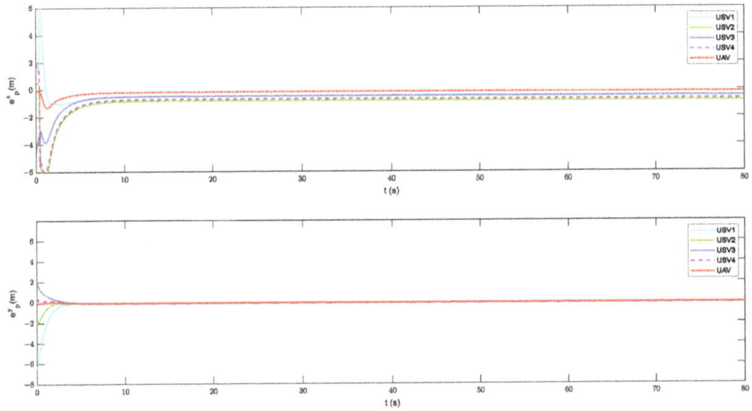

Figure 6. Position tracking errors of the UAV and USVs with collision avoidance.

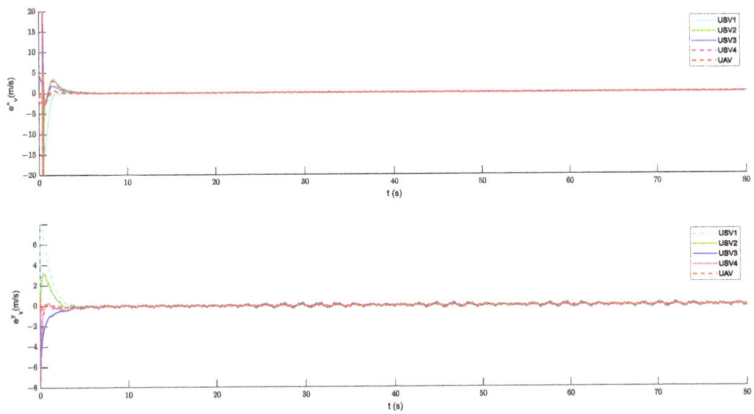

Figure 7. Velocity errors of the UAV and USVs with collision avoidance.

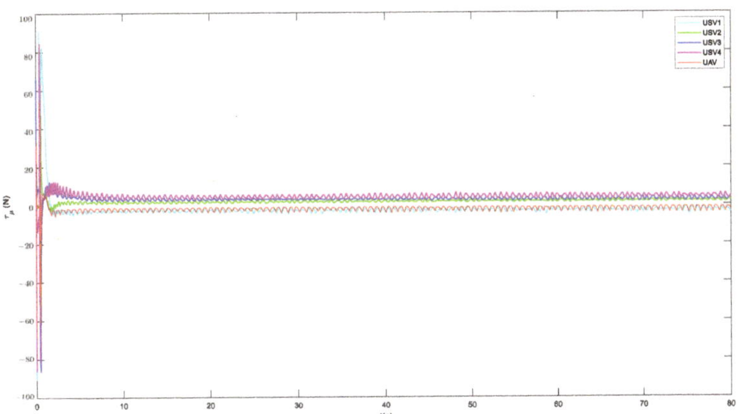

Figure 8. The surge force of heterogeneous system.

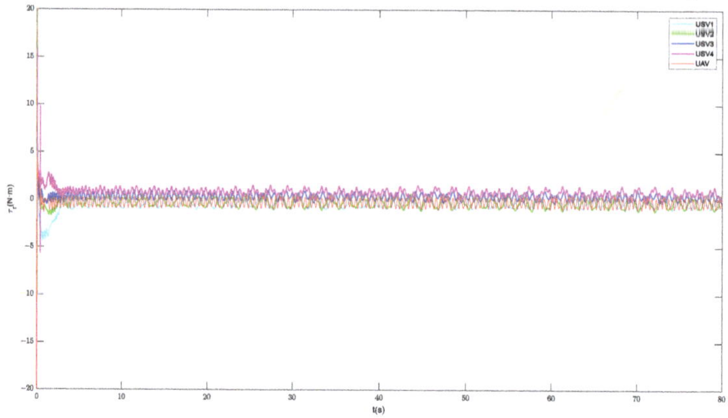

Figure 9. The yaw torque of heterogeneous system.

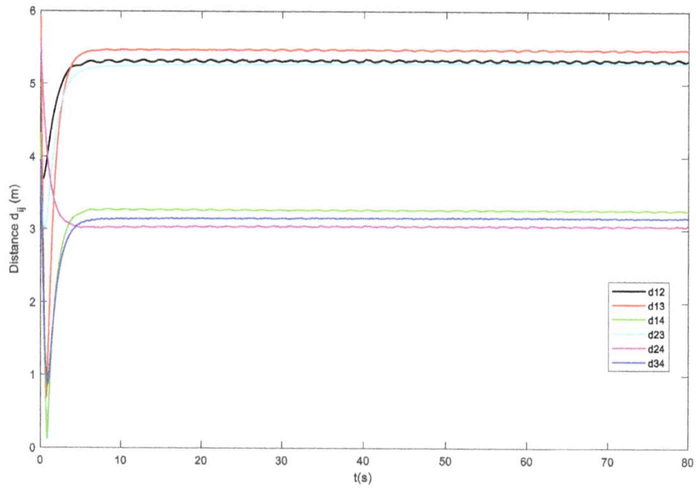

Figure 10. The distance between the USVs without collision avoidance.

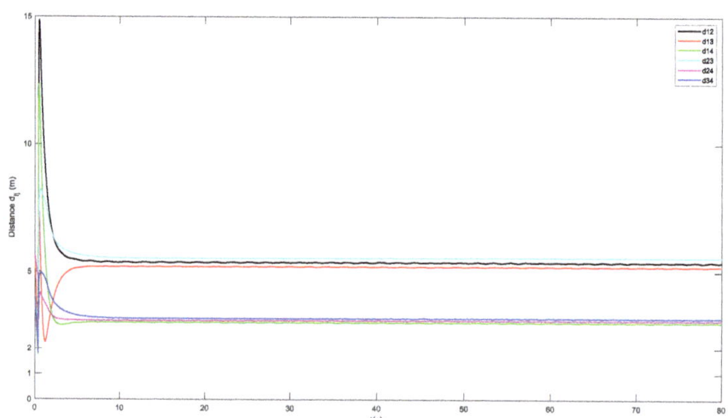

Figure 11. The distance between the USVs with collision avoidance.

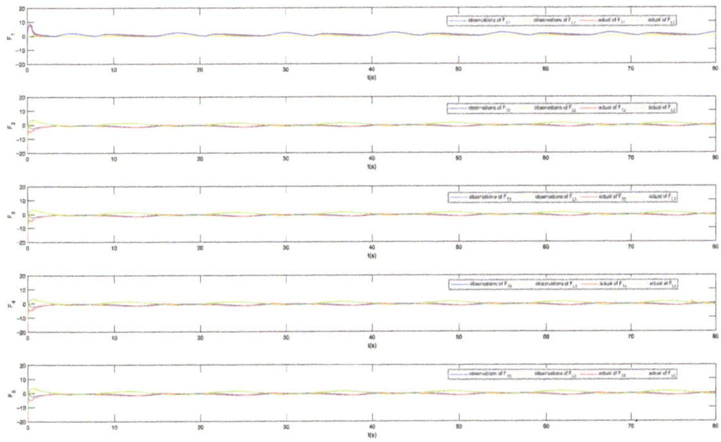

Figure 12. The estimated performance of ESO.

5. Conclusions

In this paper, the cooperative formation trajectory tracking problem for heterogeneous systems with collision avoidance performance under parameter uncertainty and external disturbance is investigated. Under the unified dynamic model of a UAV and USVs in the XY plane, a formation control protocol based on ESO is proposed. The collision avoidance control approach for USV formation is then constructed using APF theory. Furthermore, the APF approach is improved by rebuilding the repulsive potential field to bring the unmanned ship's collision avoidance behavior more in compliance with COLREGs. According to the Lyapunov stability theory, the time-varying formation errors and tracking errors are uniformly ultimately limited. Simulation results verify the effectiveness of the proposed scheme.

This paper focuses solely on heterogeneous formation and collision avoidance between a single UAV and multiple USVs. Going forward, cooperative collision avoidance involving multiple UAVs and USVs within intricate environments deserves further study. In discussions of formation control issues for multi-agent or multi-vehicle systems, it is common to assume that the controlled entities operate within an ideal network communication environment. However, in practical scenarios, such as with USVs, the communication environment is often fraught with issues like noise, time delays, and packet loss. Consequently, the problem of cooperative collision avoidance control for swarm systems under adverse communication conditions also merits thorough investigation.

Author Contributions: Conceptualization, Y.H. and J.N.; methodology, Y.H. and J.N.; software, Y.H.; validation, Y.H.; writing—original draft preparation, Y.H.; writing—review and editing, Y.H., W.L., J.N. and Z.L.; supervision, W.L. and J.N.; project administration, W.L. and Z.L.; funding acquisition, W.L. and J.N. All authors have read and agreed to the published version of the manuscript.

Funding: This work was supported in part by the National Natural Science Foundation of China under Grants 52271304,and 61976033, and in part by the Fundamental Research Funds for the Central Universities under Grant 3132023151.

Institutional Review Board Statement: Not applicable.

Informed Consent Statement: Not applicable.

Data Availability Statement: Data are contained within the article.

Conflicts of Interest: The authors declare no conflict of interest.

Abbreviations

The following abbreviations are used in this manuscript:

UAV	Unmanned Aerial Vehicle
UGV	Unmanned Ground Vehicle
USV	Unmanned Surface Vehicle
AUV	Autonomous Underwater Vehicle
ESO	Extended State Observer
APF	Artificial Potential Field
COLREG	International Regulations for Preventing Collisions at Sea
RBF	Radial Basis Function

References

1. Yan, X.; Jiang, D.; Miao, R.; Li, Y. Formation control and obstacle avoidance algorithm of a multi-USV system based on virtual structure and artificial potential field. *J. Mar. Sci. Eng.* **2021**, *9*, 161. [CrossRef]
2. Shan, Q.; Wang, X.; Li, T.; Chen, C.P. Finite-time control for USV path tracking under input saturation with random disturbances. *Appl. Ocean Res.* **2023**, *138*, 103628. [CrossRef]
3. Guerrero-Ibañez, J.; Contreras-Castillo, J.; Zeadally, S. Deep learning support for intelligent transportation systems. *Trans. Emerg. Telecommun. Technol.* **2021**, *32*, e4169. [CrossRef]
4. Li, X.; Tan, J.; Liu, A.; Vijayakumar, P.; Kumar, N.; Alazab, M. A Novel UAV-Enabled Data Collection Scheme for Intelligent Transportation System Through UAV Speed Control. *IEEE Trans. Intell. Transp. Syst.* **2021**, *22*, 2100–2110. [CrossRef]
5. Menouar, H.; Guvenc, I.; Akkaya, K.; Uluagac, A.S.; Kadri, A.; Tuncer, A. UAV-Enabled Intelligent Transportation Systems for the Smart City: Applications and Challenges. *IEEE Commun. Mag.* **2017**, *55*, 22–28. [CrossRef]
6. Sun, Y.; Zhang, D.; Wang, Y.; Zong, Z.; Wu, Z. Model Experimental Study on a T-Foil Control Method with Anti-Vertical Motion Optimization of the Mono Hull. *J. Mar. Sci. Eng.* **2023**, *11*, 1551. [CrossRef]
7. Ke, C.; Chen, H. Cooperative path planning for air–sea heterogeneous unmanned vehicles using search-and-tracking mission. *Ocean Eng.* **2022**, *262*, 112020. [CrossRef]
8. Ren, Y.; Zhang, L.; Ying, Y.; Li, S.; Tang, Y. Model-Parameter-Free Prescribed Time Trajectory Tracking Control for Under-Actuated Unmanned Surface Vehicles with Saturation Constraints and External Disturbances. *J. Mar. Sci. Eng.* **2023**, *11*, 1717. [CrossRef]
9. Peng, Z.; Wang, J.; Wang, D. Distributed maneuvering of autonomous surface vehicles based on neurodynamic optimization and fuzzy approximation. *IEEE Trans. Control Syst. Technol.* **2017**, *26*, 1083–1090. [CrossRef]
10. Zhou, Z.; Li, M.; Hao, Y. A Novel Region-Construction Method for Multi-USV Cooperative Target Allocation in Air–Ocean Integrated Environments. *J. Mar. Sci. Eng.* **2023**, *11*, 1369. [CrossRef]
11. Fu, H.; Yao, W.; Cajo, R.; Zhao, S. Trajectory Tracking Predictive Control for Unmanned Surface Vehicles with Improved Nonlinear Disturbance Observer. *J. Mar. Sci. Eng.* **2023**, *11*, 1874. [CrossRef]
12. Gu, N.; Wang, D.; Peng, Z.; Liu, L. Distributed containment maneuvering of uncertain under-actuated unmanned surface vehicles guided by multiple virtual leaders with a formation. *Ocean Eng.* **2019**, *187*, 105996. [CrossRef]
13. Tan, G.; Zhuang, J.; Zou, J.; Wan, L. Multi-type task allocation for multiple heterogeneous unmanned surface vehicles (USVs) based on the self-organizing map. *Appl. Ocean Res.* **2022**, *126*, 103262. [CrossRef]
14. Peng, Z.; Wang, D.; Chen, Z.; Hu, X.; Lan, W. Adaptive dynamic surface control for formations of autonomous surface vehicles with uncertain dynamics. *IEEE Trans. Control Syst. Technol.* **2012**, *21*, 513–520. [CrossRef]
15. Miao, R.; Wang, L.; Pang, S. Coordination of distributed unmanned surface vehicles via model-based reinforcement learning methods. *Appl. Ocean Res.* **2022**, *122*, 103106. [CrossRef]
16. Li, J.; Zhang, G.; Shan, Q.; Zhang, W. A novel cooperative design for USV-UAV systems: 3D mapping guidance and adaptive fuzzy control. *IEEE Trans. Control. Netw. Syst.* **2022**, *10*, 564–574. [CrossRef]
17. Peng, Z.; Wang, J.; Wang, D.; Han, Q.L. An overview of recent advances in coordinated control of multiple autonomous surface vehicles. *IEEE Trans. Ind. Inform.* **2020**, *17*, 732–745. [CrossRef]
18. Liu, H.; Weng, P.; Tian, X.; Mai, Q. Distributed adaptive fixed-time formation control for UAV-USV heterogeneous multi-agent systems. *Ocean Eng.* **2023**, *267*, 113240. [CrossRef]
19. Liu, W.; Ye, H.; Yang, X. Model-Free Adaptive Sliding Mode Control Method for Unmanned Surface Vehicle Course Control. *J. Mar. Sci. Eng.* **2023**, *11*, 1904. [CrossRef]
20. Huang, C.; Zhang, X.; Zhang, G.; Deng, Y. Robust practical fixed-time leader–follower formation control for underactuated autonomous surface vessels using event-triggered mechanism. *Ocean Eng.* **2021**, *233*, 109026. [CrossRef]
21. Li, J.; Zhang, G.; Li, B. Robust adaptive neural cooperative control for the USV-UAV based on the LVS-LVA guidance principle. *J. Mar. Sci. Eng.* **2022**, *10*, 51. [CrossRef]
22. Tan, G.; Zhuang, J.; Zou, J.; Wan, L. Coordination control for multiple unmanned surface vehicles using hybrid behavior-based method. *Ocean Eng.* **2021**, *232*, 109147. [CrossRef]
23. Huang, D.; Li, H.; Li, X. Formation of Generic UAVs-USVs System Under Distributed Model Predictive Control Scheme. *IEEE Trans. Circuits Syst. II Express Briefs* **2020**, *67*, 3123–3127. [CrossRef]

24. Wang, N.; Ahn, C.K. Coordinated Trajectory-Tracking Control of a Marine Aerial-Surface Heterogeneous System. *IEEE/ASME Trans. Mechatronics* **2021**, *26*, 3198–3210. [CrossRef]
25. Liu, S.; Jiang, B.; Mao, Z.; Ma, Y. Adaptive Fault-Tolerant Formation Control of Heterogeneous Multi-Agent Systems under Directed Communication Topology. *Sensors* **2022**, *22*, 6212. [CrossRef] [PubMed]
26. Li, S.; Wang, X.; Wang, S.; Zhang, Y. Distributed Bearing-Only Formation Control for UAV-UWSV Heterogeneous System. *Drones* **2023**, *7*, 124. [CrossRef]
27. Li, L.; Wu, D.; Huang, Y.; Yuan, Z.M. A path planning strategy unified with a COLREGS collision avoidance function based on deep reinforcement learning and artificial potential field. *Appl. Ocean Res.* **2021**, *113*, 102759. [CrossRef]
28. Sun, X.; Wang, G.; Fan, Y. Collision avoidance guidance and control scheme for vector propulsion unmanned surface vehicle with disturbance. *Appl. Ocean Res.* **2021**, *115*, 102799. [CrossRef]
29. Ghommam, J.; Saad, M.; Mnif, F.; Zhu, Q.M. Guaranteed Performance Design for Formation Tracking and Collision Avoidance of Multiple USVs With Disturbances and Unmodeled Dynamics. *IEEE Syst. J.* **2021**, *15*, 4346–4357. [CrossRef]
30. Dai, S.L.; He, S.; Lin, H.; Wang, C. Platoon Formation Control With Prescribed Performance Guarantees for USVs. *IEEE Trans. Ind. Electron.* **2018**, *65*, 4237–4246. [CrossRef]
31. Peng, Z.; Wang, D.; Li, T.; Han, M. Output-Feedback Cooperative Formation Maneuvering of Autonomous Surface Vehicles With Connectivity Preservation and Collision Avoidance. *IEEE Trans. Cybern.* **2020**, *50*, 2527–2535. [CrossRef] [PubMed]
32. Xue, K.; Wu, T. Distributed Consensus of USVs under Heterogeneous UAV-USV Multi-Agent Systems Cooperative Control Scheme. *J. Mar. Sci. Eng.* **2021**, *9*, 1314. [CrossRef]
33. Xu, X.; Pan, W.; Huang, Y.; Zhang, W. Dynamic Collision Avoidance Algorithm for Unmanned Surface Vehicles via Layered Artificial Potential Field with Collision Cone. *J. Navig.* **2020**, *73*, 1306–1325. [CrossRef]
34. Lyu, H.; Yin, Y. COLREGS-constrained real-time path planning for autonomous ships using modified artificial potential fields. *J. Navig.* **2019**, *72*, 588–608. [CrossRef]
35. Song, J.; Hao, C.; Su, J. Path planning for unmanned surface vehicle based on predictive artificial potential field. *Int. J. Adv. Robot. Syst.* **2020**, *17*, 172988142091846. [CrossRef]
36. Tan, G.; Zhuang, J.; Zou, J.; Wan, L.; Sun, Z. Artificial potential field-based swarm finding of the unmanned surface vehicles in the dynamic ocean environment. *Int. J. Adv. Robot. Syst.* **2020**, *17*, 172988142092530. [CrossRef]
37. Sang, H.; You, Y.; Sun, X.; Zhou, Y.; Liu, F. The hybrid path planning algorithm based on improved A* and artificial potential field for unmanned surface vehicle formations. *Ocean Eng.* **2021**, *223*, 108709. [CrossRef]
38. Chen, F.; Jiang, R.; Zhang, K.; Jiang, B.; Tao, G. Robust backstepping sliding-mode control and observer-based fault estimation for a quadrotor UAV. *IEEE Trans. Ind. Electron.* **2016**, *63*, 5044–5056. [CrossRef]
39. Song, Y.; He, L.; Zhang, D.; Qian, J.; Fu, J. Neuroadaptive Fault-Tolerant Control of Quadrotor UAVs: A More Affordable Solution. *IEEE Trans. Neural Netw. Learn. Syst.* **2019**, *30*, 1975–1983. [CrossRef]
40. Park, B.S.; Kwon, J.W.; Kim, H. Neural network-based output feedback control for reference tracking of underactuated surface vessels. *Automatica* **2017**, *77*, 353–359. [CrossRef]
41. Wen, G.; Chen, C.L.P.; Liu, Y.J. Formation Control With Obstacle Avoidance for a Class of Stochastic Multiagent Systems. *IEEE Trans. Ind. Electron.* **2018**, *65*, 5847–5855. [CrossRef]
42. Wen, G.; Chen, C.L.P.; Dou, H.; Yang, H.; Liu, C. Formation control with obstacle avoidance of second-order multi-agent systems under directed communication topology. *Sci. China Inf. Sci.* **2019**, *62*, 1–11. [CrossRef]
43. Shi, Q.; Li, T.; Li, J.; Chen, C.P.; Xiao, Y.; Shan, Q. Adaptive leader-following formation control with collision avoidance for a class of second-order nonlinear multi-agent systems. *Neurocomputing* **2019**, *350*, 282–290. [CrossRef]
44. Wen, G.; Chen, C.L.P.; Feng, J.; Zhou, N. Optimized Multi-Agent Formation Control Based on an Identifier–Actor–Critic Reinforcement Learning Algorithm. *IEEE Trans. Fuzzy Syst.* **2018**, *26*, 2719–2731. [CrossRef]
45. Kim, Y.H.; Ahn, S.C.; Kwon, W.H. Computational complexity of general fuzzy logic control and its simplification for a loop controller. *Fuzzy Sets Syst.* **2000**, *111*, 215–224. [CrossRef]

Journal of
Marine Science and Engineering

MDPI

Article

Model Experimental Study on a T-Foil Control Method withAnti-Vertical Motion Optimization of the Mono Hull

Yifang Sun [1], Dapeng Zhang [1,*], Yiqun Wang [1], Zhi Zong [2,3] and Zongduo Wu [1]

1. Ship and Maritime College, Guangdong Ocean University, Zhanjiang 524088, China
2. Liaoning Engineering Laboratory for Deep-Sea Floating Structures, School of Naval Architecture, Dalian University of Technology, Dalian 116024, China
3. Collaborative Innovation Center for Advanced Ship and Deep-Sea Exploration, Shanghai 200240, China

* Correspondence: zhangdapeng@gdou.edu.cn

Abstract: T-foils with active control systems can adjust their attack angle according to the movement of the ship in real time, providing higher lift force and improving the seakeeping performance of a ship. The optimization of the control signal and that of the control method have an important influence on the effect of active T-foils. In this paper, the control method of the T-foil's swinging angle is established and optimized on the basis of model testing in order to increase the effect of the T-foil. First, the governing equation is introduced by establishing the proportional relationship between the angular motion of the hull and the lift moment of the T-foil. On the basis of the model of the T-foil's lift force, the governing equation of the T-foil's swinging angle is deduced and simplified using the test results of the ship model with a passive T-foil and without a T-foil. Then, the active T-foil control system is established by comparing the effects of T-foils with different control signals. Finally, the efficacies of the passive and active T-foil are reported and discussed. It is found that the pitch angular velocity is a more appropriate signal than the pitch angle and pitch angular acceleration. T-foils with pitch angular velocity control can decrease the vertical motion response in the resonance region of a ship's encounter frequency by more than about 20% compared to the case of the bare ship model, while also increasing the anti-bow acceleration effect by more than 15% compared to the case of passive control. The results obtained by model testing have a certain guiding significance for specific engineering practices.

Keywords: anti-vertical motion; model test; T-foil; control method

check for updates

Citation: Sun, Y.; Zhang, D.; Wang, Y.; Zong, Z.; Wu, Z. Model Experimental Study on a T-Foil Control Method withAnti-Vertical Motion Optimization of the Mono Hull. *J. Mar. Sci. Eng.* **2023**, *11*, 1551. https://doi.org/10.3390/jmse11081551

Academic Editor: Kostas Belibassakis

Received: 2 July 2023
Revised: 25 July 2023
Accepted: 25 July 2023
Published: 4 August 2023

1. Introduction

High-performance ships have excellent comprehensive performance and have gradually been accepted in terms of reliability, security, economy, etc. With the development of the world's marine engineering and shipping market, the requirement for R&D is increasing, and this is developing actively in the ship market, which possesses great vitality. Since the 1990s, large-tonnage semi-planing ships have been widely used because they combine the advantage of large displacement, which characterizes conventional displacement-type ships, with the good rapidity of planing boats [1]. However, semi-planing ships are susceptible to waves when sailing at high speed, and the amplitude of the vertical motion is high, thus increasing the rate of seasickness. Furthermore, high-amplitude motion can easily cause slamming phenomena, and it generates a large slamming load [2]. This can easily cause fatigue damage, or even fracture, in the hull structure. Therefore, methods for reducing the motion amplitude of high-speed ships in waves represent extremely important work for improving motion performance.

Recent studies have shown that vertical motion can be reduced considerably by installing a T-foil on the bow when the ship is sailing at high speed (i.e., at a Froude number between 0.5 and 1) [3,4]. T-foils with vertical foils and horizontal foils can counteract the effect of wave disturbance force by applying a vertical force (moment) on the bow in the opposite direction to that of the ship's motion [5], and then they can play a suppressing role in the ship's motion. By introducing the automatic control program (PID control) into the T-foil system, the swing angle of the horizontal foil can be adjusted in real time with the movement of the ship. Compared to passive T-foils (i.e., without control), active T-foils can significantly increase the restoring force (moment) and improve the anti-vertical motion effect [6–8].

In PID control, the key issue is establishing the transfer function between force and motion. The longitudinal motion control is related to numerous parameters, such as heave displacement, pitch angle, vertical acceleration, hysteresis effect, heave velocity, etc. Therefore, the main problem for the motion control of semi-planing ships is finding the most important motion parameters among these motions.

Esteban et al. [9,10] used vertical acceleration (Worst Vertical Acceleration, WVA) and the rate of seasickness (Motion Sickness Incidence, MSI) at typical positions on the hull as optimization criteria, and they performed numerical simulations of the ship's motion using MATLAB's Simulink module to compare the effects of each control parameter. The simulation results showed that reasonable adjustment of each parameter in the PID control was able to effectively reduce the vertical acceleration by 26% and the rate of seasickness by 10% compared to the passive control.

Giron-Sierra [11–13] installed a T-foil and stern flaps on a high-speed ferry and conducted model tests in a towing basin. A multi-objective optimized PD control procedure was designed for the rate of seasickness, the cavitation phenomenon and mechanical efficiency. It was shown that the heave and pitch motions of the ship were important parameters for controlling the rotation of the T-foil, which was able to effectively limit the bow acceleration of the ship and improve the rate of seasickness.

Alavimehr et al. [14,15] proposed a nonlinear control method based on model tests in still water. The swing angles of the T-foil and the stern flaps were controlled using a single-signal control (pitch angular velocity or heave velocity). Model testing was conducted in regular waves to compare the anti-heave and pitch effects. The results showed that the nonlinear controlled T-foil had a better anti-vertical motion effect than the linear control. The effect of anti-heave motion is more obvious when using an active T-foil with heave velocity control. However, the effect of suppressing pitch and bow acceleration was not obvious. Accordingly, it is more suitable for reducing the pitch and bow acceleration response of the ship model by using the pitch angular velocity to control the swing angle. However, it is difficult to significantly reduce the heave, pitch, and bow acceleration of the ship at the same time using a single-signal control. The equation for controlling the swing angle should be further optimized.

Previous studies have mostly focused on different control strategies for active T-foils. However, no matter what control method is used (PID control, fuzzy control, etc.), there will be obvious different anti-vertical motion effects when using different motion control signals. Therefore, the determination of the main motion signal for the T-foil angle is extremely important, and it is also a key factor in optimizing the control method. However, there are a limited number of comparative studies on the control effect of different motion signals in the existing research, and these studies have mostly been based on numerical dynamic simulation, resulting in a lack of experimental research. Therefore, it is necessary to further optimize the control method of T-foils and optimize the master signal of T-foils through model testing.

In this paper, the experimental study of a model under high speed (Fr = 0.63) in a regular wave is carried out for a semi-planing deep-V mono-hull ship. The test principle is first introduced including control equation and motion signals for the T-foil's active control. By measuring the motion of the bare ship model with the passive T-foil (the T-foil's swing angle is 0°), the control equation of the T-foil is proposed, and the control parameters are integrated. Then, the model test was established, and three motion signals are used to control the lift force (moment) of the T-foil, respectively. Comparing the anti-vertical motion effect by the T-foil with different control signals, the control method of the T-foil was optimized. Finally, model tests were conducted based on the optimized control method to verify its anti-vertical motion effect.

2. Control Method

The swing angle of a horizontal foil is affected by factors such as sea state, real-time hull motion, profile size, etc. The design of an active control system of a T-foil can be simplified by modeling the uncertain problem and expressing it in a highly structured parametric form. For high-speed hull motion on waves, in particular, an active control system should have strong control timeliness and accuracy. This requires that the control method is not too complex, thus preventing delay in the swing angle control of the horizontal foil in practical applications and producing the desired anti-vertical motion effect on the longitudinal motion of the ship.

2.1. Mathematical Model of Active Control of T-Foil

The vertical motion of a ship is reduced by the lift force f_T and lifting moment M_T of a T-foil; the lift force (moment) can counteract the wave force (moment). The unsteady thin airfoil theory was applied to perform theoretical calculations for determining dynamic lift effects, especially in high-encounter frequencies because of the unsteady motion of hydrofoil [16]. Belibassakis et al. [17–19] established unsteady lifting models based on the integration of 2D sectional lift along a span to calculate the lift force of a hydrofoil. In this study, the T-foil's deflection amplitude is limited ($-10°$–$10°$). The deviation of the lift force's prediction between unsteady and quasi-static lift theories was acceptable in high Froude numbers and low wavelength [16]. The lift force and lifting moment are generally expressed as:

$$\begin{cases} f_T = \frac{1}{2}\rho U^2 A \frac{dC_L}{d\alpha}\alpha \\ M_T = \frac{1}{2}l_F \rho U^2 A \frac{dC_L}{d\alpha}\alpha \end{cases} \tag{1}$$

where ρ represents the fluid density (kg/m^3), A is the T-foil area (m^2), and C_L is the lift coefficient.

The attack angle of the T-foil α consists of the horizontal foil's swing angle φ (i.e., the deflection angle of the T-foil's horizontal foil with respect to the intermediate position), the pitch angle of the ship θ, and an additional angle θ_F formed by both the ship and vertical motions of the fluid particle in the flow field [20]. This is presented in Figure 1, and the attack angle α is expressed as:

$$\alpha = \varphi - \theta + \theta_F \tag{2}$$

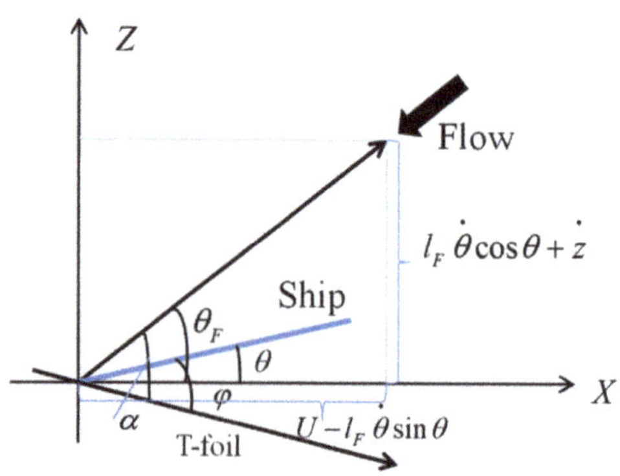

Figure 1. Graphic analysis of effective attack angle.

The additional angle θ_F is affected by the combined effect of the heave velocity $\dot{z}$, the pitch angular velocity $\dot{\theta}$, the vertical velocity of the wave particle at the hydrofoil surface $\dot{\kappa}$, and the ship's speed U. θ_F is expressed as:

$$\theta_F = \arctan\frac{l_F\dot{\theta}\cos\theta - \dot{z} - \dot{\kappa}}{U - l_F\dot{\theta}\sin\theta} \tag{3}$$

According to the micro-amplitude wave theory, the fluid particle motion is a simple harmonic motion, and its vertical velocity decreases exponentially with water depth. Great depth causes the vertical velocity $\dot{\kappa}$ of the fluid particle at the T-foil position to be extremely lower than the pitch angular velocity $\dot{\theta}$ and the heave velocity $\dot{z}$; hence, the effect of $\dot{\kappa}$ can be neglected. Furthermore, the incident wave amplitude A_0, which is significantly lower than the wavelength λ, causes the amplitude of the ship's vertical motion on the wave to be low. Equation (3) can then be simplified as:

$$\theta_F \approx \frac{l_F\dot{\theta} - \dot{z}}{U} \tag{4}$$

where l_F is the distance from the installation position of the T-foil to the longitudinal position of the ship's center of gravity (longitudinal center of gravity, LCG), as illustrated in Figure 2. Here, the lift force (f_T) and lifting moment (M_T) can be expressed as:

$$\begin{cases} f_T = \frac{1}{2}\rho U^2 A \frac{dC_L}{d\alpha}\left(\varphi - \theta + \frac{l_F\dot{\theta} - \dot{z}}{U}\right) \\ M_T = \frac{1}{2} l_F \rho U^2 A \frac{dC_L}{d\alpha}\left(\varphi - \theta + \frac{l_F\dot{\theta} - \dot{z}}{U}\right) \end{cases} \tag{5}$$

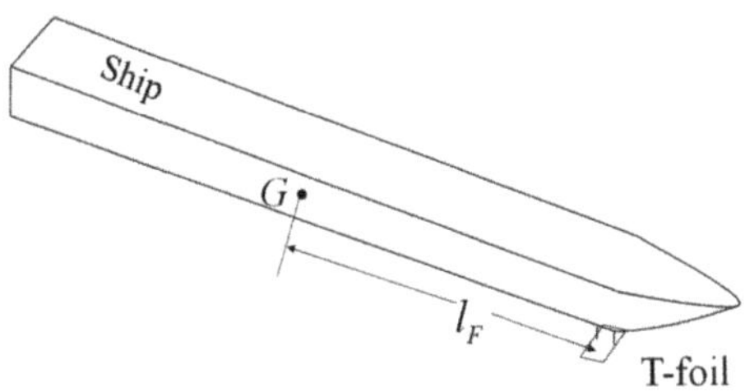

Figure 2. Distance between installation position and ship's LCG.

2.2. *Lifting Moment Control Equation*

The active T-foil control system can achieve anti-vertical motion by adjusting the attack angle in real time and then providing a greater reverse lift (moment) during significant ship motions. Following this analysis, the rotation of the T-foil is controlled in real time by the lifting moment control, i.e., the lifting moment M_T generated by the T-foil is in the opposite direction of the hull motion [21] (including heave and pitch motions). This is expressed as:

$$M_T = -C_1\theta' - C_2\dot{\theta} - C_3\ddot{\theta} - C_4 z_0 - C_5\dot{z} - C_6\ddot{z} \tag{6}$$

where C_1, C_2, C_3, C_4, C_5, and C_6 are the control parameters; these values represent the profile parameters of the T-foil, ship type, ship speed, and sea state; θ' is the adjusted pitch angle θ, i.e., relating to the equilibrium position of the ship during motion; z_0 is the adjusted heave displacement z, i.e., relating to the equilibrium position of the ship during motion.

According to previous studies, the anti-pitch and bow acceleration effects are more obvious while using the pitch angular velocity signal to control swing angle. T-foils controlled by the heave velocity signal only show a better anti-vertical motion effect on heave motion. Since bow acceleration greatly influences the extent of seasickness, this

Figure 3. Installation site of the T-foil.

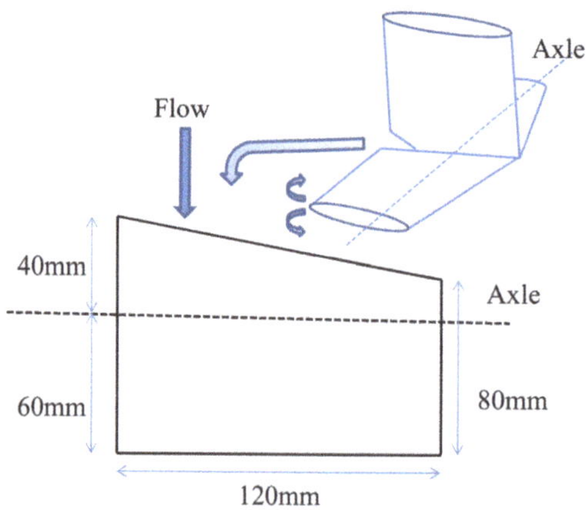

Figure 4. Dimensions of the T-foil.

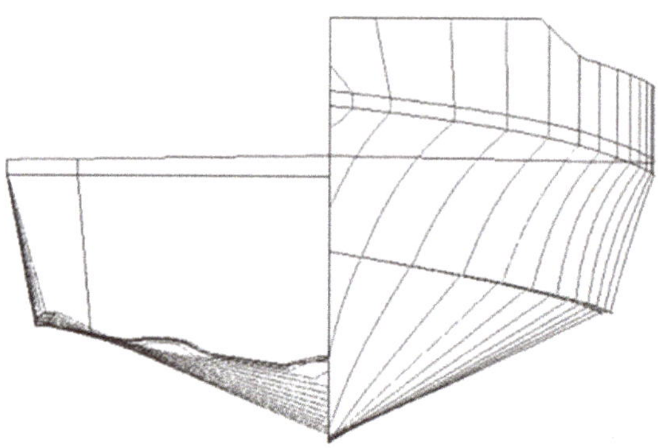

Figure 5. Model geometry.

Table 2. Main dimensions of the ship model.

Index	Value
Overall length/m	3.833
Length of Waterline length/m	3.616
Breadth/m	0.758
Draught/m	0.321
Displacement/kg	259.7
Displacement (model)/kg	152.4
Designed draft/m	0.204

3.2. T-Foil Control System

The T-foil control system is divided into the automatic control and the mechanical drive parts. The automatic control part consists of a control board with a built-in AVR microcontroller (manufacturer: Zhiwei Robotics Corps, Shanghai City, China), an inertial measurement unit (IMU), an angle sensor, and an upper computer measurement program (shown in Figure 6). The IMU has a built-in three-axis gyroscope and acceleration sensor, incorporating a Kalman filter algorithm to filter out noise during the attitude solution process. The IMU is fixed on the mid-longitudinal section of the bow above the T-foil to measure the heave velocity, pitch angle, and pitch angular velocity of the model. The mechanical drive part comprises a worm gear transmission system; an inclination sensor is fixed above the transmission system; and the rotation angle of the T-foil's horizontal foil is determined by measuring the rotation angle of the servo motor.

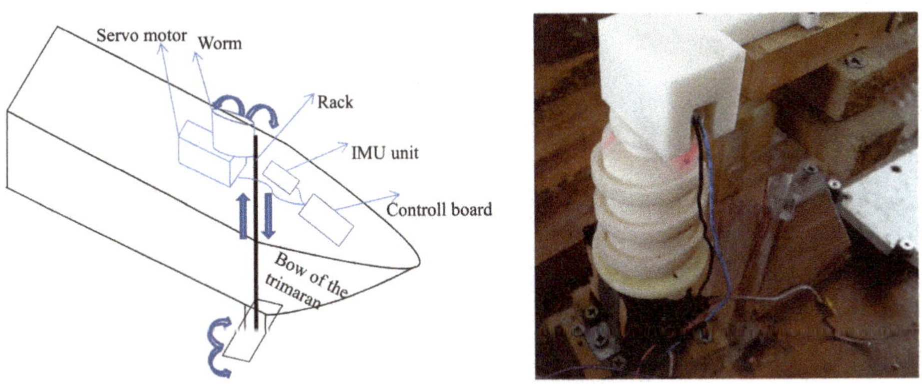

Figure 6. Automatic and mechanical control parts.

The workflow of the control system is shown in Figure 7. The head sea regular waves induce the vertical motion of ship model in corresponding frequencies. The IMU sensor installed directly above the T-foil can measure the data of heave amplitude z and pitch angle θ. Then, the motion signals are output to the control board with built-in AVR microcontroller. The control board with the governing Equation (8) processes the input signal in real time to obtain the real-time swinging angle of the T-foil. The data of motion can also send to the upper computer data acquisition system. The upper computer data acquisition system can record the motion parameters of the model in real time, including the heave amplitude, heave velocity, pitch angle, pitch angular velocity, vertical acceleration, and T-foil's swing angle, and it can export the recorded data to the computer in the form of Excel table for storage. On this basis, the swinging angle signal has been input into the mechanical system. The servo motor drives the worm gear transmission mechanism to adjust the T-foil's swing angle in real time. T-foil's deflection can induce the vertical force (moment) to reduce the vertical motion.

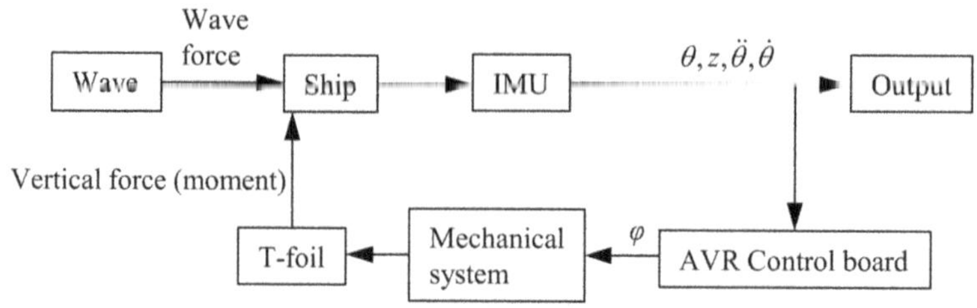

Figure 7. Flow chart of control system.

The reliability of the mechanical drive part (as shown in Figure 6) of the T-foil control system was first examined before tests began, and the differences in the actual and theoretical swing angles of the T-foil were analyzed. The actual swing angle was tested by giving the steering engine a sinusoidal swing angle signal with a frequency of 2.4 Hz; this was compared with its theoretical counterpart, and the results are illustrated in Figure 8. The mechanical clearance of the steering engine and deviation of the mechanical drive system caused the actual swing angle to lag for about 30 ms compared to its theoretical counterpart; the difference was little when compared to the encounter period and could be ignored. The test results showed that the transmission system met the test requirements.

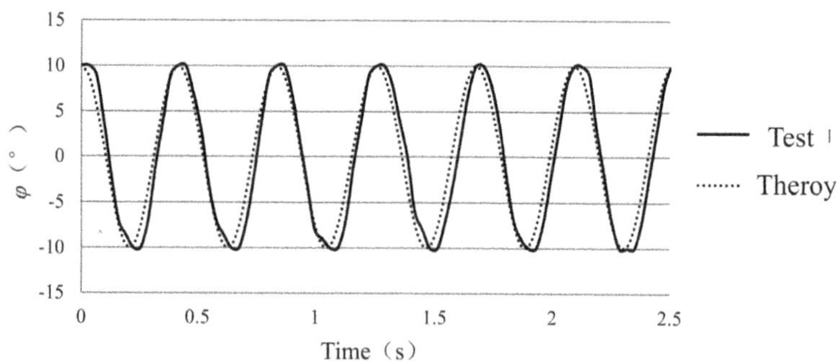

Figure 8. Differences in T-foil's swing angles (2.4 Hz).

3.3. Test Equipment

The test tank of the ship model was in the towing tank of Dalian University of Technology [23], a member of the International Towing Tank Conference (ITTC), having a total length of 160 m, width of 7 m, and depth of 3.65 m. The wave-making system of the tank used a push-plate wave-making machine, which had a maximum wave-making height of 0.4 m and high wave-making accuracy. It could produce regular and irregular waves with good repeatability. The characteristic parameters (such as wave height and period) were measured by the wave height meter and fed back to the control computer for adjustment. A wave damper was installed at the side wall of the tank. A sketch of the tank's layout of equipment is shown in Figure 9.

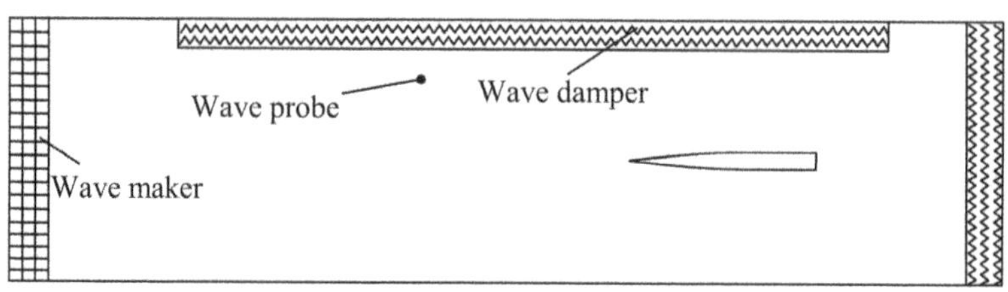

Figure 9. Experimental setup of towing tank.

In the tank, the ship model was towed forward by a Computerized Planar Motion Carriage system; the experimental setup of the towing tank is shown in Figure 10. The ship model was fixed on a seaworthy instrument on the carriage (as shown in Figure 11), so its heave amplitude, pitch and roll angles, as well as drag force, could be measured. In the test, the measurement points of the heave amplitude and pitch angle were located at the LCG of the ship model. Furthermore, acceleration sensors were installed on the bow of the ship model to measure bow acceleration, and filters were installed on each measurement device.

Figure 10. Ship model towing tank.

Figure 11. Four-degrees-of-freedom seaworthy instrument.

3.4. Model Experimental Design

The model tests focus on the vertical motion response of this semi-planing monohull ship with regular waves of different wavelengths at high speeds. Therefore, the speed chosen as its maximum design speed is U = 3.861 m/s (Fr = 0.63); the wavelength λ varies at 3–8 m; and the wave height is h = 0.046 m. The tests steps are as follows:

(1) The response to heave, pitch, and bow accelerations was measured when the ship model sailed under regular waves with different wavelengths.

(2) The motion responses of the ship model with a passive T-foil were measured under the same speed and wave conditions to calculate the value of the control parameter, C, according to the measured time-record curve.

The active control system of the T-foil was introduced to examine the response of the ship model when different motion signals are used as inputs; the optimal lifting moment control signal was determined by comparing the effects of the active T-foil.

4. Analysis of Experiment Results and Discussion

4.1. The Effect of Passive T-Foil

The model test of the passive T-foil was first conducted on the longitudinal motion of the semi-planing monohull ship with different wavelengths of regular waves to provide a basis for calculating the control parameter C in the active T-foil control system. The results of the analysis of the heave amplitude response at each encounter frequency are presented in Table 3, and the graph is presented in Figure 12. The heave response of the ship model increases as wavelength increases. The passive T-foil can reduce heave amplitude by 6–7% in higher heave response conditions (wavelength: $\lambda = 6$, 7, and 8 m). In lower-response conditions ($\lambda = 3, 4$ m), although the percentage of suppression effect improves slightly, the effect is not significant because of the low-response amplitude at that moment.

Table 3. Heave motion response (passive T-foil control).

λ (m)	Encounter Frequency	Bare Ship	Passive Control	%
3	12.86	0.0322	0.0277	13.98
4	10.07	0.1913	0.176	8
5	8.54	0.5565	0.515	7.46
6	7.39	0.9322	0.865	7.21
7	6.55	1.1496	1.07	6.92
8	5.91	1.1861	1.115	5.99

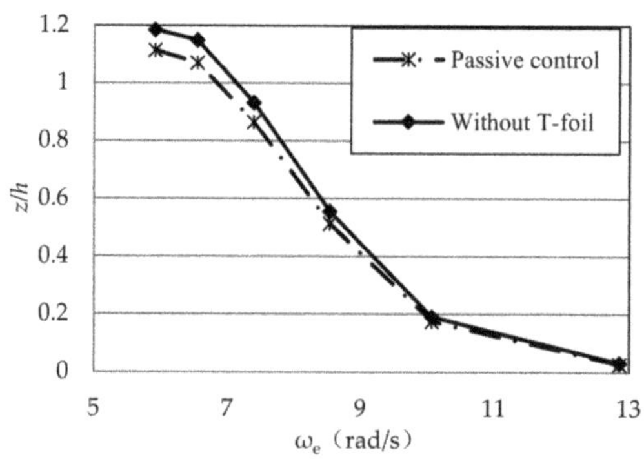

Figure 12. Heave motion reduction by passive T-foil.

The result analysis of the pitch motion is presented in Table 4, and the graph is presented in Figure 13. The variation trend of the pitch motion response with wavelength is like that of the heave motion; the pitch angle is larger at long waves than at short ones. The anti-vertical motion effect of the passive T-foil on pitch motion is between 6.3 and 7.7% in the high-pitch response conditions. This is similar to the suppression ability of heave motion. This phenomenon occurs because the passive T-foil's attack angle is small; hence, the anti-pitch motion percentage is limited.

Table 4. Pitch motion response (passive T-foil control).

λ (m)	Encounter Frequency	Bare Ship	Passive Control	%
3	12.86	0.171	0.148	13.45
4	10.07	0.54	0.475	12.04
5	8.54	1.125	1.037	7.82
6	7.39	1.605	1.482	7.66
7	6.55	1.851	1.72	7.08
8	5.91	1.905	1.785	6.30

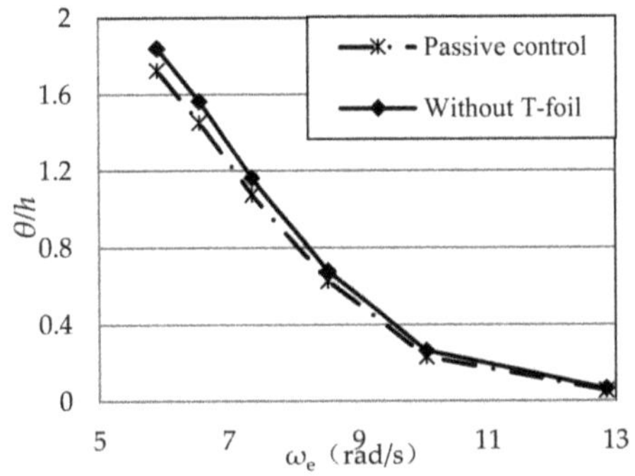

Figure 13. Pitch motion reduction by passive T-foil.

For the bow acceleration motion, the result analysis is presented in Table 5, and the graph is presented in Figure 14. Unlike pitch and heave motions, the peak response of bow acceleration is located at λ = 6 m. The vertical acceleration value of the T-foil can be reduced by 7.56% under this condition, and its anti-vertical motion ability decreases slightly as the wavelength continues to increase. This phenomenon occurs because the installation position of the T-foil is more backward, and the lifting moment it produces is relatively small; thus, the effect is not obvious.

Table 5. Bow acceleration response (passive T-foil control).

λ (m)	Encounter Frequency	Bare Ship	Passive Control	%
3	12.86	0.0467	0.04	14.35
4	10.07	0.1114	0.101	9.34
5	8.54	0.1925	0.178	7.53
6	7.39	0.2288	0.2115	7.56
7	6.55	0.217	0.202	6.91
8	5.91	0.1807	0.1705	5.64

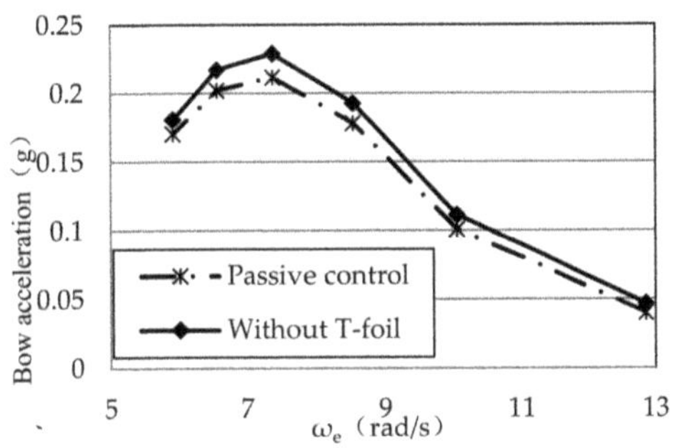

Figure 14. Bow acceleration reduction by passive T-foil.

4.2. Comparison of Motion Signals

Based on the ship model and passive control suppression tests and according to the measured ship motion results, the values of each control parameter C can be obtained using the trial and error method with Equations (8) and (9). Several typical working conditions need to be selected for motion response measurement and for comparison of the difference in the anti-vertical motion effect of the ship model using different angular

5.2. Further Work

Further work can encompass the improvement of the following aspects:

(1) The T-foil can be installed closer to the bow, and the T-foil's parameter needs to be further optimized.
(2) The governing equation of the swing angle can be optimized using a multi-signal control and control parameters adjusted in real time.
(3) The anti-roll effect when using a T-foil needs to be studied.
(4) In this study, the calculation of lift force is based on a static lift force theory, neglecting unsteady hydrofoil effects. The model can be improved by introducing an unsteady airfoil theory.

Author Contributions: Conceptualization, Y.S.; methodology, Y.S. and Z.Z.; software, D.Z.; formal analysis, Y.S. and Y.W.; data curation, Z.W.; writing—original draft preparation, Y.S.; writing—review and editing, D.Z.; supervision, D.Z.; funding acquisition, Y.S. and D.Z. All authors have read and agreed to the published version of the manuscript.

Funding: This research was funded by Program for Scientific Research Start-up Funds of Guangdong Ocean University, grant number 060302072102 and Zhanjiang Marine Youth Talent Project—Comparative Study, grant number 2021E5007.

Institutional Review Board Statement: The study did not require ethical approval.

Informed Consent Statement: The study did not involve humans.

Data Availability Statement: The data that support the findings of this study are available from the corresponding author, upon reasonable request.

Conflicts of Interest: The authors declare no conflict of interest.

References

1. Faltinsen, O.M. *Hydrodynamics of High-Speed Marine Vehicles*; Cambridge University Press: Cambridge, UK, 2005.
2. Tang, H.; Ren, H.; Wan, Q. Investigation of longitudinal vibrations and slamming of a trimaran in regular waves. *J. Ship Res.* **2017**, *61*, 153–166. [CrossRef]
3. Davis, M.R.; Watson, N.L.; Holloway, D.S. Wave response of an 86 m high speed catamaran with active T-foils and stern tabs. *Trans. R. Inst. Nav. Archit. Part A Int. J. Marit. Eng.* **2003**, *145*, 15–34.
4. Jiao, J.; Sun, S.; Li, J.; Adenya, C.A.; Ren, H.; Chen, C.; Wang, D. A comprehensive study on the seakeeping performance of high speed hybrid ships by 2.5D theoretical calculation and different scaled model experiments. *Ocean Eng.* **2018**, *160*, 197–223. [CrossRef]
5. Haywood, A.; Duncan, A.; Klaka, K.; Bennett, J. The development of a ride control system for fast ferries. *Control Eng. Pract.* **1995**, *3*, 695–702. [CrossRef]
6. Fang, M.C.; Shyu, W.J. Improved prediction of hydrodynamic characters of SWATH ships in wave. *Proc. Natl. Sci. Counc.* **1994**, *18*, 495–507.
7. De la Cruz, J.; Aranda, J.; Giron-Sierra, J.; Velasco, F.; Esteban, S.; Diaz, J.; de Andres-Toro, B. Improving the comfort of a fast ferry. *IEEE Control Syst.* **2004**, *24*, 47–60.
8. Mehr, J.A.; Ali-Lavroff, J.; Davis, M.R.; Holloway, D.; Thomas, G. An experimental investigation of ride control algorithms for high-speed catamarans Part 1: Reduction of ship motions. *J. Ship Res.* **2017**, *61*, 35–49.
9. Esteban, S.; Giron-Sierra, J.M.; De Andres-Toro, B.; De La Cruz, J.M. Development of a control-oriented model of the vertical motions of a fast ferry. *J. Ship Res.* **2004**, *48*, 218–230. [CrossRef]
10. Esteban, S.; Giron-Sierra, J.; De Andres-Toro, B.; Cruz, J.D.; Riola, J.M. Fast ships models for seakeeping improvement studies using flaps and T-foil. *Math. Comput. Model.* **2005**, *41*, 1–24. [CrossRef]
11. Giron-Sierra, J.M.; Andres-Toro, B.; Esteban, S.; Recas, J.; Besada, E.; Cruz, J.M.; Maron, A. First principles modelling study for the development of a 6 DOF motions model of a fast ferry. *IFAC Proc. Vol.* **2004**, *36*, 73–78. [CrossRef]
12. Giron-Sierra, J.; Andres-Toro, B.; Esteban, S.; Recas, J.; Besada, E.; De la Cruz, J.; Riola, J.M. Model based analysis of seasickness effects in a fast ferry. *IFAC Proc. Vol.* **2003**, *36*, 103–108. [CrossRef]
13. Giron-Sierra, J.M.; Esteban, S.; Cruz, J.M.; Andres, B.D.; Riola, J.M. Fast ship's longitudinal motion attenuation with T-Foil and flaps. In Proceedings of the Novel Vehicle Concepts and Emerging Vehicle Technologies Symposium, Ottawa, ON, Canada, 18–21 October 1999; The RTO Applied Vehicle Technology Panel: Neuilly-sur-Seine, France, 2003; pp. 26–34.
14. Alavimehr, J.; Davis, M.R.; Lavroff, J.; Holloway, D.S.; Thomas, G.A. Response of a high-speed wave-piercing catamaran to an active ride control system. *Trans. R. Inst. Nav. Archit. Part A Int. J. Marit. Eng.* **2016**, *158*, A325–A335. [CrossRef]

15. Alavimehr, J.; Lavroff, J.; Davis, M.R.; Holloway, D.; Thomas, G. An experimental investigation of ride control algorithms for high-speed catamarans Part 2: Mitigation of wave impact loads. *J. Ship Res.* **2017**, *61*, 51–63. [CrossRef]
16. Alavimehr, J. The Influence of Ride Control System on the Motion and Load Response of a Hydroelastic Segmented Catamaran Model. Ph.D. Thesis, University of Tasmania, Hobart, Australia, 2016.
17. Belibassakis, K.A.; Politis, G.K. Hydrodynamic performance of flapping wings for augmenting ship propulsion in waves. *Ocean Eng.* **2013**, *72*, 227–240. [CrossRef]
18. Filippas, E.; Papadakis, G.; Belibassakis, K.A. Free-surface effects on the performance of Flapping-Foil thruster for augmenting ship propulsion in waves. *J. Mar. Sci. Eng.* **2020**, *8*, 357. [CrossRef]
19. Ntouras, D.; Papadakis, G.; Belibassakis, K.A. Ship bow wings with application to trim and resistance control in calm water and in waves. *J. Mar. Sci. Eng.* **2022**, *10*, 492. [CrossRef]
20. López, R.; Santos, M. Neuro-Fuzzy system to control the fast ferry vertical acceleration. In Proceedings of the 15th IFAC World Congress, Barcelona, Spain, 21–26 July 2002; pp. 319–324.
21. Giron-Sierra, J.M.; Esteban, S. Frequency domain study of longitudinal motion attenuation of a fst ferry using a T-Foil. In Proceedings of the 17th World Congress the International Federation of Automatic Control, Seoul, Republic of Korea, 6–11 July 2008; pp. 15004–15009.
22. Wang, W.; Zong, Z.; Ni, S.; Zhang, L.; Chen, Z. Model tests of effect of interceptor on resistance of a semi-planing ship. *Chin. J. Res.* **2012**, *7*, 18–22.
23. Jia, J.; Zong, Z.; Shi, H. Model experiments of a trimaran with transom stern. *Int. Shipbuild. Prog.* **2009**, *56*, 119–133.

incorporates ship maneuverability, human experience, and navigation rules, and designed a restricted water test method to effectively test the capabilities of intelligent ships in a limited time frame [15]. Sawade et al. proposed a collision avoidance algorithm based on proximal policy optimization (PPO), which improves the obstacle zone by target (OZT) and enables the control of the rudder angle in continuous action space [16]. Zhao et al. proposed a DRL algorithm for ship collision avoidance based on Actor-Critic (AC), which divides the target ship area into four regions based on COLREGs and solves the case of different numbers of target ships by fixing the neural network input dimensions [17]. However, the above methods based on the single-agent concept deal with ship collision avoidance from the perspective of the own ship and do not describe the interaction behavior relations among ships directly, which is inconsistent with reality. The individual behaviors will have an impact on the overall collision avoidance result, and collision avoidance measures need to be decided in coordination with each other, especially in multi-ship collision avoidance scenarios.

Therefore, experts and research scholars have gradually extended the research direction from the single-agent system to the multi-agent system (MAS). Groups of agents within the MAS share the same environment, use sensors to perceive the environment, and take actions by using actuators. MAS usually adopts a distributed structure, which allows control authority to be distributed to the individual agents [18]. It has high reliability and robustness by using MAS to solve practical problems. However, MAS has difficulty dealing with high-dimensional continuous environments because of its concurrency. On the contrary, DRL is able to deal with high-dimensional inputs and learn to control complex actions.

MADRL combines the advantages of DRL and MAS and overcomes their inherent disadvantages. Specifically speaking, DRL models often require a large number of samples for training, and the inherent concurrency of the MAS system enables agents to generate a large amount of data concurrently, which greatly increases the number of samples, accelerates the learning process, and achieves better learning effects. At the same time, the internal structure of the neural network can solve the communication problem in MAS by using a shared policy network that exhibits implicit coordination to overcome the problem of inadequate artificially defined communication methods.

MADRL is an effective method for solving the multi-ship autonomous collision avoidance problem, which is a typical sequential decision-making process. Zhao et al. proposed a DRL-based algorithm to address the multi-ship collision avoidance problem. The algorithm adopts policy network sharing, i.e., eight ships are trained simultaneously, which improves the efficiency of policy convergence and obtains higher returns [17]. Luis et al. proposed a centralized convolutional Deep Q-network. Each agent has an ultimately independent dense layer to handle scalability [19]. Chen et al. proposed a multi-ship cooperative collision avoidance method based on the MADRL algorithm. By designing different reward weights to vary the degree of cooperation among the agents, the impact of agents in different cooperation modes on their collision avoidance behavior is discussed [20]. However, the above DRL algorithms are constructed and trained by pure simulation data. As a result, even if these models perform well in simulation environments, there is no guarantee that they will be able to make equally effective and safe decisions in real waters. Compared with simulation data, models trained by real data can not only better cope with real navigational challenges but also more deeply absorb human experience and wisdom to ensure the ship's safety and reliability in various scenarios.

The shipborne navigation aid systems, which include RADAR/ARPA, AIS, and ECDIS (Electronic Chart Display and Information System), provide the source and real data of ship collision avoidance scenarios at sea. As a requirement (part of the International Convention for Safety of Life at Sea), AIS, which should be carried for all ships from 2002, shall provide information including the ship's identity, type, position, course, speed, navigational status, and other safety-related information—automatically to appropriately equipped shore stations, other ships, and aircraft. Meanwhile, the reporting interval of AIS messages is from 2 s to 6 min, depending on the message types and the ship's dynamic conditions [21].

Growing ships have been equipped with AIS devices in the past twenty years, so a huge amount of marine traffic scenarios that are useful to develop ship autonomous collision avoidance algorithms have been recorded and accumulated in shore-based systems.

Motivated by all of the above, this paper proposes A multi-ship distributed collision avoidance algorithm with MADRL by real AIS data-driven, taking into consideration mixed traffic scenarios and uncoordinated scenarios in real waters. In this paper, the overall framework and its constituent units follow the principle of "reality as primary and simulation as supplementary", which determines that real AIS data-driven model structure occupies a dominant position. Then, we combine the statistical results of the real water traffic data to guide and design the MADRL framework and select the representative influencing factors to be designed into the collision avoidance decision-making algorithm's reward function. Next, based on the idea of "reality as primary and simulation as supplementary", the proportion of practical significance is selected to use real-AIS data and simulation data for model training, respectively. Finally, the simulation tests the collision avoidance effect of this algorithm in a library of complex and difficult ship encounter scenarios based on the idea of the Imazu problem.

3. Multi-Ship Collision Avoidance Decision-Making Algorithm Design

In this section, we will describe COLREGs, ship coordinated and uncoordinated behaviors, and design the flow chart, observation state, action space, reward function, and neural network model in the proposed algorithm.

3.1. COLREGs

In the sight of one another, overtaking situations, head-on situations, and crossing situations are three situations of encounters or three positional relationships that are constituted when two ships meet during navigation. Chapter two of COLREGs defines the conditions that constitute these three situations and also the rights and obligations of the ship in them. The situations defined by COLREGS are also the environment in which the ship's autonomous collision avoidance decision system operates as the agent. The specific definitions are shown below [22,23]:

- Rule 13 (Overtaking): If a vessel is deemed to be overtaking when coming up with another vessel from a direction more than 22.5° above her beam, the situation is considered to be overtaking. Notwithstanding anything contained in the Rules of Part B, Sections I and II, any vessel overtaking any other shall keep out of the way of the vessel being overtaken.
- Rule 14 (Head-on situation): Each ship should turn to the starboard and pass on the port side of the other ship when there is a risk of collision.
- Rule 15 (Crossing situation): If the courses of two vessels cross, the situation is considered as crossing situation; When two power-driven vessels are crossing so as to involve risk of collision, the vessel which has the other on her own starboard side shall keep out of the way and shall, if the circumstances of the case admit, avoid crossing ahead of the other vessel.

As is shown in Figure 1, the yellow region indicates the head-on situation, the red region indicates the port crossing situation, the green region indicates the starboard crossing situation, and the white region indicates the overtaking situation in which the agent ship is the overtaken vessel. In addition, the own ship (OS) is pink, and the target ship (TS) is blue.

3.2. Ship Coordinated and Uncoordinated Behaviors

The following situation may occur during the process of maned-vessel collision avoidance in the real waters: one or more vessels do not take coordinated communication or take collision avoidance actions based on COLREGs, resulting in uncoordinated collision avoidance behaviors [24]. Meanwhile, there will be a mixed traffic scenario in which manned ships and autonomous ships coexist for a certain period in the future [25]. Therefore, the possible uncoordinated behavior of all ships from the global perspective is one of the factors that MASS collision avoidance algorithms need to focus on when designing.

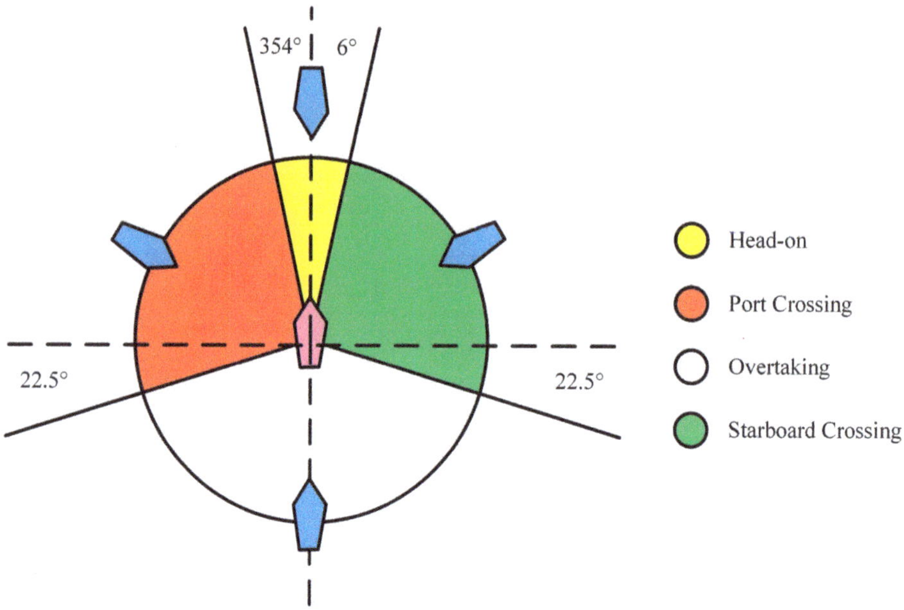

Figure 1. The typical diagram of different encounter situation.

Based on this, we define "coordinated collision avoidance behaviors" in this paper as those taken by the ship, which has the attribute of the trained agent. Specifically, the ship can take safe and rule-compliant collision avoidance decision-making measures when it recognizes a collision risk. Likewise, "uncoordinated collision avoidance behaviors" are defined as those taken by the ship which does not have the attribute of the trained agent, such as keeping speed and course without taking collision avoidance actions or taking non-rule-compliant actions.

We adopt the MAS framework, i.e., all ships within the scenario are default set as positive and rational agents that adopt coordinated collision avoidance behaviors. In order to simulate the uncoordinated scenarios in real waters, as well as to consider the sampling flexibility and enhance the model robustness factors, this paper selects the Weighted Random Sampling (WRS) method. The interval [0, 1] is divided into equal parts at interval intervals of 0.2 by the WRS method. Each interval is assigned a weight value, as shown in Table 1. A higher weight value means a higher probability that the interval will be selected.

Table 1. Selection probability of the random number generation based on WRS.

The Interval for Random Number Generation	Probability of Selecting the Interval
[0.0, 0.2]	0.10
[0.2, 0.4]	0.20
[0.4, 0.6]	0.40
[0.6, 0.8]	0.20
[0.8, 1.0]	0.10

Based on the above method, we set that there are n ships within the encounter scenario. When the $i-th$ ship decides whether to perform the coordinated collision avoidance action or not, a random number $R_i(i = 1, 2, \dots n)$ that falls within the [0, 1] probability interval will be generated. R_i represents the probability of whether the $i - th$ ship performs a collision-avoidance action or not, which can also be interpreted as the probability that the ship is given the attributes of a positive and rational agent.

In order to effectively manage the non-coordination behaviors and improve the system's overall performance, this paper proposes a flexibly adjustable non-coordination avoidance factor θ. When $R_i > \theta$, the $i - th$ ship is regarded as having the attribute of the trained agent in the collision avoidance scenario and follows the reward function design concept to positively take avoidance measures in Section 3.8. On the other hand, when $R_i < \theta$, the $i - th$ ship will no longer have the attribute of the trained agent. Specifically, the ship may keep speed and course without taking collision avoidance actions or taking non-rule-compliant actions. We set the ship's hazard recognition switch and the agent attribute switch to be mutually exclusive. When the ship recognizes a hazard, the algorithmic model will extract the failure experience or worse experience from the training experience pool. And the action space corresponding to the selected experience will be used as the action measure. This may create a more dangerous situation within the whole scenario. At this time, ships with uncoordinated behaviors will follow the new reward function, as detailed in Section 3.8.

At the same time, we can control the proportion of uncoordinated scenarios appearing by adjusting the weights of the WRS intervals and the size of the non-coordination avoidance factor θ to make the generated test scenarios as close as possible to real water. This increases the diversity and authenticity of the training data set.

3.3. Flow Chart

Figure 2 shows the flow chart of the algorithm. At the start of each cycle, state parameters are obtained, and the values of DCPA (distance of the closest point of approach), TCPA (time to the closest point of approach), distance, and bearing are calculated to obtain the current status information. Then, the risks of encounter situations are calculated during each state transfer. If there are no risks and the ship has passed and cleared the target ship, the ship will return to the planned route. If there are no risks and the ship has not passed and cleared the target ship, the ship will keep the original course and speed. If there are risks, the observation state will be calculated and input to the DDQN (Double Deep Q-Network) to make the decision. The corresponding action information is then transferred to the ship motion control system, which updates the current status information in conjunction with the ship motion model. The cycle ends when the ship reaches the planned route point or when a collision occurs with the target ship. Otherwise, the cycle will continue.

3.4. Definition of Ship Collision Avoidance Problem Based on MDP

Markov chain is a random process with Markov property, i.e., the future state depends only on the current state and is unrelated to the past state. In the ship collision avoidance problem, we can use factors such as the ship's position and speed as input states. The actions of the ship in each state are affected by certain probabilities, which can be expressed as state transfer probabilities. These describe the probability of transferring to another state in a given state.

However, the actions of ships are not only affected by the states but also by the other ships' actions in the environment, as well as the ship's desired goals. Therefore, we need to introduce the Markov Reward Process (MRP) to consider these factors. MRP is an extension of the Markov chain. It combines the probability of each state transfer with an immediate reward to take into account the effect of the particular behavior in a given state. In the ship collision avoidance problem, we can define the reward function. For example, the smaller the deviation distance, the larger the reward that the agent receives to encourage the ship to choose the appropriate actions to avoid the collision.

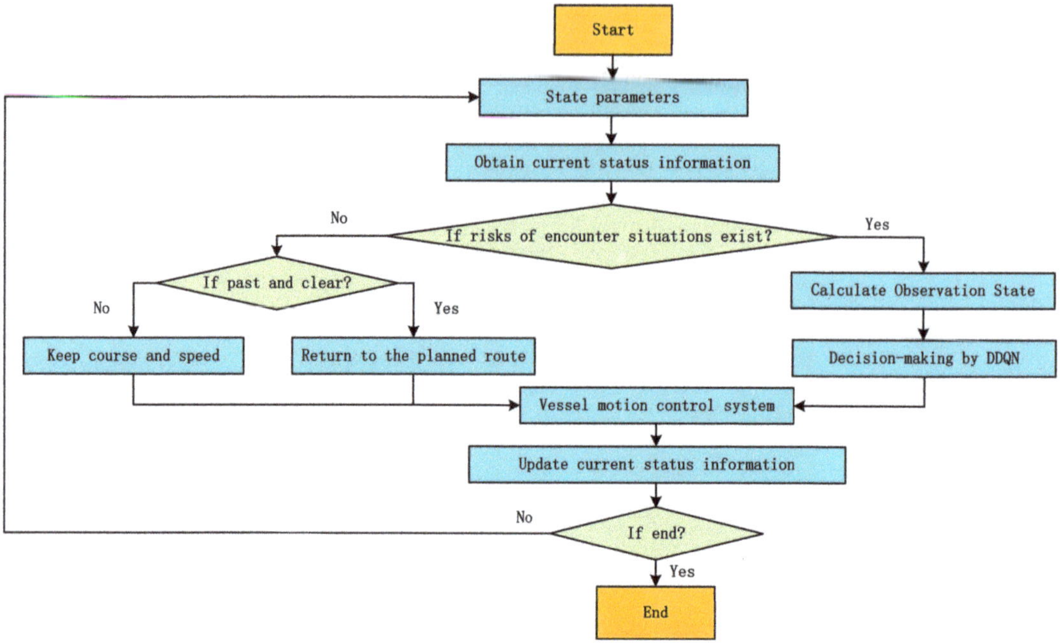

Figure 2. The flow chart of autonomous collision avoidance decision-making algorithm.

On this basis, we continue to introduce decision variables that allow the ship to choose the actions under each state, thereby forming a complete MDP. At the same time, by considering all possible actions that can be taken in each state, we can establish decision rules or policies to guide the ships' actions so that the overall reward is maximized or a specific objective function is optimal.

Therefore, when applying the MADRL framework to solve the multi-ship collision avoidance decision-making problem, we describe this problem as an MDP. This method can help us to solve the ship collision avoidance problem systematically and provide guidance for decision-making. In the MDP, the agent obtains the observation state from the current environment and decides to perform the action based on it. The chosen action, in turn, indirectly affects the update of the environment and the size of the reward value. Based on the above, this paper represents the MDP as an 8-tuple $(S, O, A, \pi, P, R, \gamma, \alpha)$ as follows:

1. S is a finite set of environment states; s is the current environment state, which mainly includes ships, dynamic obstacles, static obstacles, etc.
2. O is the set of observed states of the agents; o_t is the observation state obtained by the agent in the environment at the moment t.
3. A is the action space set of the agents; a_t is the action performed by the agent at the moment t, generated by the policy function $\pi(a \mid o) = \mathbb{P}(A = a|O = o)$.
4. P is the state transfer function and $P \in [0,1]$; $P(s' \mid s, a) = \mathbb{P}(S'_{t+1} = s'|S_t = s,\ A_t = a)$ is the probability that the state is transferred from s to $s\prime$ after the agent performs the action a_t at the moment t.
5. R is the reward function; r_t is the reward that the agent receives from the environment at the moment t.
6. γ is the decay value for future reward; α is the learning rate of the agent.

3.5. PER-DDQN

In 2015, V. Mnih's team proposed the concept of target neural networks, which officially marked the birth of DQN (Deep Q-network) [26]. Compared with traditional

Q-Learning, DQN no longer records the Q-value but uses a neural network $Q(s,a;w)$ to approximate the optimal action-value function $Q^*(s_t,a_t)$. The DQN algorithm's main advantage is its ability to deal with high-dimensional state spaces. Meanwhile, the algorithm's generalization ability can be improved through deep neural network learning to ensure scalability and applicability.

However, DQN does not guarantee that the network will always converge because DQN suffers from the maximum operator and bootstrap problems. To solve this problem, DDQN (Double Deep Q-Network) was proposed by the DeepMind team in 2016 [27]. DDQN works by setting up two independent Q-networks. One is the main neural network for selecting the maximum value action, and the other is the target neural network for evaluating this action's Q-value. The target neural network is usually a duplicate of the main network, but its parameter θ^- is not updated with each training iteration. Instead, it is copied from the main network at a certain frequency. Specifically, when we use the target network to compute the target's Q-value, the parameter θ^- is only updated once every certain number of steps so as to maintain the stability of the objective function. This results in less variation in the target value during the training process and allows for more efficient training of the primary network. At the same time, it reduces the noise and volatility in the learning process and improves the stability of training and convergence speed.

We compare the neural network performance of Nature-DQN, Target Network, and DDQN by the process of computing TD-target, as shown in Table 2.

Table 2. Comparison of three neural network constructions.

Type	Action Selection	Value Evaluation
Nature-DQN	DQN: $a^* = argmaxQ(s_{t+1},a;\theta)$	Target Network: $y_t = r_t + \gamma Q(s_{t+1},a^*;\theta)$
Target Network	Target Network: $a^* = argmaxQ(s_{t+1},a;\theta^-)$	Target Network: $y_t = r_t + \gamma Q(s_{t+1},a^*;\theta^-)$
DDQN	DQN: $a^* = argmaxQ(s_{t+1},a;\theta)$	Target Network: $y_t = r_t + \gamma Q(s_{t+1},a^*;\theta^-)$

DDQN not only alleviates the high-estimate problem but also improves usability and makes training more stable and efficient. In addition, Schaul's team proposed the Prioritized Experience Replay (PER) method in 2016 [28]. It is an enhanced experience replay method for learning by agents for training deep neural networks. It introduces the priority concept based on the traditional experience replay, i.e., it prioritizes the more important experiences for learning and makes more efficient use of the samples in the experience pool to improve the training efficiency and performance.

Based on the above, this paper adopts the PER-DDQN algorithm. It extracts all the transfer information in the experience pool that can be used for experience replication and then selects and gives priority to the transfers with a larger TD error. These experiences are more worthy of agent learning, so they are given greater priority. The model of PER-DDQN is shown in Figure 3.

Overall, the combination of DDQN and PER amplifies its intelligence advantage on a macro level, which can be understood as the agent paying more attention to failed experiences and choosing the learning order according to the experience priority. This can greatly reduce the trial-and-error process, make the network converge more quickly, and use the samples in the experience pool more efficiently to avoid experience waste. At the same time, using PER can eliminate the correlation between transitions and improve the performance of the DRL algorithm, making it more efficient and stable in dealing with complex tasks.

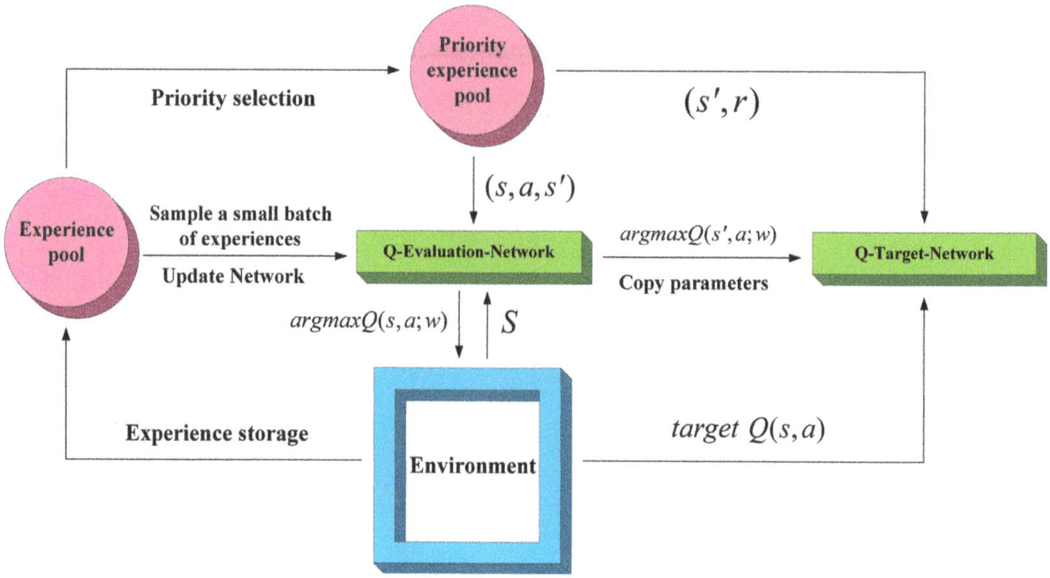

Figure 3. The model of PER-DDQN.

3.6. Observation State

In this paper, we define the distribution of MAS to constitute the set of environments as follows:

$$S = \begin{bmatrix} \psi_1 & v_1 & x_1 & y_1 \\ \psi_2 & v_2 & x_2 & y_2 \\ \cdots & \cdots & \cdots & \cdots \\ \psi_{n-1} & v_{n-1} & x_{n-1} & y_{n-1} \\ \psi_n & v_n & x_n & y_n \end{bmatrix}$$

where ψ_n is the ship's course or the dynamic obstacle's moving direction; v_n is the ship's speed or the dynamic obstacle's moving speed; x_n and y_n are the latitude and longitude of the ship or obstacle, respectively; n is the number of targets in the environment.

In past studies, research scholars have proposed many methods for predicting the hazard area of ship collision. For example, the obstacle zone by target (OZT) method based on the risk evaluation circle (REC) [29], the avoidance of bow crossing detection method [30], the predicted area of danger (PAD) model, the collision probability model, fuzzy logic and rule-based reasoning, and digital simulation. Comprehensively considering factors such as the real-time nature of environmental changes and the uncertainty of ship navigation, this paper will use an improved method based on OZT to predict the collision hazard area of each ship in the MAS.

The core idea and design principle of OZT is to "enlarge obstructions" and "advance avoidance". Specifically, ships use sensors such as LiDAR and cameras to capture information about their surroundings, including the location, size, and shape of obstacles, which is fed into the OZT algorithm. The OZT algorithm "enlarges" the obstacle at the system's decision level; namely, the size of the obstacle is virtually magnified. Therefore, the ship's perception system will consider the obstacle to be closer than its actual distance when the ship is in close proximity to the obstacle. Ships will start to change course or slow down when they are still a certain distance away from the targets and take avoidance action in advance.

Although the OZT can allow ships to achieve certain results in avoidance actions, the method has some practical application problems. Firstly, the correct execution of OZT relies heavily on the sensors' performance. If the sensor data are inaccurate (sensor malfunction, ambient noise, obstacle occlusion, etc.), the OZT may not be able to correctly "enlarge" the obstacle, resulting in reduced avoidance performance. Secondly, OZT requires real-time environmental analysis and decision-making, which may require significant computational resources. For some unmanned systems with limited hardware resources, there may be a trade-off between OZT and other navigation tasks. Thirdly, since the design principle of OZT is "avoidance in advance", there may be the possibility of over-avoidance, which reduces the operational efficiency of the ship and the unreasonable avoidance behaviors.

Considering the above possible problems, the OZT method is improved in this paper to enhance the method's ability to cope with emergencies because the CPA (closest point of approach) is the point where two ships are closest to each other when they meet at sea. As a result, the high probability of collision in real waters is near the CPA [31]. In addition, DCPA and TCPA are CPA-derived physical quantities. DCPA is the distance between the closest approach of two ships. TCPA is the time required for a ship to reach the CPA. These parameters are very important concepts in ship collision avoidance and core indexes for developing navigation policies and assessing ship safety [32]. Therefore, the target ship's CPA is taken as the center of the circle, and the speed navigation distance (SND) R_{SND} is taken as the radius (The diameter $D_{SND} = 2R_{SND}$) to create a circular area C_1. When the ship sails to the moment t, based on the speed v of the target ship, the system calculates the distance $D_{calculation}$ that the target ship will travel in the next k set time steps (kh), and the calculation equation is shown in Equation (1).

$$D_{calculation} = khv \tag{1}$$

We extend C_1 along the direction of the target ship's course at the moment t by a distance $D_{calculation}$ to form a new circular area C_2, which is the target ship's CPA area after k time steps. As shown in Figure 4, the capsule-shaped area formed by geometrically connecting C_1 and C_2 is the collision hazard prediction area C_{OZT} set up in this paper. The length of this geometric area is $D_{Length} = D_{calculation} + D_{SND}$ and the width is $D_{width} = D_{SND}$, and all ships in the MAS should avoid entering this area. At the same time, according to the speeds of different target ships, they will be given different prediction time steps. The purpose is to control the extension distance $D_{calculation}$ unchanged so that all target ships form a collision hazard area of equal size. In this paper, we set $D_{calculation} = 1.5$ NM, $R_{SND} = 0.5$ NM, $D_{Length} = 2.5$ NM. By this way, it can balance the differences of the target ship with different features such as course, speed and size, which can reduce the algorithm's computation and facilitate the scene clustering. At the same time, the method can deal with emergencies when the sensors are faulty and prevent the observation space from generating chaos.

Considering that the input to a neural network can only be a tensor of fixed dimension, this paper designs the observation state space as an observable discretized environment and quantifies the predicted hazard area by using the grid method. This ensures that the dimension of the observation state does not change with the number of target ships in the environment. In order to be closer to the real navigational environment at sea, this grid environment uses its own perspective as the center and establishes a field of view (FOV) to detect the environment's state. At the same time, taking itself as the center of the circle, it extends outwards with a fixed value of distance interval and angle interval to form a certain number of concentric circles. In addition, we set the due north direction as the course 0°, the clockwise as the positive direction, and the angle range as 360°. The whole circumference is evenly divided by a 15° interval with a detection radius distance of 8 NM and a distance interval of 0.5 NM, as shown in Figure 5.

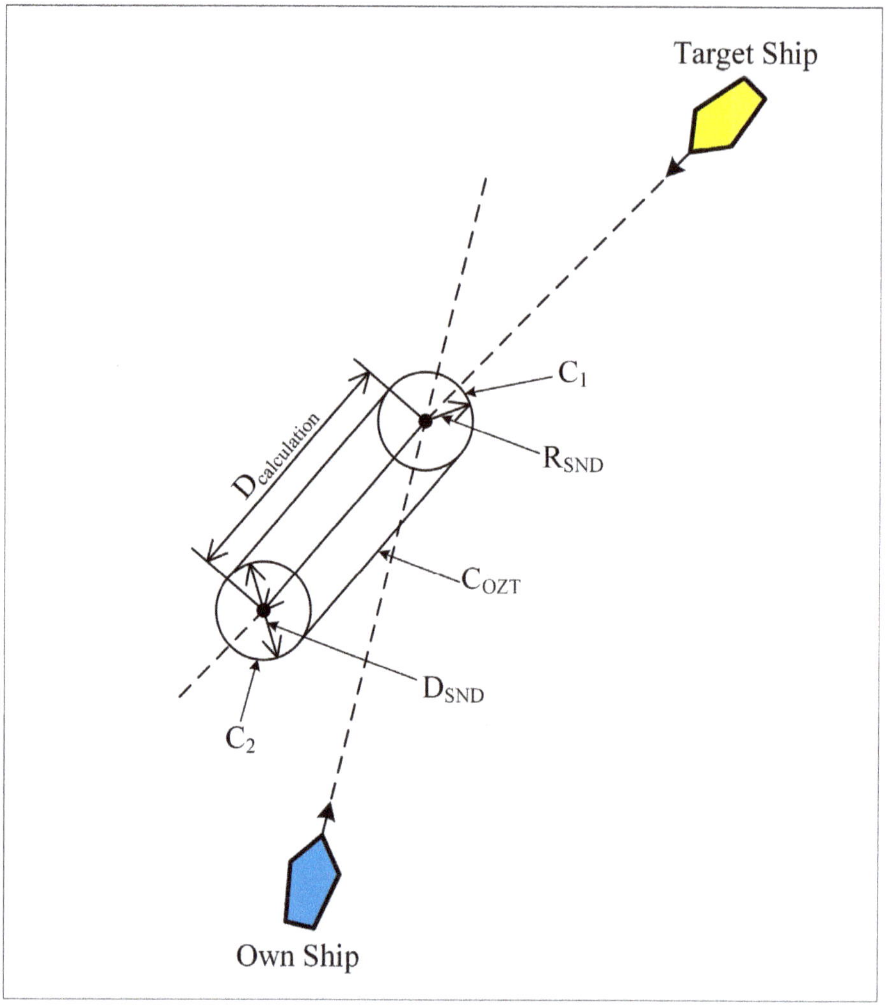

Figure 4. Improved OZT collision hazard area.

In addition, this paper defines the observation state by Boolean Operators: When the predicted collision hazard area of a ship is not in the FOV range, the ship's observation state o_t is 0; When the ship's predicted collision hazard area crosses the FOV range, the ship's observation state o_t changes to one and the collision avoidance decision-making switch is turned on. During the process of taking collision avoidance actions, the observation state o_t remains at one. The collision avoidance decision-making switch is turned off after the ship has passed and cleared the target ship. And the ship's observation state o_t becomes 0, which means the current collision avoidance task is completed.

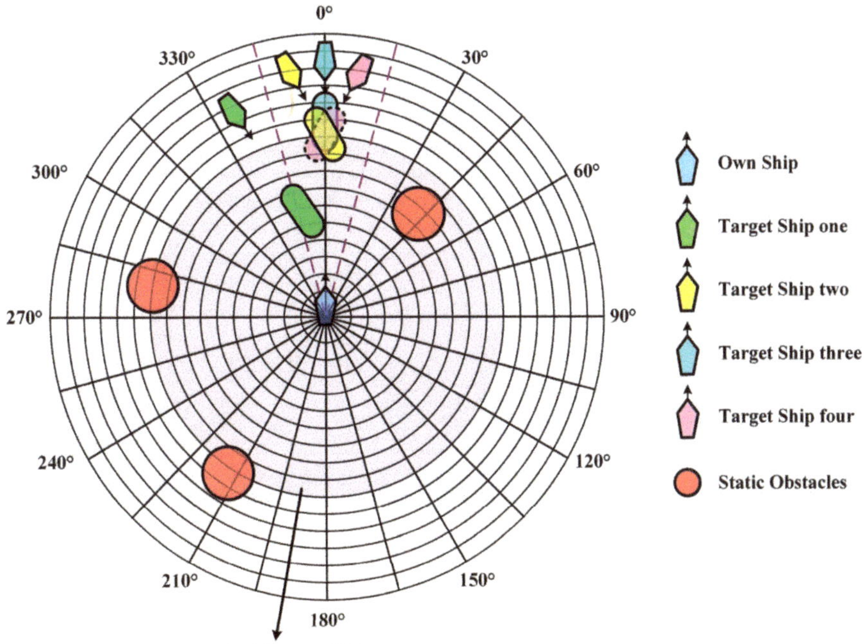

Figure 5. Observation state from the agent's own perspective.

Meanwhile, in order to reduce the input dimension of the neural network and reduce the risk of overfitting, we fixed the FOV's range and set the observation range of the agent to the environment to within 5 NM, which is helpful for us to better evaluate the generalization ability of the model. We believe that considering the partially observable perspective is an important step in the application of intelligent ships to real marine environments. At the same time, it is an effective means of replacing the state of the marine environment with areas that predict the possible risk of future collisions when we are dealing with a class of similar scenarios. In this way, similar encounter situations can be clustered and can lead to more stable decisions made by the model. By adopting the above method, the computation amount of the algorithm can be greatly reduced, and the size of the observation state space can be effectively reduced. It also prevents the observation space from generating chaotic superposition or wrong recognition of the external environment.

3.7. Action Space

Ship collision avoidance usually consists of four parts: environmental perception, taking collision avoidance action, keeping on course and speed, and returning to the planned route. In the entire collision avoidance process, the time spent on the collision avoidance decision-making (taking collision avoidance action and returning to the planned route) is much less than that spent on keeping course and speed, but it is the core part of the whole action. If the RL algorithm is used in the whole process, it will greatly increase the number of state transfers in the decision-making process, causing difficulties in model convergence. Therefore, the algorithm in this paper will only be used in the collision avoidance decision part, meaning that the agent interacts with the environment only in the collision avoidance decision-making phase, effectively shortening the number of state transitions in the MDP and substantially improving the efficiency of the algorithm. According to the above and Rule 8 [22,23]: If there is sufficient sea room, alteration alone, of course, may be the most effective action to avoid a close-quarters situation provided that it is made in good time, is substantial, and does not result in another close-quarters situation.

In collision avoidance, the pilot usually takes steering avoidance measures, including controlling the rudder angle and the course of a ship. The rudder angle change is different for different ships in the same encounter scenario. It is worth noting that the ship's course is the same at this point. Therefore, this paper will adopt the second avoidance measure as the action space, through a series of discrete course angle commands to continuously adjust the course and finally complete the ship collision avoidance. In other words, the discrete course change angle range is set as this algorithm's action space [20].

The six-degrees-of-freedom (6-DOF) model is widely used in the field of ship motion, but we usually adopt the three-degrees-of-freedom (3-DOF) model in ship collision avoidance. The 3-DOF mathematical model of a ship is shown in Figure 6.

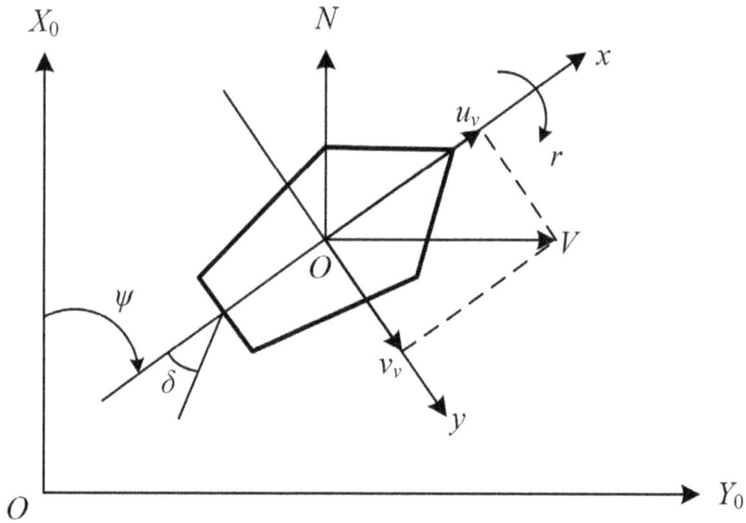

Figure 6. The 3-DOF mathematical model of a ship.

In this paper, the ship motion parameters are calculated by using Nomoto Equation [33], as expressed in Equation (2).

$$\begin{bmatrix} \dot{\psi} \\ \dot{r} \\ \dot{\delta} \end{bmatrix} = \begin{bmatrix} r \\ (K\delta - r)/T \\ (\delta_E - \delta)/T_E \end{bmatrix} \tag{2}$$

At the same time, the rudder angle is calculated by the PD controller and solved by the differential equation, as expressed in Equations (3)–(5).

$$r(t) = K\delta\left(1 - e^{-\frac{t}{T}}\right) \tag{3}$$

$$\psi(t) = K\delta\left(t - T + e^{-\frac{t}{T}}\right) \tag{4}$$

$$\delta(t) = K_p[\psi_c - \psi(t)] + K_d r(t) \tag{5}$$

The formula for the agent position at any moment $t_2 = t_1 + \Delta t$ is as follows:

$$x(t_2) = x(t_1) + \int_{t_1}^{t_2} v \cdot sin\psi(t_2)\, d(t) \tag{6}$$

$$y(t_2) = y(t_1) + \int_{t_1}^{t_2} v \cdot cos\psi(t_2)\, d(t) \tag{7}$$

where ψ is the course of the ship; ψ_c is the target course of the ship; r is the yaw rate; δ is the real rudder angle; δ_E is the command rudder angle; T_E is the time constant of the steering gear; K and T are the index parameters of ship maneuverability in clam water; K_p is the controller gain coefficient; K_d is the controller differential coefficient.

This algorithm discretizes the action space and executes a series of discrete course change angle a_t commands to complete ship collision avoidance based on the collision degree hazard identification results. This paper defines that the agent turns to the left as a negative angle and the right as a positive angle. The range of discrete course change angle is $[-10°, +10°]$. The calculation formula of a ship's new course is expressed in Equation (8), and the discrete interval a_t is expressed in Equation (9).

$$\psi = \psi_{last} + a_t \tag{8}$$

$$a_t \in [-10°, -8°, -6°, -4°, -2°, 0°, +2°, +4°, +6°, +8°, +10°] \tag{9}$$

3.8. Reward Function

The agent in the RL algorithm learns by acquiring rewards through interaction with the environment and decides the appropriate action by the amount of reward value. Therefore, the reward function becomes the key to how well the agent learns. It is also the core part of the RL algorithm, which directly affects the effectiveness of the collision avoidance decision.

In order to construct a meaningful and effective reward function, this paper invests a lot of time, resources, and effort in the preliminary data collection. At the same time, considering the uncertainty of the marine environment and the diversity of navigation situations, this paper collects a large amount of relevant historical data under various types of ship navigation situations, including sailing trajectories, radar information, sensor data, and so on. By processing and integrating the collected real data, this paper analyses and clusters the data of real ship collision avoidance scenarios to reveal the correlations and trends.

Therefore, in the process of designing the reward function in this paper, the statistical results of real water traffic data are fully integrated. This is an important theoretical basis to guide the construction of the reward function so that the decisions made by the agent are closer to the results of navigation in real waters.

Combined with the COLREGs of Rule 8, Rule 16, good seamanship, expert advice, practical experience and other factors [22,23], the reward function has six main parts, as follows:

- Failure Reward: When the distance between ships is less than 0.5 NM, the algorithm defines it as a collision occurs, i.e., collision avoidance fails. Then, it will receive a larger negative reward from the environment.
- Warning Reward: When the ship moves into the collision hazard area, it will receive a small negative reward from the environment.
- Out-of-bounds Reward: When the ship enters the unplanned sea area because of taking collision avoidance actions, it will receive a medium negative reward from the environment.
- Ship Size-Sensitivity Reward: The ship's size and sensitivity can affect the ship's collision avoidance strategy and decision-making. Larger ships typically require a larger turning radius and longer braking distances, so ship size can be considered for

inclusion in the reward function. For example, larger ships could be given more success rewards based on their size and sensitivity to emphasize their collision avoidance difficulty. This can guide different types of intelligent ships to make appropriate collision avoidance decisions for themselves.

- Success Reward: When the ship successfully avoids other ships, i.e., there is no risk of collision with any other ship at the next moment, it will receive a positive reward from the environment. This reward is refined into six components by considering all factors, i.e., rule compliance, the deviation distance at the end of the avoidance, the total magnitude of ship course changes during the avoidance process, the amount of the cumulative rudder angle during the avoidance process, the DCPA when clear of the other ship and the number of rudder operations.
- Other Reward: Except for the four cases mentioned above, the agent will not receive a reward from the environment, i.e., the reward is 0.

To sum up, the definition of the reward function used in this algorithm is specified in Equations (10) and (11).

$$Reward = \begin{cases} -20 & \textit{ship collision} \\ -2 & \textit{enter the collision hazard waters} \\ -5 & \textit{enter the unscheduled waters} \\ w_1 \times w_2 \times L \times B \times D & \textit{size-sensitivity impact extent} \\ [k_1\ k_2\ k_3\ k_4\ k_5\ k_6][R_{COLREGs}\ R_{deviation}\ R_{\Delta\psi}\ R_{\delta}\ R_{DCPA} R_{rudder}]^T & \textit{reach the destination} \\ 0 & \textit{other} \end{cases} \quad (10)$$

$$\begin{aligned} R_{COLREGs} &= \begin{cases} +5, \textit{ rule compliance} \\ -5, \textit{ rule noncompliance} \end{cases} \\ R_{deviation} &= \tfrac{d_{deviation}}{2} = \tfrac{d_{avoid}+d_{resumption}-d_{planned}}{2} \\ R_{\Delta\psi} &= \tfrac{|\Delta\psi|}{10} \\ R_{\delta} &= \sum_i |\delta_i| \\ R_{DCPA} &= \tfrac{1}{(n-1)} \sum_i^{n-1} DCPA_i \\ R_{rudder} &= (7 - n_{rudder}) \end{aligned} \quad (11)$$

where $d_{deviation}$ is the deviation distance at the end of the avoidance; $\Delta\psi$ is the ship's course angle during the avoidance process; δ_i is the magnitude of the $i-th$ rudder angle; n is the total number of ships in the current encounter situation; $DCPA_i$ is the distance to closest point of approach when passing and clearing the $i-th$ target ship; n_{rudder} is the total number of rudder operations; w_1 is maneuver difficulty coefficient based on the ship size; w_2 is ship maneuver sensitivity coefficient; L is the ship's length between perpendiculars; B is the ship's breadth; T is the ship's draft; k_i is the weight of each successful collision avoidance reward and $\sum_1^6 k_i = 1$, where $k_1 = 0.3$, $k_2 = 0.15$, $k_3 = 0.15$, $k_4 = 0.1$, $k_5 = 0.2$, $k_6 = 0.1$.

By selecting an action based on the above reward function, the ship is given the attribute of the trained agent and takes a coordinated collision avoidance action. However, ships with uncoordinated behaviors, as elaborated in Section 3.2, will no longer fully follow this reward function. We modify the reward function in terms of safety, rule compliance, and deviation distance, as shown in Equations (12) and (13).

$$Reward = \begin{cases} -20 & ship\ collision \\ +5 & enter\ the\ collision\ hazard\ waters \\ -5 & enter\ the\ unscheduled\ waters \\ w_1 \times w_2 \times L \times B \times D & size\text{-}sensitivity\ impact\ extent \\ [k_1\ k_2\ k_3\ k_4\ k_5\ k_6][R_{COLREGs}\ R_{deviation}\ R_{\Delta\psi}\ R_{\delta}\ R_{DCPA} R_{rudder}]^T & reach\ the\ destination \\ 0 & other \end{cases} \quad (12)$$

$$\begin{array}{l} R_{COLREGs} = \begin{cases} -5, & rule\ compliance \\ +5, & rule\ noncompliance \end{cases} \\ R_{deviation} = \frac{d_{deviation}}{2} = \frac{d_{avoid} + d_{resumption} - d_{planned}}{2} \\ R_{\Delta\psi} = \frac{|\Delta\psi|}{10} \\ R_{\delta} = \sum\limits_{i} |\delta_i| \\ R_{DCPA} = \frac{1}{(n-1)} \sum\limits_{i}^{n-1} DCPA_i \\ R_{rudder} = (7 - n_{rudder}) \end{array} \quad (13)$$

where k_i is the weight of each successful collision avoidance reward and $\sum_1^6 k_i = 1$, where $k_1 = 0.35, k_2 = 0.25, k_3 = 0.1, k_4 = 0.1, k_5 = 0.15, k_6 = 0.05$.

4. Training and Testing of Algorithm Model

In this paper, CPU (12th Gen Intel® Core™ i5-12400, Santa Clara, CA, USA) and GPU (Intel® UHD Graphics 730) are the equipment configurations for training and testing the algorithmic model. At the same time, pycharm software (Runtime version: 17.0.4.1) with python 3.10 is used to develop the algorithmic model.

4.1. Training Set

4.1.1. Real-Data Training Set

In this paper, the real encounter situation scenario data obtained from the literature [24] are used as the real-data training set for the algorithm. The specific approach is to screen out five groups of encounter information with different ship numbers, which are used as five units in the training set to serve the model training. And the ship's longitude and latitude information are converted to coordinate parameters in the XY coordinate system of this paper by Mercator projection so as to reproduce the real encounter scene in the training set.

We define a complete training cycle to consist of a single training session of its five constituent units. This paper completes a total of 10 training cycles and records the success rate of collision avoidance for each unit under each training cycle. We treat each unit of single training in each training cycle as an epoch, with each epoch containing n_1 iterations, and each iteration containing n_2 episodes. Each epoch trains all encounter situations (episodes) in its scene and records its training data at approximately equal intervals. At the same time, the initial value of $\varepsilon - greedy$ is defined as 0.90, increasing by 0.005 for every n_3 episodes; the neural network parameter θ_t^- is updated once for every n_4 episodes. The data information for each part of the training set is shown in Table 3.

In this paper, East is set as the positive X-axis direction, and North is set as the positive Y-axis direction in NM. The course is set using a circular representation. The results of all training cycles are shown in Figure 7. The curves represent the collision avoidance success rate of each unit in each training cycle driven by real data.

Table 3. The information of real-data training set.

Number of Ships	Episodes	n_1	n_2	n_3	n_4
Two	67,849	118	575	400	1000
Three	17,940	65	276	200	500
Four	4316	26	166	100	200
Five	951	19	50	50	50
Six	248	31	8	8	25

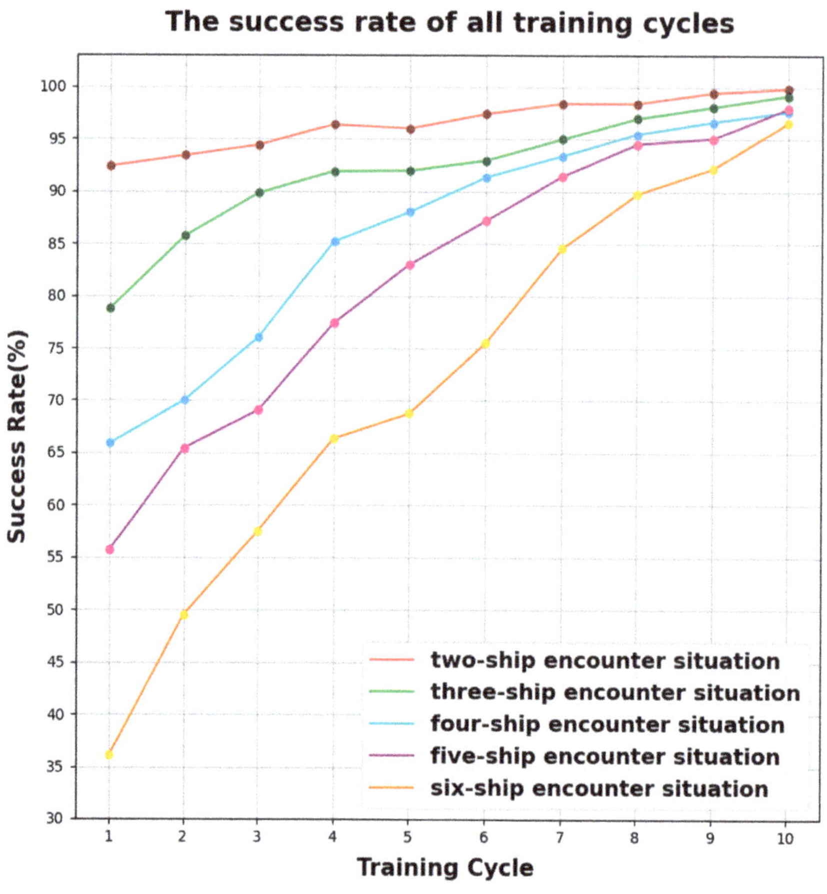

Figure 7. The real-data success rate for all training cycles.

From Figure 7, the model gradually and steadily converges in the success rate of collision avoidance with increasing training. Although there are very few cases of regression in the success rate, the success rate still shows an overall increasing trend. For situations where the number of ships is less, the success rate can increase at a steady pace with each training cycle. For situations with a high number of ships, the success rate is usually not high in the first training cycle. However, after a certain number of training cycles, the failure experience is focused on in the next learning. Therefore, the success rate of collision avoidance shows a significant increase. The greater the number of ships in the encounter situation, the faster the success rate improves.

In summary, the learning ability of the agent is gradually improved through the accumulation of training volume, and its abilities to deal with complex situations are becoming more and more strong. At the same time, the model trained by real data-driven training can ensure a high success rate when dealing with multi-ship situations. This shows that the model originated from reality and can be applied to it, which has a certain practical significance.

4.1.2. Simulation Data Training Set

The collected real-data-driven training sets do not cover all possible encounter scenarios because of high economic and time costs. Alternatively, the ship encounter scenarios are endless, and any slight change in the ship parameters will form new scenarios. And it may have an impact on the decision-making and the collision avoidance result. Although the agent's learning ability had been trained very well by real data, it may have insufficient coping ability when the agent faces unfamiliar and complex situations in the future.

Based on the above, we can conclude that it requires us to continue to enrich a large number of brand-new training scenarios so as to obtain more efficient and better training models. According to the COLREG definition of the encounter situation, we could "virtually" break the situation down into several single-ship situations under the perspective of any one ship. Therefore, we put 12 ships into the MAS. By designing the ship's course, speed, position, and destination, we make these ships constitute a variety of encounter situations, including head-on situations, port crossing situations, starboard crossing situations, overtaking situations, and overtaken situations. At the same time, considering the realism and uncertainty of the traffic flow, the weights are assigned to the integers within the interval [2,12] by the WRS method before starting the training of each episode. The larger weight value means the higher probability that the number is selected in the sample, as shown in Table 4.

Table 4. Selection probability of the ship number in the encounter scenario based on WRS.

Integer Interval Indicating the Number of Ships	Probability of Each Element in the Interval Being Selected
[2, 3]	0.05
[4, 5, 6]	0.15
[7, 8, 9]	0.10
[10, 11, 12]	0.05

Based on the above real-data training, the model can ensure a high success rate when dealing with situations with a relatively small number of ships. Therefore, situations with fewer ships will be given less weight when training on the simulation data in this subsection. This can improve learning efficiency and reduce the learning of similar experiences. At the same time, we also give less weight to encountering situations with excessive ships, such as 10, 11, and 12 ships. Although it is also achievable to successfully complete all ship collision avoidances with a certain amount of training, the real traffic flow is seldom so complex with such a large number of ships.

After selecting and determining the number of encounter situation ships i ($i = 1, 2, 3, \ldots, 12$) in the above way, we further select the ships corresponding to the number i in the MAS with 12 ships set up by complete randomization. In this way, the initial position of the ship and the training scenario are determined. The encounter scenarios set by double random selection of the ship number and ship position can greatly enrich the diversity of the simulation data training set, which is conducive to improving the model's coping ability and learning ability.

At the same time, considering that the ship's course is not constant in the real traffic situation, it will be affected by external factors such as wind, waves, currents, etc. Therefore, this paper sets that the course of each agent will be randomly determined within $\pm 5°$ of the set value. The trajectory mapping interval in the collision avoidance decision-making phase is set to 30 s, i.e., the time step of decision-making is 30 s. The information on ship navigation in the simulation-data training set is shown in Table 5. In this paper, the intelligent ship "YU KUN" is selected as the experimental model [34], and its parameters are shown in Table 6.

Table 5. Navigation information of MAS.

Ship No.	ψ (0°)	X (NM)	Y (NM)
Ship 1	[355, 5]	0.000	0.000
Ship 2	[25, 35]	−2.500	−4.330
Ship 3	[55, 65]	−6.062	−3.500
Ship 4	[85, 95]	−10.000	0.000
Ship 5	[115, 125]	−6.062	3.500
Ship 6	[145, 155]	−5.000	8.660
Ship 7	[175, 185]	0.000	8.000
Ship 8	[205, 215]	3.000	5.196
Ship 9	[235, 245]	8.660	5.000
Ship 10	[265, 275]	8.000	0.000
Ship 11	[295, 305]	6.062	−3.500
Ship 12	[325, 335]	4.500	−7.794

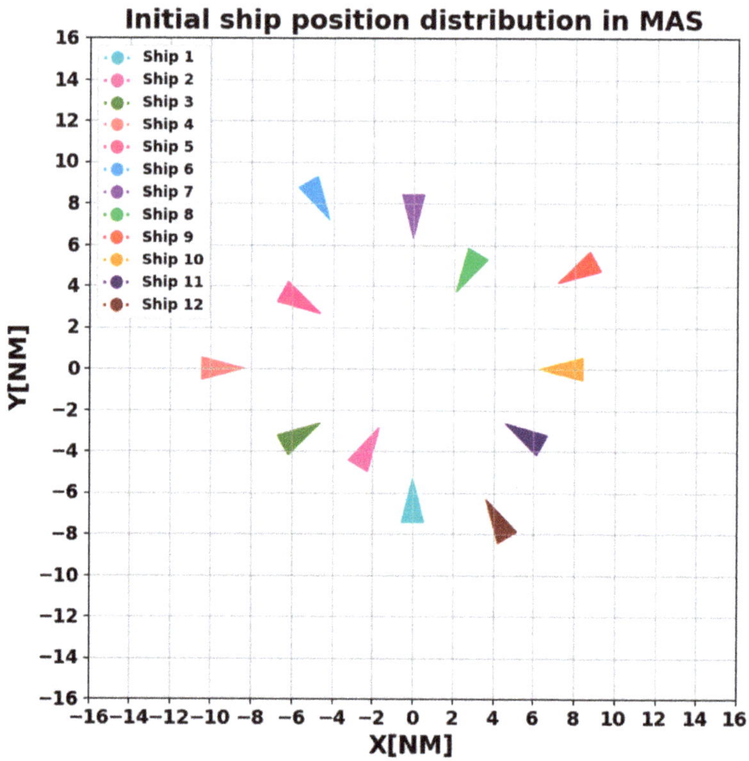

Figure 8. The initial ship position distribution in MAS.

Table 6. Ship parameters of the "YU KUN".

Physical Quantity	Symbol	Numerical Value
Length between perpendiculars (m)	L	105
Breadth (m)	B	18
Speed (kn)	V	12
Draft (m)	D	5.4
Turning ability index (1/s)	K	−0.2257
Following index (s)	T	86.8150
Controller gain coefficient (-)	K_p	2.2434
Controller differential coefficient (-)	K_d	35.9210

$LBVDKTK_pK_d$The initial ship position distribution in MAS is shown in Figure 8.

In addition, this paper follows the principle of "reality as primary and simulation as supplementary" to set up the total training set, and its content composition is shown in Figure 9.

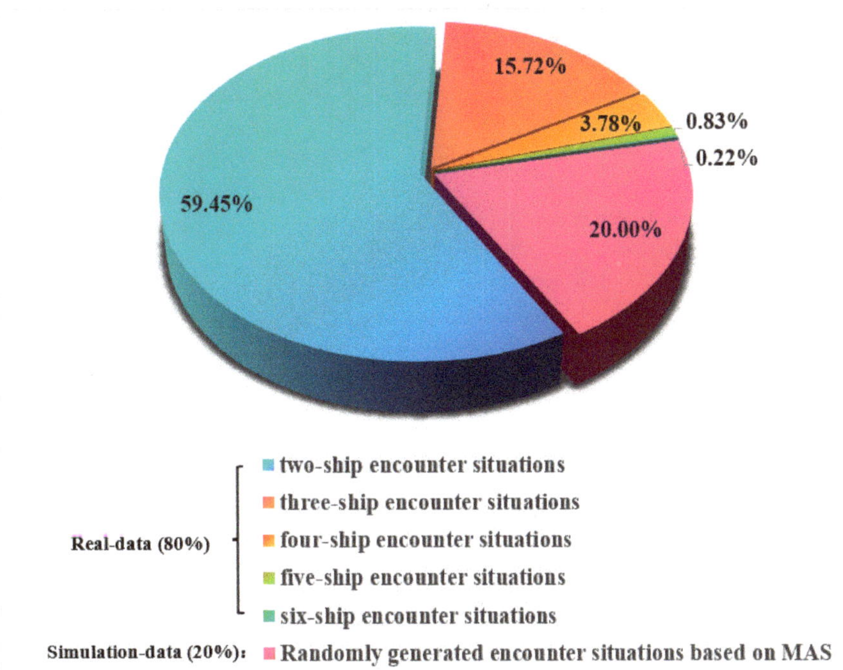

Figure 9. Content composition of the total training set.

From Figure 9, it can be seen that the real-data set in the previous subsection occupies 80% of the total training set, with a total of 913,040 episodes trained in 10 training cycles. And the remaining 20% has the simulation-data training set of this subsection constituting. In this part of the training set, each multi-ship encounter scenario has randomly generated ships in the MAS. We randomly generated 22,826 episodes by the method of WRS described above. The information and collision avoidance success rate of each episode is recorded and used as a complete training cycle (training subset). After that, this training subset was continued to repeat nine times without changing any of the training parameters, and the success rate of collision avoidance was recorded. Because each training cycle contains a sufficiently large number of episodes, and they are all generated in a random manner with a certain level of complexity. The resulting large number of random training samples is meaningful for both the improvement of the model generalization ability and the applicability extension.

Likewise, the idea of parameter setting in this subsection is consistent with the real data set. We treat each unit of single training in each training cycle as an epoch, with each epoch containing 115 iterations and each iteration containing 200 episodes. Each epoch trains all encounter situations (episodes) in its scene and records its training data at approximately equal intervals. At the same time, the initial value of $\varepsilon - greedy$ is defined as 0.90, increasing by 0.005 for every 1000 episodes; the neural network parameter θ_t^- is updated once every iteration. The results of all training cycles are shown in Figure 10. The curves represent the collision avoidance success rate of each unit in each training cycle driven by simulation data.

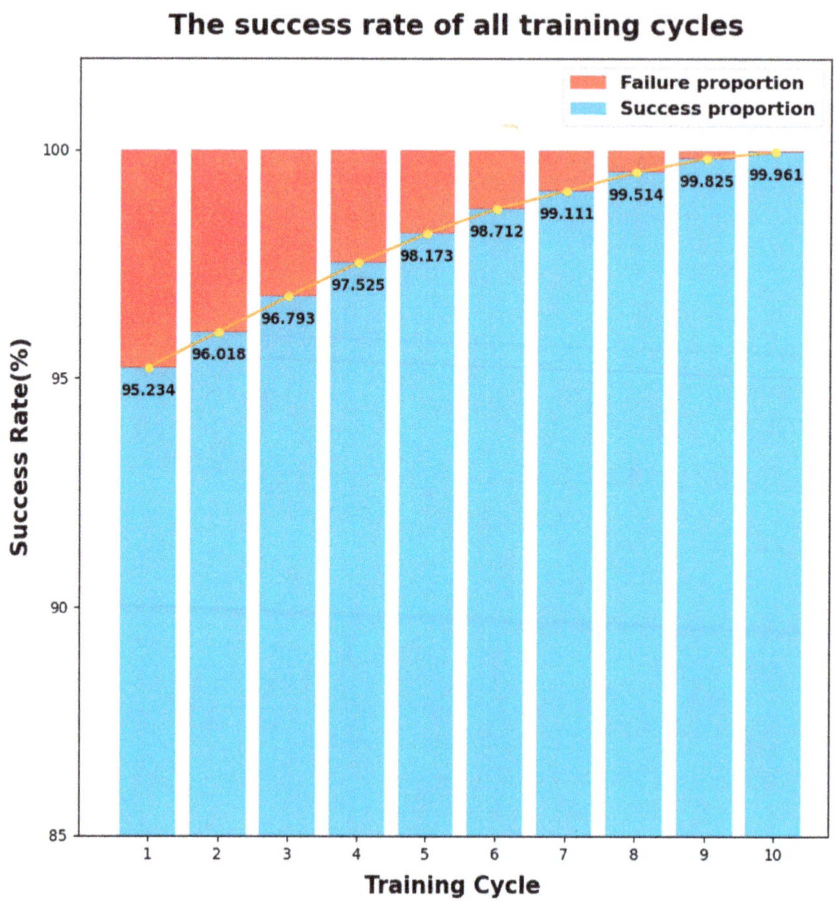

Figure 10. The simulation-data success rate for all training cycles.

In combination with the model training process and Figure 10, we can find that the model may fail the first few times in complex encounter situations. However, the model uses the PER technique and always follows the principle of "scenario adaptation" when constructing encounter scenarios. Therefore, after continuous focused learning, the agent can make the model converge quickly and stably in situations where the number of ships is "moderate", such as four ships, five ships ... eight ships, etc. At the same time, the model performs excellently and can be trained successfully for all episodes in most iterations. It was even able to gradually optimize the navigation process based on successful collision avoidance.

4.2. Testing Set

In the autonomous ship navigation field, the Imazu problem is widely considered a series of navigational collision avoidance challenges. In order to verify the algorithm's effectiveness and the model's generalization ability, this section designs and extends 40 scenarios as the encounter scenario library based on the Imazu problem's idea. The encounter scenario library includes relatively difficult and very difficult scenarios as a way to verify the model's expressiveness and usefulness in complex environments. The idea of building this scenario library mainly stems from the following aspects:

- Comprehensiveness extension: By testing to include a variety of possible real-world sailing scenarios, we can ensure that the algorithm is able to cope with the challenges in various aspects of actual sailing;
- Improving the model's generalization ability: Diversified scenarios can help the model learn richer data, thus making its performance more stable and reliable in unknown environments;
- Simulating extreme situations: The particularly difficult scenarios in the encounter scenario library can simulate extreme situations that might be encountered in reality, which is essential for assessing the model's performance under stress;
- Enhancing verification credibility: By verifying the model's performance in various scenarios, we can more confidently ensure its safety and effectiveness in real-world applications.

Overall, the encounter scenario library has been built to provide a comprehensive, practical, and challenging test environment to ensure the wide applicability of the model. By verifying in such a scenario library, the model not only demonstrates its excellent performance in complex environments but also further ensures its usefulness and safety. The initial information of the scenario library is shown in Table 7. Where Cases 1–4 are two-ship encounter situations, Cases 5–14 are three-ship encounter situations, Cases 15–31 are four-ship encounter situations, Cases 32–36 are five-ship encounter situations, and Cases 37–40 are six-ship encounter situations. The schematic of each scenario is shown in Figure 11. The agent model is still set to the "YU KUN" with a speed of 12 kn, and the overtaken ship's speed is set to 8 kn. The non-coordination avoidance factor θ is set to 0.5. Like Section 4.1.2, the trajectory mapping interval in the collision avoidance decision-making phase is set to 100 s.

Considering the large number of figures in the test results, we structured the article by including the figures in Appendix A. The model test results are shown in Figures A1–A3. Figure A1 shows the ship's trajectory, where the initial position of each agent is represented by a different triangle, the destination is represented by the circle of the corresponding color, and the trajectory's color is the same as that of the agent in the legend. It is assumed that Agent ship one is the perspective of the own ship, and Figure A2 shows the distance change between the own ship (ship one) and the target ships under this perspective, where the dotted line 0.5 NM represents the minimum encounter distance for the urgent situation specified in this paper. Figure A3 shows the minimum passing distance of each agent ship from other agent ships.

In order to better analyze the process of collision avoidance actions of each ship, Case 35 is used as an example to elaborate the whole sequential decision-making process in detail. Figure 12 shows the ships' trajectories.

Table 7. The initial information of the scenario library.

Case No.	Ship 1 X	Ship 1 Y	Ship 1 ψ (°)	Ship 2 X	Ship 2 Y	Ship 2 ψ (°)	Ship 3 X	Ship 3 Y	Ship 3 ψ (°)	Ship 4 X	Ship 4 Y	Ship 4 ψ (°)	Ship 5 X	Ship 5 Y	Ship 5 ψ (°)	Ship 6 X	Ship 6 Y	Ship 6 ψ (°)
1	0.000	−6.000	000	0.000	6.000	180	-	-	-	-	-	-	-	-	-	-	-	-
2	0.000	−6.000	000	6.000	0.000	270	-	-	-	-	-	-	-	-	-	-	-	-
3	0.000	−6.000	000	−6.000	0.000	090	-	-	-	-	-	-	-	-	-	-	-	-
4	0.000	−6.000	000	0.000	10.000	000	-	-	-	-	-	-	-	-	-	-	-	-
5	0.000	−6.000	000	0.000	6.000	180	6.000	0.000	270	-	-	-	-	-	-	-	-	-
6	0.000	−6.000	000	−6.000	0.000	090	0.000	6.000	180	-	-	-	-	-	-	-	-	-
7	0.000	−6.000	000	0.000	−10.000	000	5.657	5.657	315	-	-	-	-	-	-	-	-	-
8	0.000	−6.000	000	6.000	0.000	270	3.000	−5.196	330	-	-	-	-	-	-	-	-	-
9	0.000	−6.000	000	−1.553	−5.796	015	6.000	0.000	270	-	-	-	-	-	-	-	-	-
10	0.000	−6.000	000	−5.000	0.000	090	3.000	−5.196	330	-	-	-	-	-	-	-	-	-
11	0.000	−6.000	000	0.000	7.000	180	1.553	−5.796	345	-	-	-	-	-	-	-	-	-
12	0.000	−6.000	000	−6.000	0.000	090	6.000	0.000	270	-	-	-	-	-	-	-	-	-
13	0.000	−6.000	000	4.243	−4.243	315	1.553	−5.796	345	-	-	-	-	-	-	-	-	-
14	0.000	−6.000	000	−1.553	−5.796	015	3.000	−5.196	330	-	-	-	-	-	-	-	-	-
15	0.000	−6.000	000	−1.553	−5.796	015	0.000	6.000	180	4.243	−4.243	315	-	-	-	-	-	-
16	0.000	−6.000	000	−1.553	−5.796	015	−4.243	−4.243	045	0.000	6.000	180	-	-	-	-	-	-
17	0.000	−6.000	000	−1.553	−5.796	015	6.000	0.000	270	4.243	4.243	315	-	-	-	-	-	-
18	0.000	−6.000	000	0.000	−10.000	000	6.000	0.000	270	4.243	−4.243	315	-	-	-	-	-	-
19	0.000	−6.000	000	−4.243	−4.243	045	−6.000	0.000	090	6.000	0.000	270	-	-	-	-	-	-
20	0.000	−6.000	000	0.000	−10.000	000	−1.553	−5.796	015	4.243	−4.243	315	-	-	-	-	-	-
21	0.000	−6.000	000	4.243	4.243	225	3.000	−5.196	330	1.553	−5.796	345	-	-	-	-	-	-
22	0.000	−6.000	000	−1.553	−5.796	015	4.243	4.243	225	1.553	−5.796	345	-	-	-	-	-	-
23	0.000	−6.000	000	0.000	−10.000	000	6.000	0.000	270	3.000	−5.196	345	-	-	-	-	-	-
24	0.000	−6.000	000	−1.553	−5.796	015	6.000	0.000	270	1.553	−5.796	345	-	-	-	-	-	-
25	0.000	−6.000	000	0.000	−10.000	000	6.000	0.000	270	3.000	−5.196	330	-	-	-	-	-	-
26	0.000	−6.000	000	−1.553	−5.796	015	0.000	4.000	180	1.553	−5.796	345	-	-	-	-	-	-
27	0.000	−6.000	000	2.000	−4.000	000	0.000	6.000	180	2.000	8.000	180	-	-	-	-	-	-
28	0.000	−6.000	000	−3.000	5.196	150	6.000	0.000	270	3.000	−5.196	330	-	-	-	-	-	-
29	0.000	−6.000	000	−3.000	5.196	150	6.000	0.000	270	1.553	−5.796	345	-	-	-	-	-	-
30	0.000	−6.000	000	−1.553	−5.796	015	−3.000	−5.196	030	3.000	−5.196	330	-	-	-	-	-	-
31	0.000	−6.000	000	0.000	6.000	180	3.000	−5.196	330	1.553	−5.796	345	-	-	-	-	-	-

Table 7. *Cont.*

Case No.	Ship 1			Ship 2			Ship 3			Ship 4			Ship 5			Ship 6		
	X	*Y*	ψ (°)	*X*	*Y*	ψ (°)	*X*	*Y*	ψ (°)	*X*	*Y*	ψ (°)	*X*	*Y*	ψ (°)	*X*	*Y*	ψ (°)
32	0.000	−6.000	000	−4.243	−4.243	045	0.000	6.000	180	6.000	0.000	270	4.243	−4.243	315	-	-	-
33	0.000	−6.000	000	−4.243	−4.243	045	−6.000	0.000	090	0.000	6.000	180	4.243	−4.243	315	-	-	-
34	0.000	−6.000	000	−4.243	−4.243	045	−6.000	0.000	090	0.000	6.000	180	6.000	0.000	270	-	-	-
35	0.000	−6.000	000	−6.000	0.000	090	0.000	6.000	180	6.000	0.000	270	4.243	−4.243	315	-	-	-
36	0.000	−6.000	000	−6.000	0.000	090	−4.243	4.243	135	0.000	6.000	180	6.000	0.000	270	-	-	-
37	0.000	−6.000	000	4.243	−4.243	045	−6.000	0.000	090	0.000	6.000	180	6.000	0.000	270	4.243	−4.243	315
38	0.000	−6.000	000	−4.243	−4.243	045	−6.000	0.000	090	−4.243	4.243	135	0.000	6.000	180	6.000	0.000	270
39	0.000	−6.000	000	−6.000	0.000	090	−4.243	4.243	135	0.000	6.000	180	6.000	0.000	270	4.243	−4.243	315
40	0.000	−6.000	000	−6.000	0.000	090	0.000	6.000	180	4.243	4.243	225	6.000	0.000	270	4.243	−4.243	315

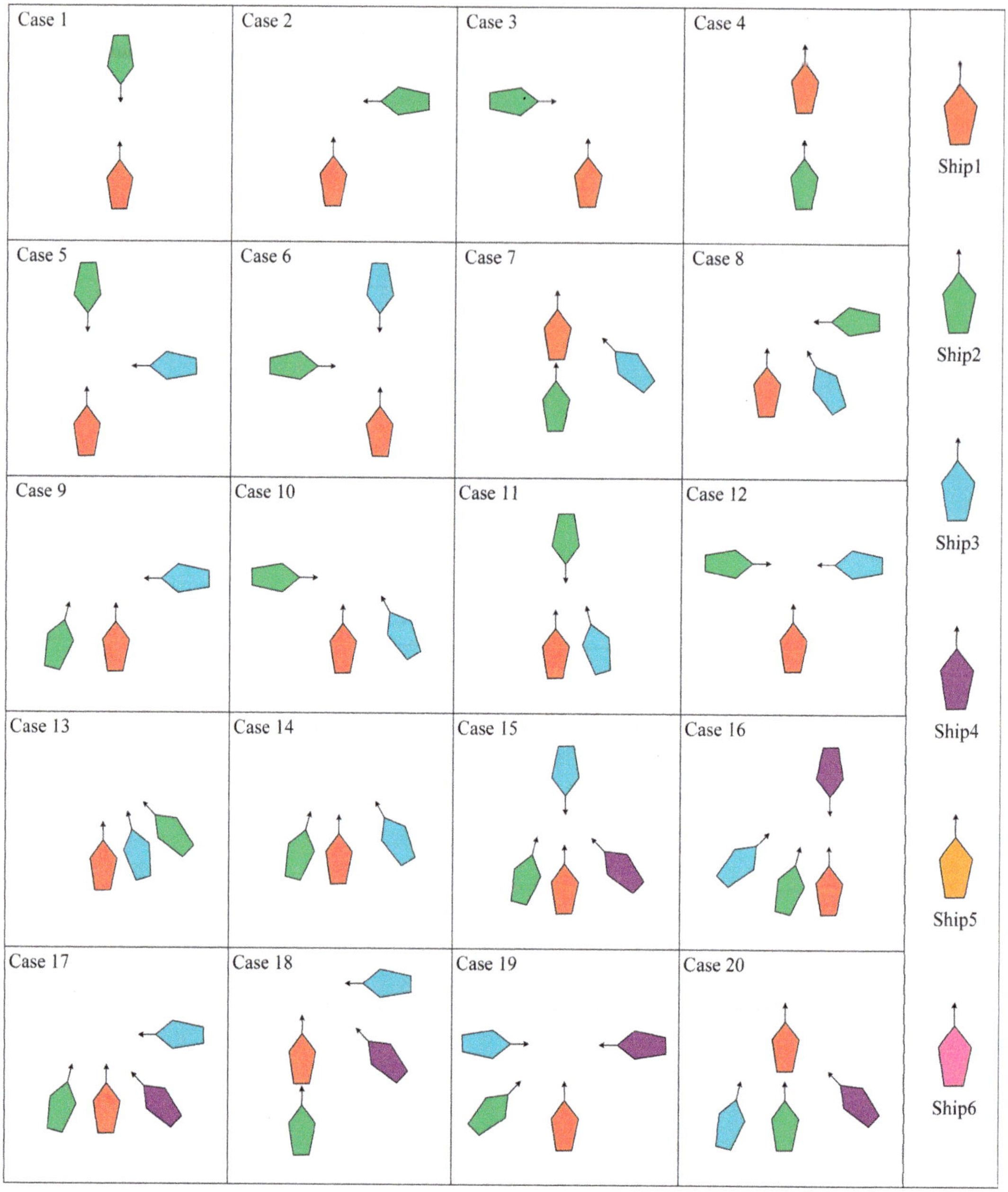

(a)

Figure 11. *Cont.*

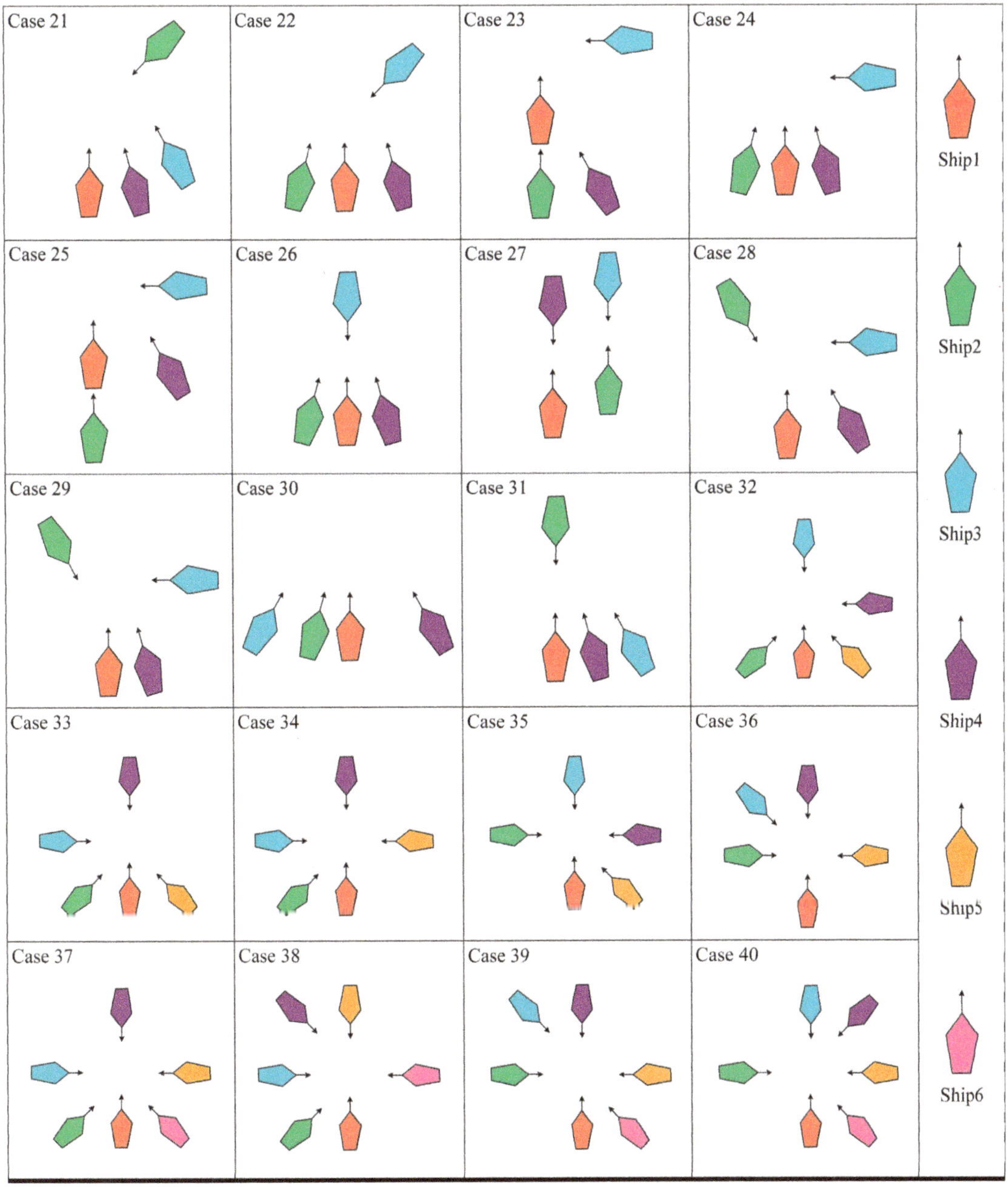

(b)

Figure 11. The extended encounter scenario library based on Imazu problem. (**a**) Case 1–20 of the extended encounter scenario library; (**b**) Case 21–40 of the extended encounter scenario library.

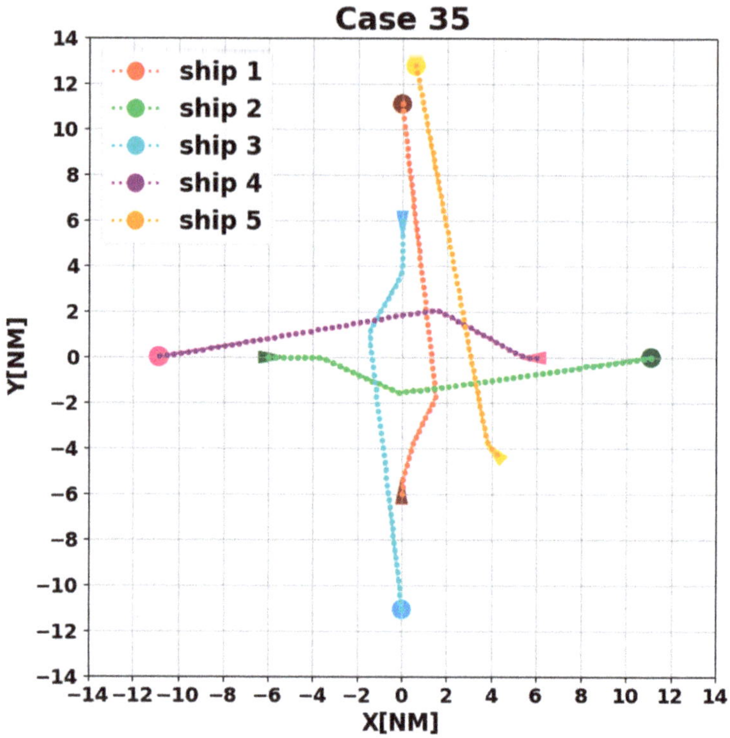

Figure 12. Ships' trajectories of Case 35.

At the initial moment, the five ships in the scenario constitute a relatively complex collision hazard situation. We split the current situation according to COLREGs and found that each ship has more than one encounter situation with other ships. For example, ship three forms the head-on situation with ship one and the crossing situation with the remaining ships, respectively.

We illustrate the working principle of the algorithmic MDP tuple by the motion process of ship three as follows. The collision avoidance algorithm model successively generates four MDP transitions (S,O,A,π,P,R,γ,α) for ship three.

1. At t = 0–600 s, ship three does not perceive a hazard in the environment, the observed state o_t is 0, and the ship is sailing towards its destination on the prescribed course;
2. At t = 600 s, ship three recognizes the hazard in the environment, at which time the observation state o_t changes to one, and collision avoidance action is started. The algorithmic model selects $a_1 = +10°$, $a_2 = +10°$, $a_3 = +10°$, $a_4 = +4°$ sequentially as actions in the action space based on the policy function π;
3. Until t = 1700 s, the ship removes the collision hazard by four course changes. At the same time, the observation state o_t becomes 0. The collision avoidance decision-making switch is turned off, and the ship starts to return to the planned route;
4. At t = 5400 s, all ships arrive at their destinations, and the sailing missions are over. The minimum passing distances of each ship from other ships are respectively 1.86 NM, 2.11 NM, 2.14 NM, 2.09 NM, and 1.86 NM. All ships are guaranteed to complete the collision avoidance decision-making beyond the safe distance.

At the same time, we can observe the agent attributes of ships in Figure 12. For example, ship five has chosen to sail around to the right instead of crossing the possible routes of the other four ships. In addition, ship two and ship four constitute the head-on situation. They are able to complete the collision avoidance task in a rule-compliant situation and do not generate extreme collision avoidance options. And the ships can resume navigation in time to avoid generating excessive deviation distances. All of the above fully reflects the core design ideas of the algorithm's reward function to focus on safety and high efficiency.

4.3. Analysis of Experimental Results

In order to clearly observe the ship's collision avoidance, we have made the colors of the ship's trajectory, distance change, and the minimum passing distance in the above figures the same as the ship's colors set in the encounter scenario library.

Among them, Figure A1 shows the navigation position and motion trajectory of each ship at different moments. We can see each ship's navigation process, including recognizing the collision risk, taking collision avoidance action, sailing to clear and past, returning to the planned route, and continuing to the destination. This is also a complete collision avoidance decision-making process. However, a ship does not completely eliminate the collision hazard through a single collision avoidance decision-making process. In general, many ships need to take several continuous steering actions in order to remove the current hazard. While some ships will still face new collision hazards during the resumption process, thus starting a new collision avoidance decision-making process. In Figure A2, this paper takes the perspective of ship one as an example. We can see that the overall trend of distance change between ships at different moments is to gradually become closer and then further away. This shows that ships can take timely collision avoidance actions after recognizing the risk so that the distances between ships are constantly moving towards higher safety. In Figure A3, we count the minimum passing distance of each agent ship from other agent ships in the complete time step. We can find that the minimum passing distance is usually presented in pairs. And this algorithm can ensure that each ship completes collision avoidance beyond the safety distance at different moments.

The results of 40 group simulation experiments show that, on the one hand, the algorithm shows sufficient coordination in unknown, diverse, and complex environments; on the other hand, it is demonstrated that the algorithm's model can be trained to full convergence through a shared policy network. Meanwhile, the trained model can be copied to the MAS with different numbers of ships to complete the collision avoidance decision-making.

5. Conclusions

This paper proposes a multi-agent collision avoidance algorithm based on DDQN and incorporates the PER technique.

Firstly, the research idea of this paper is established. The overall framework of this paper and its components follow the principle of "real data-driven as primary, simulation-driven as supplementary", so real AIS data-driven dominates the model construction. Secondly, the agent's observation state is determined by quantifying the hazardous area. Identifying the external environment from the perspective of any ship, scene clustering of target ships with similarly predicted collision hazard areas within a certain range can obtain the same observation state, effectively reducing the size of the observation state space. Then, ship-coordinated and uncoordinated behaviors are defined. In order to simulate uncoordinated scenarios in real waters, this paper proposes a non-coordination avoidance factor to decide whether to give attributes to ship intelligence or not. Thereby, the idea of multi-ship distributed collision avoidance considering the uncoordinated behaviors of the target ship is added to this paper. Next, based on a full understanding of COLREGs and the preliminary data collection, this paper combines the statistical results of the real water traffic data to guide and design the MADRL framework and selects the representative influencing factors to be designed into the collision avoidance decision-making algorithm's reward function. Subsequently, we divide the total training set of this model into two parts: one is the real data training set, and the other is the simulation data training set. Based on the idea of "reality as primary and simulation as supplementary" in this paper, the former consists of five parts of real water data, and its proportion is set to be 80% of the total training set; at the same time, this paper adopts the model of "YU KUN" for simulation and designs a MAS with 12 ships based on the ship encounter scenarios classified by COLREGs. Before each training model, the MAS will select the number of ships and their positions to complete the scenario construction by double randomization. The proportion of this part is set to be 20% of the total training set. Finally, 40 encounter scenarios are designed and extended to verify the algorithm performance based on the idea of the Imazu problem. The experimental results show that the algorithm proposed in this paper can solve the multi-ship collision avoidance problem in multiple scenarios quite efficiently. The algorithm improves the safety of autonomous ship navigation and provides a reference idea for the research of autonomous ship collision avoidance.

At present, the MADRL application in the ship collision avoidance field is still in its infancy, and the applicable conditions of this algorithm still need to be further improved. For example, the agent uses the recognition function in a way that treats other agents more as part of the environment. Such a way of coordination is obviously implicit, and the communication is not sufficient. This may lead to an unstable learning state of agents, slow convergence of the algorithms, etc. Therefore, in the next research, we will focus on achieving a more specific and efficient recognition function of agents, i.e., we will delve into the explicit method of coordinated communication among multiple agents. Meanwhile, a self-supervision mechanism can be added to the original algorithm. The aim is to better supervise the decision-making behaviors made by the agents themselves, as well as to continuously further improve the algorithm's practicality.

Author Contributions: Conceptualization, Y.N. and F.Z.; funding acquisition, F.Z.; methodology, Y.N., F.Z. and P.Z.; software, Y.N.; validation, Y.N., F.Z., M.W., Y.D. and P.Z.; formal analysis, Y.N.; investigation, F.Z.; resources, F.Z.; data curation, Y.N. and F.Z.; writing—original draft preparation, Y.N. and F.Z.; writing—review and editing, Y.N., F.Z., M.W., Y.D. and P.Z.; visualization, Y.N.; supervision, F.Z.; All authors have read and agreed to the published version of the manuscript.

Funding: This work was supported in part by the National Key Research and Development Program of China under Grant 2018YFB1601505, the National Natural Science Foundation of China under Grant 52231014, and the Liaoning Provincial Shipping Joint Fund under Grant 2020-HYLH-28.

Institutional Review Board Statement: Not applicable.

Informed Consent Statement: Not applicable.

Data Availability Statement: Not applicable.

Conflicts of Interest: The authors declare no conflict of interest.

Appendix A

As stated in Section 4.2, the model test results are shown in Figures A1–A3. Panel (a) and Panel (b) of each figure, respectively, represent the test results of Case 1–20 and Case 21–40 in the extended encounter scenario library.

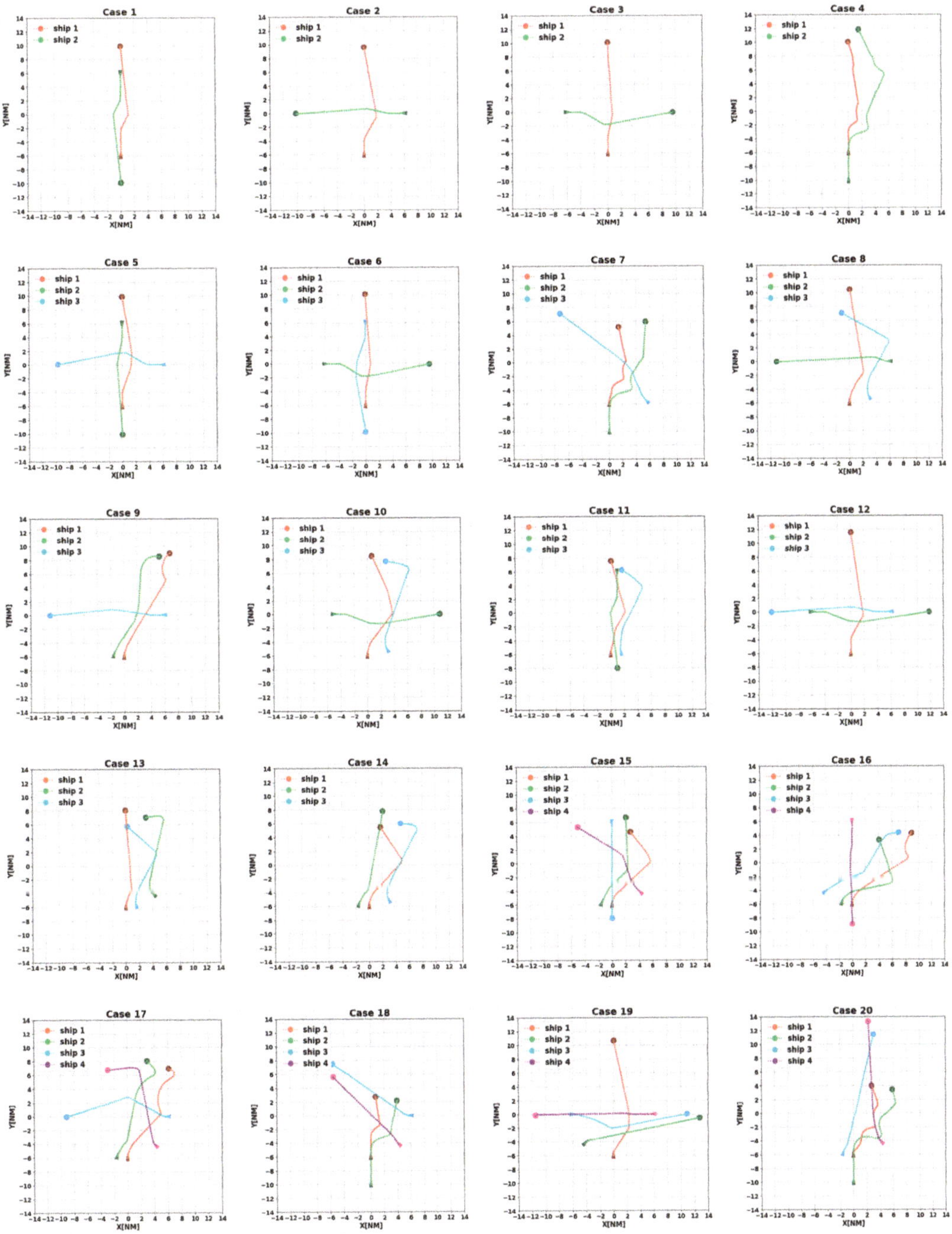

(**a**)

Figure A1. *Cont.*

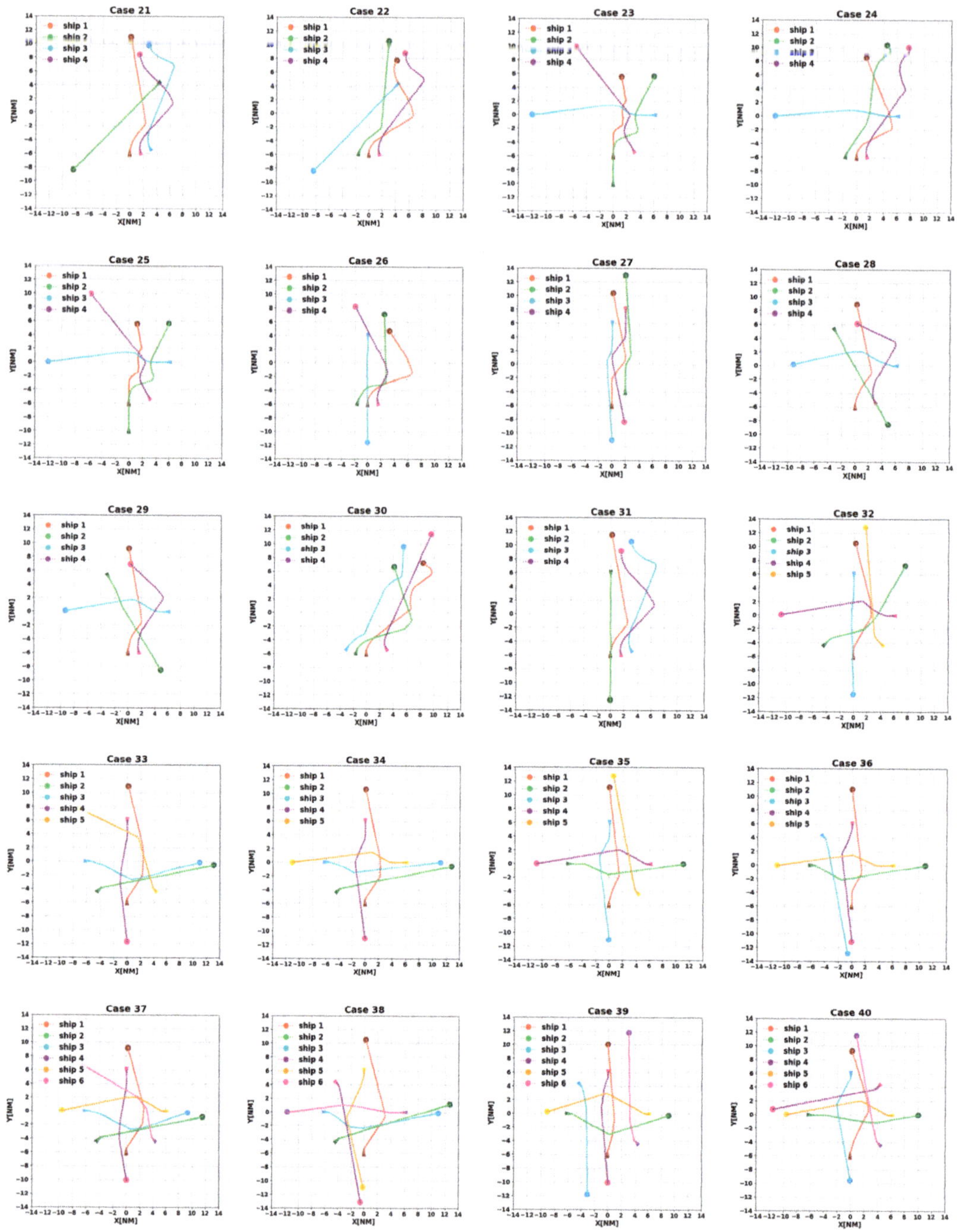

(**b**)

Figure A1. Ships' trajectories in the encounter scenario library. (**a**) Case 1–20; (**b**) Case 21–40.

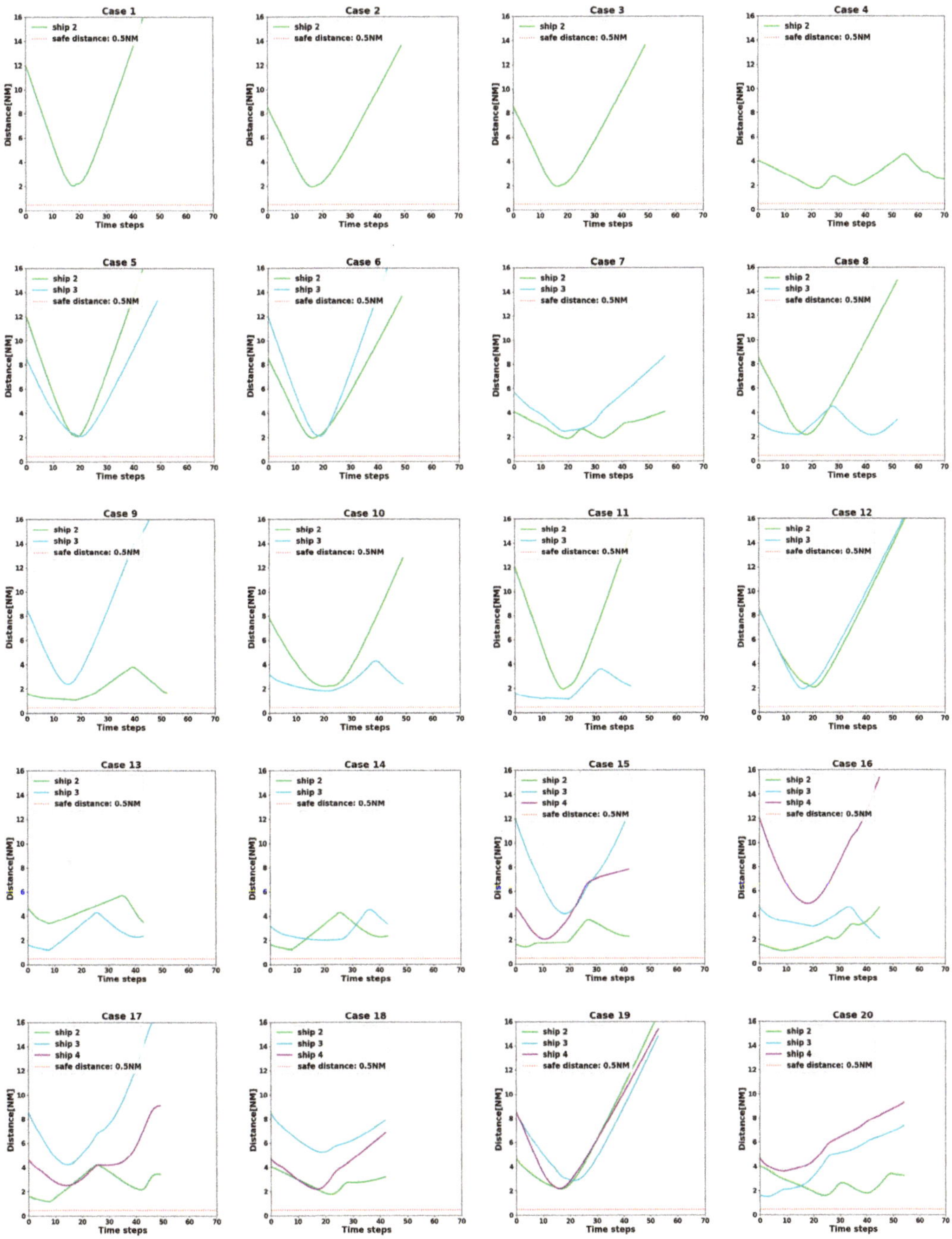

(a)

Figure A2. *Cont.*

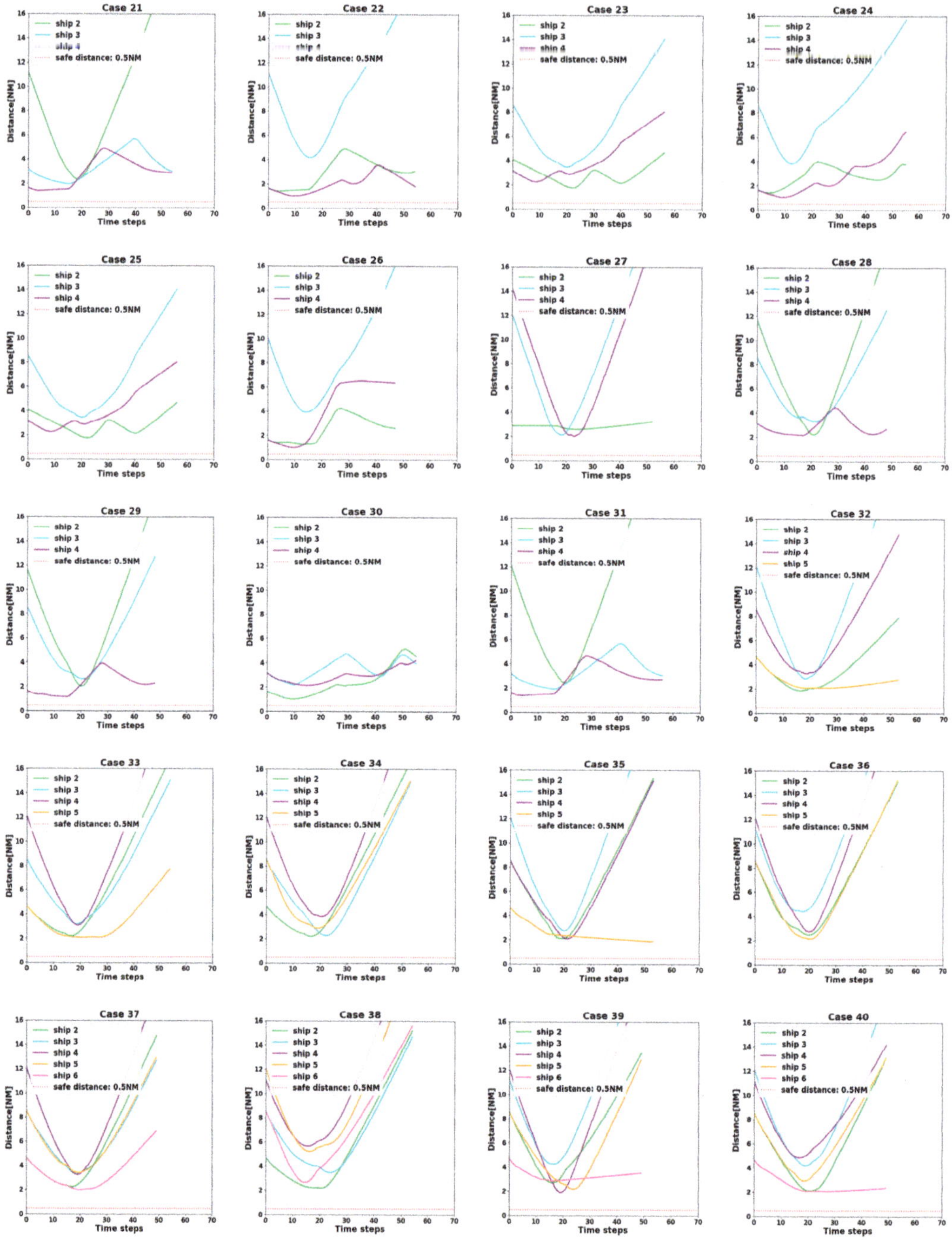

(**b**)

Figure A2. The changes in distance between the own ship (ship 1) and the target ships under this perspective. (**a**) Case 1–20; (**b**) Case 21–40.

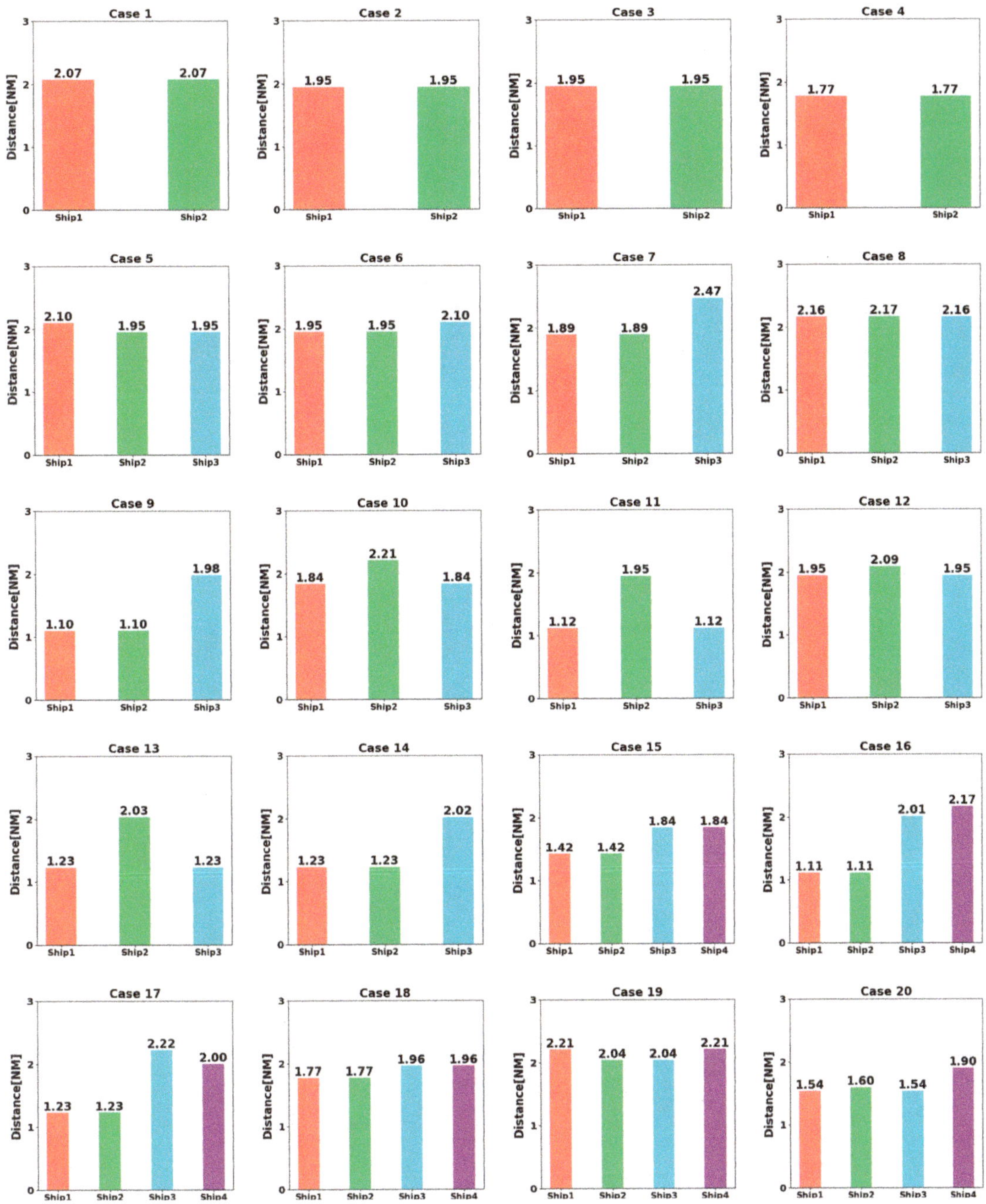

(a)

Figure A3. *Cont.*

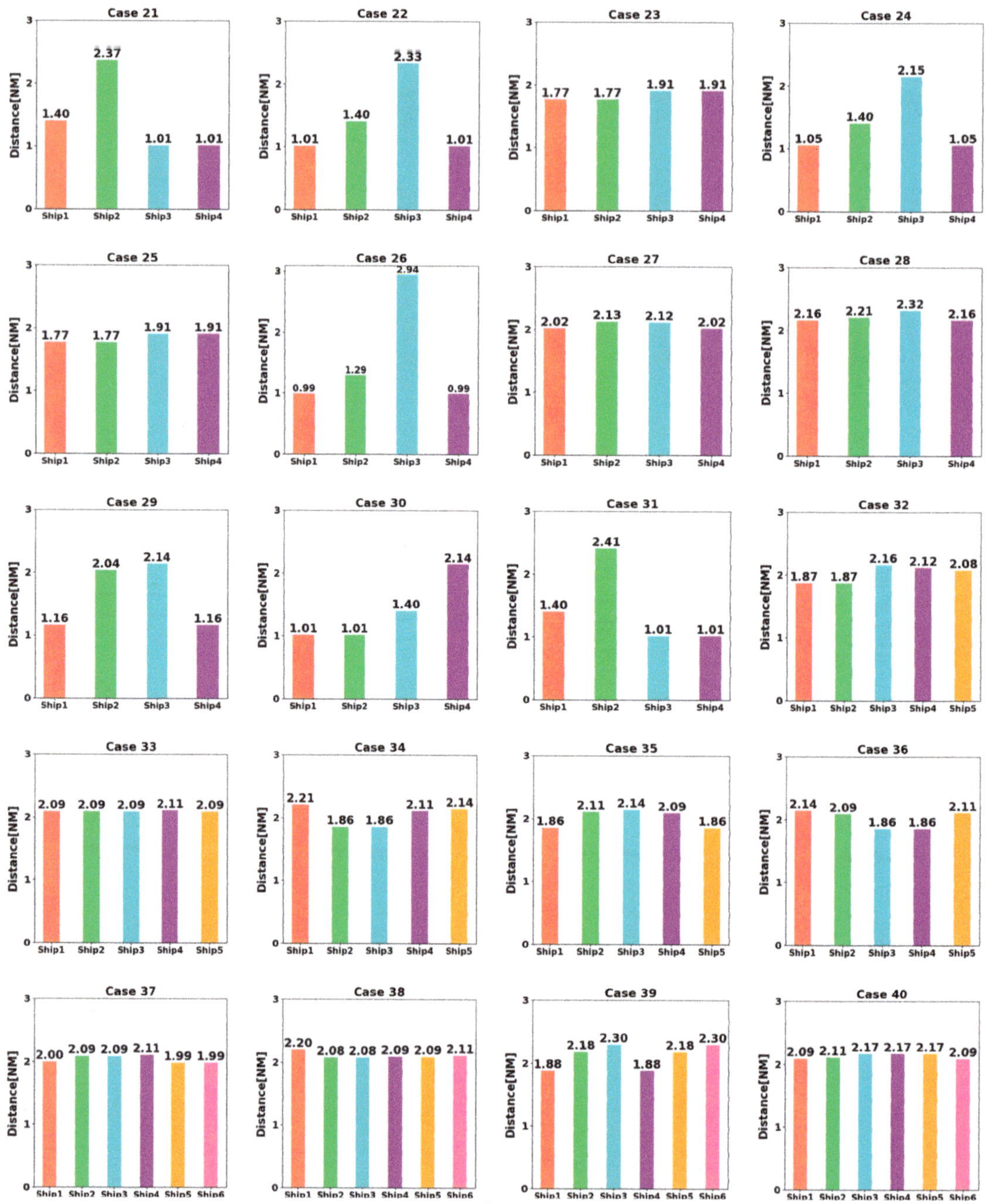

(**b**)

Figure A3. The minimum passing distance of each agent ship from other agent ships. (**a**) Case 1–20; (**b**) Case 21–40.

References

1. European Maritime Safety Agency. *Annual Overview of Marine Casualties and Incidents 2021*; EMSA: Lisbon, Portugal, 2021; Available online: https://www.emsa.europa.eu/newsroom/latest-news/item/4266-annual-overview-of-marine-casualties-and-incidents-2020.html (accessed on 11 August 2023).
2. Maritime Safety Committee. *Report of the Maritime Safety Committee on Its Ninety-Ninth Session*; IMO: London, UK, 2018; Available online: https://www.imo.org/en/MediaCentre/MeetingSummaries/Pages/MSC-99th-session.aspx (accessed on 16 August 2023).
3. Wei, G.; Kuo, W. COLREGs-Compliant Multi-Ship Collision Avoidance Based on Multi-Agent Reinforcement Learning Technique. *J. Mar. Sci. Eng.* **2022**, *10*, 1431. [CrossRef]
4. Zhang, Y.; Zhai, P. Research progress and trend of autonomous collision avoidance technology for marine ships. *J. Dalian Marit. Univ.* **2022**, *48*, 1–11.
5. Papadimitrakis, M.; Stogiannos, M.; Sarimveis, H.; Alexandridis, A. Multi-Ship Control and Collision Avoidance Using MPC and RBF-Based Trajectory Predictions. *Sensors* **2021**, *21*, 6959. [CrossRef]
6. Shaobo, W.; Yingjun, Z.; Lianbo, L. A collision avoidance decision-making system for autonomous ship based on modified velocity obstacle method. *Ocean Eng.* **2020**, *215*, 107910. [CrossRef]
7. Huang, Y.; Chen, L.; van Gelder, P.H.A.J.M. Generalized velocity obstacle algorithm for preventing ship collisions at sea. *Ocean Eng.* **2019**, *173*, 142–156. [CrossRef]
8. Ma, J.; Su, Y.; Xiong, Y.; Zhang, Y.; Yang, X. Decision-making method for collision avoidance of ships in confined waters based on velocity obstacle and artificial potential field. *China Saf. Sci. J.* **2020**, *30*, 60–66. [CrossRef]
9. Singh, Y.; Sharma, S.; Sutton, R.; Hatton, D.; Khan, A. A Constrained A* Approach towards Optimal Path Planning for an Unmanned Surface Vehicle in a Maritime Environment Containing Dynamic Obstacles and Ocean Currents. *Ocean Eng.* **2018**, *169*, 187–201. [CrossRef]
10. Ahn, J.-H.; Rhee, K.-P.; You, Y.-J. A study on the collision avoidance of a ship using neural networks and fuzzy logic. *Appl. Ocean Res.* **2012**, *37*, 162–173. [CrossRef]
11. Szłapczyński, R.; Ghaemi, H. Framework of an evolutionary multi-objective optimisation method for planning a safe trajectory for a marine autonomous surface ship. *Pol. Marit. Res.* **2019**, *26*, 69–79. [CrossRef]
12. Statheros, T.; Howells, G.; Maier, K.M.D. Autonomous ship collision avoidance navigation concepts, technologies and techniques. *J. Navig.* **2008**, *61*, 129–142. [CrossRef]
13. Wang, C.; Zhang, X.; Cong, L.; Li, J.; Zhang, J. Research on Intelligent Collision Avoidance Decision-Making of Unmanned Ship in Unknown Environments. *Evol. Syst.* **2019**, *10*, 649–658. [CrossRef]
14. Sun, Z.; Fan, Y.; Wang, G. An Intelligent Algorithm for USVs Collision Avoidance Based on Deep Reinforcement Learning Approach with Navigation Characteristics. *J. Mar. Sci. Eng.* **2023**, *11*, 812. [CrossRef]
15. Shen, H.; Hashimoto, H.; Matsuda, A.; Taniguchi, Y.; Terada, D.; Guo, C. Automatic collision avoidance of multiple ships based on deep Q-learning. *Appl. Ocean Res.* **2019**, *86*, 268–288. [CrossRef]
16. Sawada, R.; Sato, K.; Majima, T. Automatic Ship Collision Avoidance Using Deep Reinforcement Learning with LSTM in Continuous Action Spaces. *J. Mar. Sci. Technol.* **2021**, *26*, 509–524. [CrossRef]
17. Zhao, L.; Roh, M.-I. COLREGs-compliant multiship collision avoidance based on deep reinforcement learning. *Ocean Eng.* **2019**, *191*, 106436. [CrossRef]
18. Sutton, R.S.; McAllester, D.; Singh, S.; Mansour, Y. Policy gradient methods for reinforcement learning with function approximation. *Adv. Neural Inf. Process. Syst.* **1999**, *12*, 1057–1063.
19. Luis, S.Y.; Reina, D.G.; Marin, S.L.T. A Multiagent Deep Reinforcement Learning Approach for Path Planning in Autonomous Surface Vehicles: The Ypacaraí Lake Patrolling Case. *IEEE Access* **2021**, *9*, 17084–17099. [CrossRef]
20. Chen, C.; Ma, F.; Xu, X.; Chen, Y.; Wang, J. A Novel Ship Collision Avoidance Awareness Approach for Cooperating Ships Using Multi-Agent Deep Reinforcement Learning. *J. Mar. Sci. Eng.* **2021**, *9*, 1056. [CrossRef]
21. Zhu, F.; Ma, Z. Ship trajectory online compression algorithm considering handling patterns. *IEEE Access* **2021**, *9*, 70182–70191. [CrossRef]
22. The International Maritime Organization (IMO). Convention on the International Regulations for Preventing Collisions at Sea (COLREGs). 1972. Available online: https://www.imo.org/fr/about/Conventions/Pages/COLREG.aspx (accessed on 21 August 2023).
23. Belcher, P. A sociological interpretation of the COLREGS. *J. Navig.* **2002**, *55*, 213–224. [CrossRef]
24. Zhu, F.; Zhou, Z.; Lu, H. Randomly Testing an Autonomous Collision Avoidance System with Real-World Ship Encounter Scenario from AIS Data. *J. Mar. Sci. Eng.* **2022**, *10*, 1588. [CrossRef]
25. Wang, X.; Zhang, Y.; Liu, Z.; Wang, S.; Zou, Y. Design of Multi-Modal Ship Mobile Ad Hoc Network under the Guidance of an Autonomous Ship. *J. Mar. Sci. Eng.* **2023**, *11*, 962. [CrossRef]
26. Mnih, V.; Kavukcuoglu, K.; Silver, D.; Rusu, A.A.; Veness, J.; Bellemare, M.G.; Graves, A.; Riedmiller, M.; Fidjeland, A.K.; Ostrovski, G.; et al. Human-level control through deep reinforcement learning. *Nature* **2015**, *518*, 529–533. [CrossRef] [PubMed]
27. Van Hasselt, H.; Guez, A.; Silver, D. Deep Reinforcement Learning with Double Q-Learning. In Proceedings of the 30th Association-for-the-Advancement-of-Artificial-Intelligence (AAAI) Conference on Artificial Intelligence, Phoenix, AZ, USA, 12–17 February 2016; pp. 2094–2100. [CrossRef]

28. Schaul, T.; Quan, J.; Antonoglou, I.; Silver, D. Prioritized Experience Replay. In Proceedings of the 4th International Conference on Learning Representations, San Juan, PR, USA, 2–4 May 2016. [CrossRef]
29. Fukuto, J.; Imazu, H. New Collision Alarm Algorithm Using Obstacle Zone by Target (OZT). *IFAC Proc. Vol.* **2013**, *46*, 91–96. [CrossRef]
30. Zhang, W.; Feng, X.; Qi, Y.; Shu, F.; Zhang, Y.; Wang, Y. Towards a model of regional vessel near-miss collision risk assessment for open waters based on AIS data. *J. Navig.* **2019**, *72*, 1449–1468. [CrossRef]
31. Yoo, Y.; Lee, J.-S. Evaluation of ship collision risk assessments using environmental stress and collision risk models. *Ocean Eng.* **2019**, *191*, 106527. [CrossRef]
32. Zhai, P.; Zhang, Y.; Shaobo, W. Intelligent Ship Collision Avoidance Algorithm Based on DDQN with Prioritized Experience Replay under COLREGs. *J. Mar. Sci. Eng.* **2022**, *10*, 585. [CrossRef]
33. Fossen, T.I. *Guidance and Control of Ocean Vehicles*; John Wiley & Sons Inc.: Hoboken, NJ, USA, 1994.
34. Liu, J.; Zhao, B.; Li, L. Collision Avoidance for Underactuated Ocean-Going Vessels Considering COLREGs Constraints. *IEEE Access* **2021**, *9*, 145943–145954. [CrossRef]

Journal of
Marine Science and Engineering

MDPI

Article

Development of a Graph-Based Collision Risk Situation Model for Validation of Autonomous Ships' Collision Avoidance Systems

Taewoong Hwang [1] and Ik-Hyun Youn [2,*]

1 Industry Academic Cooperation Foundation, Mokpo National Maritime University, Mokpo 58628, Republic of Korea; hwangtw6539@gmail.com
2 Division of Navigation & Information Systems, Mokpo National Maritime University, Mokpo 58628, Republic of Korea
* Correspondence: iyoun@mmu.ac.kr; Tel.: +82-61-240-7283

Abstract: In the maritime industry, the systematic validation of collision avoidance systems of autonomous ships is becoming an increasingly important issue with the development of autonomous ships. The development of collision avoidance systems for autonomous ships faces inherent risks of programming errors and has mostly been tested in limited scenarios. Despite efforts to verify these systems through scenario testing, these scenarios do not fully represent the complex nature of real-world navigation, limiting full system verification and reliability. Therefore, this study proposed a method for analyzing collision risk situations extracted from AIS data through graph-based modeling and establishing validation scenarios. This methodology categorizes collision risk scenarios according to their centrality and frequency and demonstrates how simple collision risk situations gradually evolve into harsh situations.

Keywords: autonomous ships; collision risk situation; graph-based model; validation scenario

check for updates

Citation: Hwang, T.; Youn, I.-H. Development of a Graph-Based Collision Risk Situation Model for Validation of Autonomous Ships' Collision Avoidance Systems. *J. Mar. Sci. Eng.* **2023**, *11*, 2037. https://doi.org/10.3390/jmse11112037

Academic Editors: Carlos Guedes Soares, Xianbo Xiang, Lúcia Moreira and Haitong Xu

Received: 23 September 2023
Revised: 19 October 2023
Accepted: 19 October 2023
Published: 24 October 2023

1. Introduction

Collision avoidance is of paramount importance in the development of autonomous ships given their significant impact on human safety, the environment, and the economy. Therefore, ensuring the reliability of their collision avoidance systems is critical [1].

Although various approaches, from kinematics to artificial intelligence, have been utilized in developing collision avoidance systems, their human design origin presents risks of programming errors due to factors such as inadequate training or program bugs. This underlines the necessity of rigorous system testing and validation [2–18].

The concerns and challenges in collision avoidance systems development are validated by incidents caused by system defects across various industrial domains. Issues arising from internal defects necessitate a systematic flaw elimination and thorough understanding of the collision avoidance algorithm's critical system states to ensure safety, particularly in complex, unanticipated scenarios during operation [19,20].

The systematic development and testing of potential scenarios is inherently complex and multifaceted. Though research has identified methods for comprehensive scenario development for testing collision avoidance systems, their effectiveness in thoroughly understanding and replicating the multifaceted and varied collision risks in real navigation settings is limited, sometimes due to the increased number and complexity of the parameters involved [21–32].

Therefore, this study aims to develop a methodology for the systematic verification of autonomous ship collision avoidance systems by generating realistic collision risk situations. This methodology, while not primarily focused on the direct validation of autonomous ship collision avoidance systems, emphasizes the understanding and development of collision risk situation scenarios through the objective analysis of graph networks. It encompasses multiple collision risk situations that have occurred in the past, and thereby seeks to extract and systematize realistic scenarios of varying levels of complexity that can be useful in the evaluation of such systems in hindsight.

2. Methodology

In this section, we propose a data-driven approach to analyze ship collision risk situations (CRSs) using a graph-based model. First, CRSs satisfying certain conditions are extracted from the AIS data. Each situation is transformed into a categorical vector by a combination of "unit scenarios" that form the context of each ship. The similarity between these vectors is then translated into a matrix computed using a modified version of the Jaccard similarity. This similarity matrix is then visualized as a network graph. A detailed view of the whole process is shown in Figure 1. The ultimate goal of these steps is to gain a systematic understanding of CRSs, with more specific methods described in the following subsections.

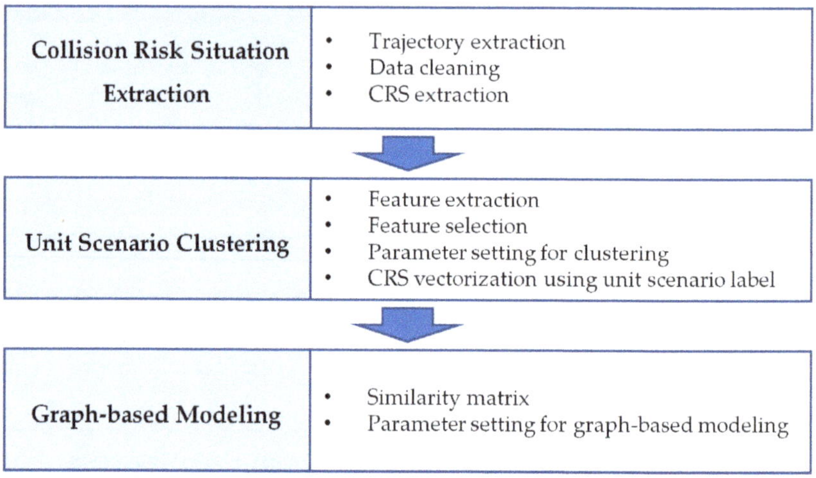

Figure 1. Workflow of the proposed methodology.

2.1. Extraction of CRSs

This section describes the extraction of collision risk situations from AIS data in the Southern Sea of Korea from September to November 2019. The procedure commences with defining CRSs, followed by a thorough data pre-processing phase to cleanse and synchronize the AIS data, and then extracting CRSs based on specific CPA and TCPA criteria. The extracted collision risk situations are converted into a relative Cartesian coordinate system to facilitate precise comparative analysis and ensure consistency across various encounter scenarios.

2.1.1. Definition of CRSs

For this study, we define a "Collision Risk Situation" as a scenario in which one or more vessels are involved and at least one vessel meets the criteria of DCPA (distance to the closest point of approach) and TCPA (time to the closest point of approach). Although the criteria for evaluating the collision risk between vessels based on DCPA and TCPA are rather ambiguous and vary with the size of the traffic area and the ships themselves, the purpose of this study is to suggest a methodology. Therefore, we identified situations with a considerable risk of collision by using a DCPA of 0.1 nautical miles and a TCPA of 6 min. Specifically, if multiple vessels are involved and the ith vessel meets a condition where

DCPA is 0.1 nautical miles or less and TCPA is 6 min or less, the given scenario involving that vessel and others is classified as a collision risk situation (according to Equation (1)):

$$\text{CRS} = \{ \text{ Exists } i \in \text{(encountered vessels) such that } (\text{DCPA\_}i \leq 0.1) \wedge \text{TCPA\_}i \leq 6) \} \quad (1)$$

2.1.2. AIS Data

We have employed automatic identification system (AIS) data, which provides information on a vessel's identity, location, course, and speed, and is commonly used to analyze maritime traffic. The spatial range of the data spans the Southern Sea of the Republic of Korea, a region chosen for its lack of terrestrial interference and ability to extract various vessel encounter scenarios as shown in Figure 2. The temporal range of the utilized data spans from 1 September 2019 to 30 November 2019, containing a total of 12,139,052 position data points, which were generated by 4216 vessels.

$$\begin{gathered} N\ 33°21'00'' < x_i < N\ 34°48'00'',\ i = 1, 2, \cdots, n-1, n \\ E\ 127°27'00'' < y_i < E\ 128°30'00'' \end{gathered} \quad (2)$$

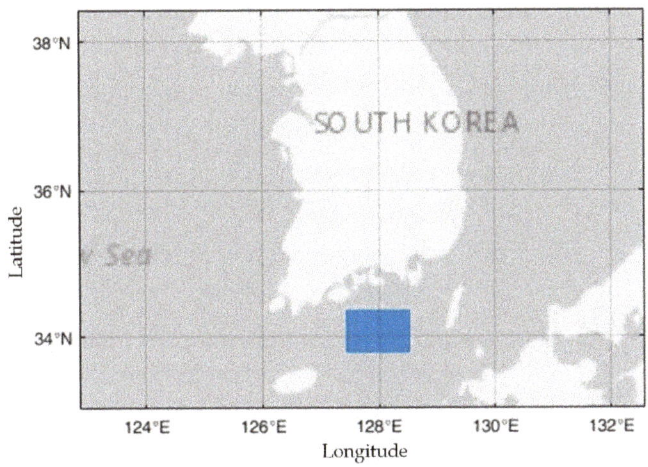

Figure 2. Spatial information of AIS data.

2.1.3. Trajectory Extraction and Data Cleaning (Pre-Processing)

AIS data, comprising navigational information collected from various vessels, form a time-series dataset. Its composition is interspersed over time and inherently contains certain inaccuracies. In this phase of the research, considering this complexity, our objective was to extract the trajectories of a specific target vessel and its surrounding vessels. To achieve this, we underwent a comprehensive data-cleaning process, which included error correction and synchronization of the time-series data.

The extraction of the subject ships and the encountering vessels was carried out via the following process. As the ongoing project supporting this study plans to develop vessels between 100 and 130 m in length, and the same length range was chosen for the subject vessels in the AIS data analysis. Subject ships were initially identified in the static data and subsequently extracted from the dynamic data. To focus on vessels actively navigating, only those with a speed exceeding 5 knots were selected. Encounter vessels were determined to be those located within 3 nautical miles of the subject ship's trajectory during the same time frame.

The extracted ships have inconsistent time series frequencies. Due to this inconsistency, the time stamps of the own ship were regularized to one-minute intervals. For the time-varying values of position, course, speed, and heading, linear interpolation was applied during the time series data normalization process. The choice of linear interpolation was driven by a need to select an algorithm with low computational demands due to the handling of large-scale data. Additionally, excluding vessels with very high variations in speed and course—which are rare cases—linear interpolation did not exhibit notable disadvantages in results compared to other algorithms (such as spline, pchip, and makima), substantiating its application. In addition, the data of the relative ships were synchronized to match the timestamps of the subject ship, and interpolation was also applied to position, course, speed, and heading. Notably, although course and speed are key variables in discriminating the navigational relationship between the subject ship and the encountering ships, they were recalculated based on the interpolated positions due to discrepancies between them.

2.1.4. CRSs Extraction

The pre-processing performed up to this stage has synchronized the trajectory data for both the subject ship and the other ships on a minute-by-minute basis. These data serve as the basis for the extraction of CRSs, which are used to assess the risk of collision on a minute-by-minute basis. In this phase, the CPA and TCPA values were derived from each trajectory using the position, course, and speed information of the target vessel and the encountering vessel. Moments that met specific criteria for CPA and TCPA were identified, and the situation involving the encountering vessel and surrounding vessels at that time was defined and recorded as a CRS. This specific scenario was referred to in our research as a "moment situation", which can be understood as a snapshot capturing navigational circumstances that meet certain collision risk criteria. Specifically, a "moment situation" is the specific moment satisfies the CPA and TCPA conditions between the subject and the encountering vessels, and is considered a CRS.

2.1.5. Converting CRSs to a Relative Cartesian Coordinate System

The extracted CRSs were detailed based on true north, describing both the ship's trajectory and the encounter situations. To refine the comparison and analysis of these encounters, the CRSs were adjusted to a course-up orientation centered on the subject ship's course. In addition, the subject ship was fixed at the origin (0,0) by a Cartesian coordinate system transformation, while the positions of the other ships were defined relative to it by x and y coordinates. The purpose of this transformation was not only to avoid clustering errors induced by the periodicity of the angles in the polar coordinate system and to consistently represent different collision risk situations (CRSs) under conditions relative to the subject ship; it was also to effectuate data reduction by expressing collision risk situations—previously articulated through the latitude, longitude, speed, course, and heading of the subject and encounter ships—succinctly in terms of their current and future positions.

2.2. *Unit Scenario*

This section introduces the concept of "unit scenarios" within CRSs, which allows for a systematic categorization of individual collision risk situations. This involves the extraction and normalization of relevant features, followed by k-means clustering to formulate distinct unit scenario groups. Finally, the CRSs are transformed into categorical vectors using unit scenario clusters to provide a compact and consistent representation that allows for detailed analysis and understanding of collision scenarios.

2.2.1. Concept of Unit Scenario

The extracted CRS not only reflects the collision risk relationship; it also comprehensively describes the maritime situation at a given time, including various parameters such as the distribution of nearby ships' positions, draught, speed, and ship length. Taking this diversity into account, this study attempted to systematically categorize the encounter relationships of all vessels that comprise the CRS and endeavored to represent the CRS effectively by combining these interactions. To achieve this, individual encounters between ships were defined as "unit scenarios". These scenarios were then clustered and labeled to provide a systematic categorization of encounter relationships within the CRS.

The definition of the unit scenario is as follows: it represents an individual ship's encounter situation as separate components of the CRS. In Figure 3, the ship shown in black represents the subject ship, while the ships shown in red represent the opposing ships. The position indicated by the triangle represents the initial point that was captured as a CRS at the specified instance. In addition, the length and direction of the arrow indicate the speed and course of each ship, respectively. The endpoint of the arrow, referred to as the "terminal point," illustrates the estimated position of the vessel after a given time interval by utilizing a vector composed of the course and speed from the AIS data of the vessel at the "initial point." In this study, a six-minute period was used to calculate the ship's terminal point. The CRS presented includes a total of four opposing vessels, including vessel (b), which poses a collision risk. Each of these vessels can be classified into individual unit scenarios, labeled (a), (b), and (c). According to this concept, the CRS is described as being composed of two ships corresponding to unit scenario (a), one ship to unit scenario (b), and one ship to unit scenario (c). This concept of unit scenario is utilized to describe the different types of ship encounters that compose the CRS as fine-grained categorical variables, which are finally applied as elements to vectorize the CRS.

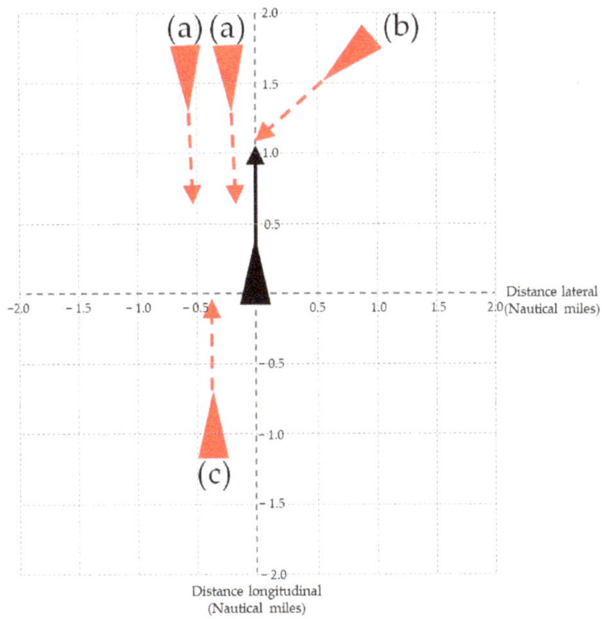

Figure 3. Concept of unit scenario.

2.2.2. Feature Engineering for Unit Scenario Clustering

For a systematic clustering of the unit scenario, it is imperative to extract and select features that effectively represent individual encounter situations. In this study, as shown in Table 1, we extracted various features related to the relative position, course, distance, and speed of the encounter ship, such as the Cartesian coordinates at both the initial and terminal points based on the reference of the subject ship, the distance between the ships at the initial point, the change in distance from the initial to the terminal point, and the course direction of the encounter ship at both points. These features reflect the relative position and its changes based on the subject ship and serve as representative values for the encounter ship's movement. After extracting these features, min–max normalization was applied to adjust for the importance bias caused by differences in the units of each feature.

Table 1. Extracted feature list for unit scenario clustering.

Number	Features	Description
F1	Distance (Initial)	Distance between subject ship and each encounter ship at initial point
F2	Distance (Change)	Difference between distance from terminal point and distance from initial point
F3	Relative position at initial point (Cartesian coordinate x)	The Cartesian coordinate (***x*-axis**) of the encounter ship at **initial point**: based on the subject ship as a reference, the port side is denoted by a negative value (−) and the starboard side by a positive value (+)
F4	Relative position at initial point (Cartesian coordinate y)	The Cartesian coordinate (***y*-axis**) of the encounter ship at the **initial point**: based on the subject ship as a reference, forward is denoted by a positive value (+) and aft by a negative value (−)
F5	Relative position at terminal point (Cartesian coordinate x)	The Cartesian coordinate (***x*-axis**) of the encounter ship at **terminal point**: based on the subject ship as a reference, the port side is denoted by a negative value (−) and the starboard side by a positive value (+)
F6	Relative position at terminal point (Cartesian coordinate y)	The Cartesian coordinate (***y*-axis**) of the encounter ship at the **terminal point**:bBased on the subject ship as a reference, forward is denoted by a positive value (+) and aft by a negative value (−)
F7	Encounter ship's course	Encounter ship's course at initial point
F8	Encounter ship's speed	Encounter ship's speed at initial point

The extracted and normalized features were applied to the Laplacian feature selection algorithm, and their importance was ranked based on their impact on clustering. While most features showed potential for clustering effectiveness with an importance score of 0.9 or higher, specific features were chosen to ensure computational efficiency and to facilitate intuitive interpretation of the clustering results. In particular, the features with the highest importance, F7 (initial course), F8 (initial speed), F3 (initial relative position—Cartesian coordinate x), and F4 (initial relative position—Cartesian coordinate y), were distinguished within two main domains: relative motion and position of the vessel. Due to their intuitive nature in interpreting the clustering results, these features were selected as the final features applied in clustering.

2.2.3. Unit Scenario Clustering

The feature matrix of the unit scenario was used in k-means clustering. To preserve the independent characteristics of each feature dimension while reducing the sensitivity to outliers, the city block distance was chosen as the distance measurement method [33]. In order to determine the optimal number of clusters, a gap analysis was conducted. Figure 4 shows a bar graph illustrating the results of this analysis. The x-axis represents the number of clusters, while the y-axis represents the gap analysis results for each cluster number. The height of the bar graph does not significantly increase beyond 66 clusters, indicating that 66 is the optimal number of clusters. Thus, the unit scenarios that make up the entire CRS were divided into 66 distinct clusters. These were then used as elements in vectors to describe each encounter situation in the next stage.

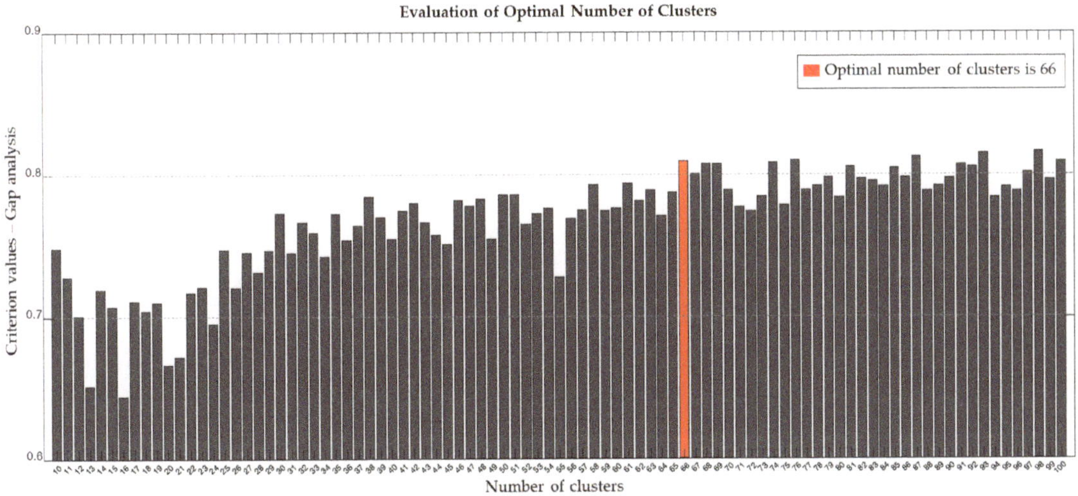

Figure 4. Result of gap analysis to evaluate optimal number of clusters.

2.2.4. Vectorization of CRSs

The categorized unit scenarios are used as components of each CRS vector to effectively represent collision risk situations. This approach provides a significant advantage in minimizing information loss during the aggregation process, as it allows for the consistent representation of all encounter ships within a CRS, from single encounters to complex situations involving multiple ships, in a uniform categorical vector format. Each CRS is transformed into a categorical vector using the labels of the encounter ships that comprise it. To mitigate the loss of similarity between vectors with the same labels but in different sequences, the order of the vectors was reorganized in ascending order.

2.3. Graph-Based Modeling

In this section, a modified Jaccard similarity measure is utilized to create a similarity matrix between CRS vectors, which is then applied in graph-based modeling to visually represent the relation of CRSs. The layout and edge thickness within the graph are adjusted to intuitively convey the degree of similarity between nodes.

2.3.1. Similarity Matrix

Computing a similarity matrix between the generated vectors is an essential step for graph analysis. In this study, we used a modified Jaccard similarity measure, tailored for specific purposes, to assess the similarity between vectors. The Jaccard similarity index traditionally calculates the similarity between two vectors based on the ratio of their intersection to their union, as described by the following formula:

$$\text{Jaccard Similarity}(\text{A}, \text{B}) = \frac{|A \cap B|}{|A \cup B|}. \quad (3)$$

However, this method is limited by its inadequate consideration of the total length of the arrays and their overlapping elements. To overcome these limitations, the modified Jaccard similarity was designed, which computes the similarity by dividing the number of identical elements, including duplicates, by the length of the longer of the two arrays. Thus, this modified method considers the duplicate elements that were previously ignored by the conventional Jaccard index, allowing for a more granular evaluation of similarity. The following formula represents the modified Jaccard similarity:

$$S = \frac{\min\left(\sum_{k=1}^{|B|} 1_{B_k \in A \cap B}, \sum_{k=1}^{|A|} 1_{A_k \in A \cap B}\right)}{\max(|A|, |B|)}, \quad (4)$$

where A and B are two vectors, S is the modified Jaccard similarity between them, and $1_{B_k \in A \cap B}$ and $1_{A_k \in A \cap B}$ are indicator functions that return 1 if B_k and A_k are in $A \cap B$, respectively, and 0 otherwise. This formula calculates the modified Jaccard similarity by dividing the minimum number of elements in A and B that are also in their intersection by the maximum length of A and B. This calculation allows us to measure the similarity between two vectors.

2.3.2. Parameter Setting for Graph-Based Modeling

The similarity matrix was used as an input variable for graph-based modeling, and the CRS was visualized as a graph-network. The layout of the nodes in the graph was configured by applying a force direction methodology to adjust the distance between nodes in an inverse proportion to their similarity. Based on these configurations, the length of the edges connecting the nodes was also set to be inversely proportional to the similarity between the nodes. Additionally, the thickness of the edges was set to be directly proportional to the similarity, so that the connections between nodes with high similarity were thicker. Centrality indices, utilized to portray the connectivity among nodes, encompass "Degree Centrality," which indicates the number of direct connections a node maintains; 'Closeness Centrality," representing the average shortest path length across all nodes; and "Betweenness Centrality," expressing the frequency at which a node appears on the shortest paths between every pair of nodes [34]. For this study, "Degree Centrality" was employed to determine the direct connections each CRS node has with other nodes.

3. Result

This section describes the results of the application of the methodology presented in this study. The results section consists of results and examples of extracted CRSs, results of vectorization of CRSs, and results of graph-based modeling.

3.1. Extracted CRSs

In this study, we analyzed ship traffic data collected over a three-month period and identified a total of 1205 collision risk situations (CRS). When the subject ship encountered various other ships during its sailing, if even one other ship met the conditions for a collision risk, the entire interaction involving the subject ship and its surrounding ships at that moment was extracted as a CRS. According to this methodology, each CRS necessarily incorporates at least one vessel with a collision risk. Figure 5 presents an analysis based on the number of vessels involved in the encounter, illustrating situations ranging from encounters with a relatively small number of other vessels to situations with multiple vessels, exemplified by cases such as CRS no.957 and no.1099. From the analysis, in each situation, at least one vessel posed a collision risk. These data were then processed and visualized centering on the subject ship using a course up display.

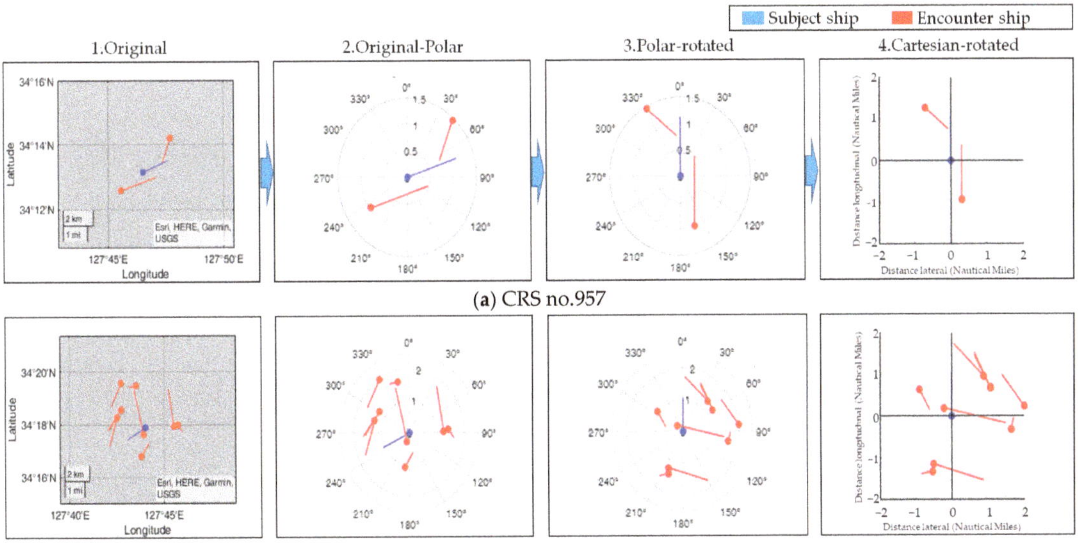

(**a**) CRS no.957

(**b**) CRS no.1099

Figure 5. Sample of extracted trajectory.

3.2. Vectorization of CRSs

The extracted 1205 CRSs comprise a total of 5110 unit scenarios. Each scenario was clustered based on the relative position, speed, and course of the opposing vessels. The optimal number of clusters was determined to be 66 unit scenario clusters using gap analysis. The 66 distinct unit scenarios identified through this clustering process are depicted in Figure 6. To illustrate the representative characteristics of each cluster, the unit scenarios were visualized using the centroids of the initial and terminal point distributions. Unit scenarios systematically break down various ship encounter situations that are not explicitly defined in the COLREGs, including vessels that pose a collision threat, stationary vessels, and vessels that do not interfere with the navigation of the subject ship. As an example of a categorization, unit Scenarios "29" and "27," highlighted by the broken line, clearly demonstrate the benefits of unit Scenarios. Both encounter ships in unit scenarios are similarly located about 40 degrees to the port side from the subject ship, but they were categorized differently due to their individual course and speed. To gain a comprehensive understanding of the unit scenarios, which are divided into 66 cluster groups, Figure 6 interprets them into three risk groups based on the distribution of DCPA (distance to closest point of approach). The first group is a high-risk group, which includes vessels approaching the subject ship's path or so close (DCPA less than 0.5) that there is a high risk of collision. The second group is of moderate risk (DCPA between 0.5 and 2), affecting the decision-making process of the subject ship but not presenting an immediate collision hazard. The third group is a low-risk group (DCPA 2 or more), which includes situations without explicit collision risk where the other ship is either not on the subject ship's course, stopped, or moving away. However, these groupings are only meant to facilitate a comprehensive understanding of the overall structure of the unit scenario clusters, and it is the cluster numbers, numerically illustrated in Figure 6, that are directly employed as categorical elements in the vectorization of the CRSs in Equation (3).

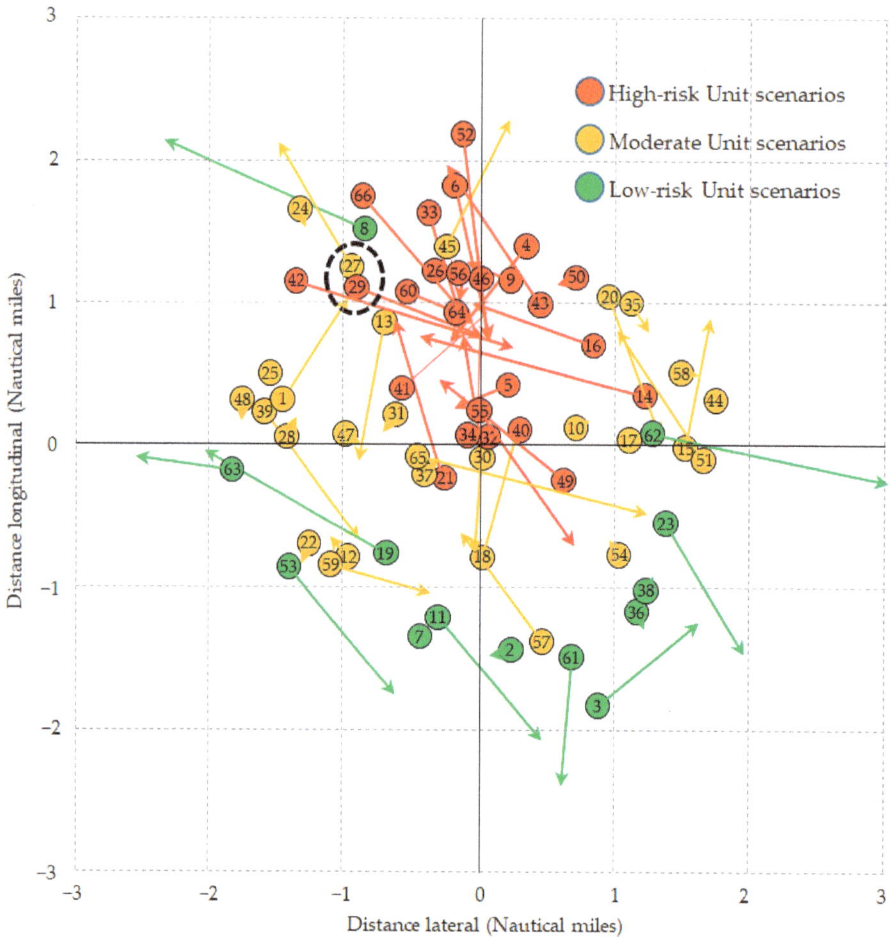

Figure 6. Result of unit scenario clustering.

This equation represents a vector in which the trajectories in Figure 5 are transformed using the categorical labels of the unit scenarios. CRS no. 957 and CRS no. 1099 consist of two and eight encounter ships, respectively, and have been transformed into vectors with two and eight elements, depending on the number of encounters. Each element represents the categorical label of the unit scenario, which means that they have been transformed into an array of vectors containing the relative position, course, and speed of the encounter ships.

$$\begin{array}{c}(\mathrm{a})\ \mathrm{CRS\ no.957} = \{``6", ``18"\} \\ (\mathrm{b})\ CRS\ no.1099 = \{``7", ``11", ``13", ``15", ``16", ``43", ``51", ``65"\}\end{array} \quad (5)$$

3.3. Graph-Based Model

The similarity between the vectorized CRSs was converted into a similarity matrix using a modified Jaccard similarity measurement. This similarity matrix was used as an input to the graph model, and the total of 1205 CRSs formed a graph network as shown in Figure 7. The dots in this graph represent CRS nodes, which are connected by edges to nodes with similarity to each other. The color of a node indicates its centrality, and it is categorized by increasing centrality, as shown in the color bar on the right. Centrality is an indicator of how many connections a node has to other nodes and is a key factor in assessing the complexity of a CRS node. The thickness of an edge is proportional to the similarity between nodes, with higher similarity represented by thicker edges. The layout of the graph adopts the force-direction method, which means that the higher the similarity between nodes, the more attractive they are, and the lower the similarity, the more repulsive they are. Therefore, the relative positions of nodes can be used to intuitively understand the similarity between them.

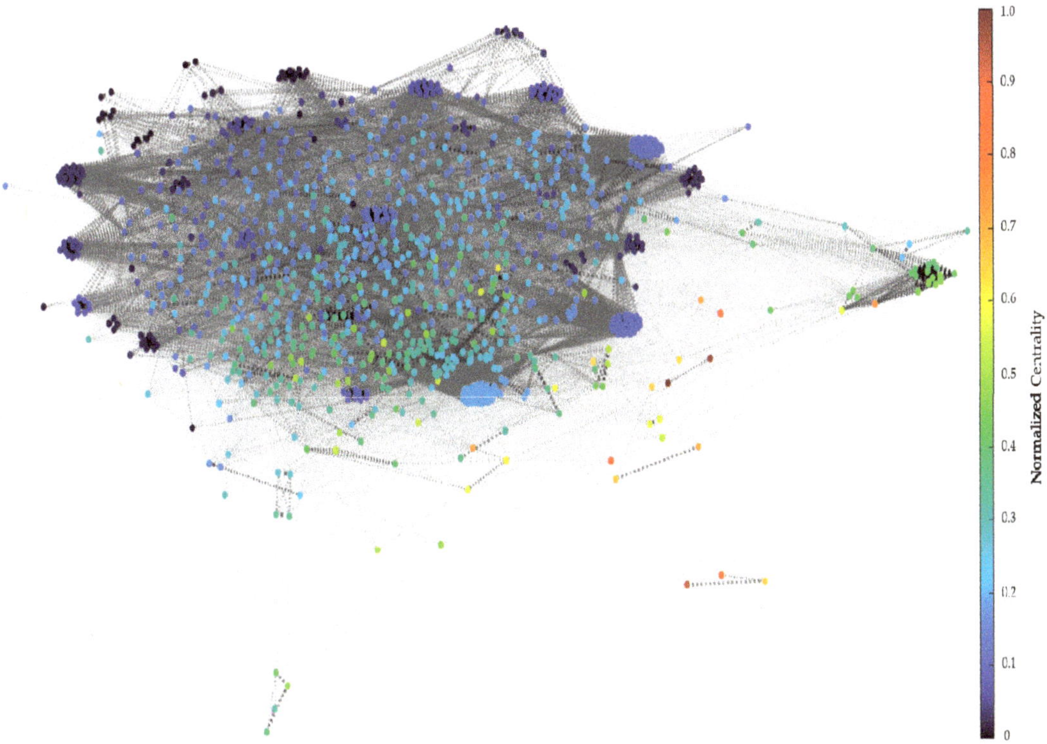

Figure 7. Graph network result.

Within the graph, nodes can be divided into three main categories based on the centrality and structure of graph network. First, nodes that have relatively less centrality and are colored in the blue range form a peripheral region, called "peripheral nodes." Second, nodes that are located in the central part of the graph, surrounded by "peripheral nodes," and are colored from green to yellow are defined as "central nodes." Finally, nodes with high centrality, which have a strong repulsion to other nodes in the graph and are located in the outer regions of the peripheral nodes, can be classified as "outlier nodes." The scenarios corresponding to "peripheral nodes," "central nodes," and "outlier nodes" are exemplified in Appendices A–C, though they are not limited to these examples. These scenarios are classified according to the main category of the aforementioned nodes and are presented alongside the normalized centrality (refer to the color bar in Figure 7).

When analyzing this graph from the perspective of scenario development for autonomous ships' collision avoidance algorithm and enhancing the understanding of CRS, it can be interpreted through three primary dimensions: centrality, frequency, and evolution.

3.3.1. Centrality and Frequency

Centrality is a metric that indicates the extent to which a specific node is connected to others. Therefore, a node's centrality value indicates the complexity of the corresponding CRS node.

Figure 8 displays the distribution of centrality across nodes via a box plot and bar graph. The median centrality value is 0.16, with an inter-quartile range (IQR) of 0.09 to 0.27. Overall, the distribution demonstrates a bias towards the left side, leaning towards 0 when 0.5 is considered a reference point. Values that are above 1.25 times the IQR in the tail end of the distribution are classified as outliers.

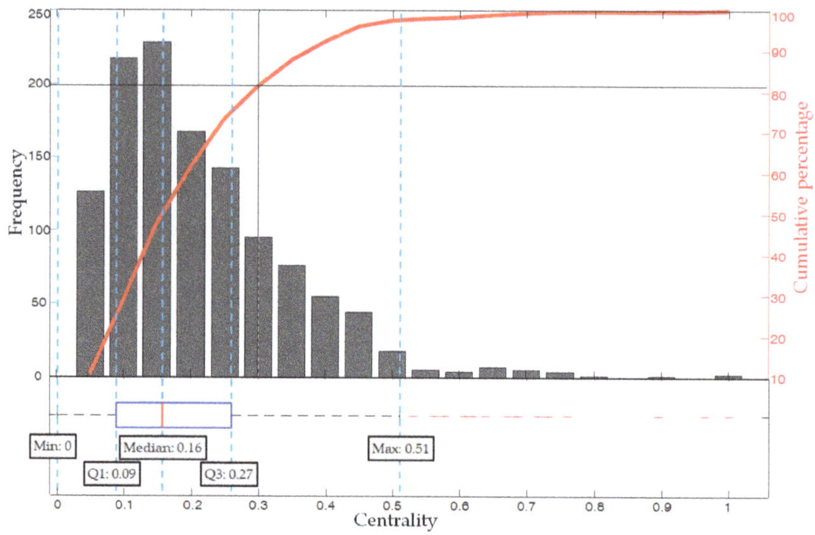

Figure 8. Distribution of centrality in box plot and Pareto chart.

The cumulative percentage also reveals that the majority of the distribution is concentrated in CRSs with low centrality. The left y-axis indicates the frequency of the bar chart, while the right y-axis shows the cumulative percentage of the bar charts. This visualization clarifies the application of the Pareto principle to CRS, which was not easily discernible from the centrality distribution alone. CRS nodes with a centrality value ranging from 0 to 0.3, comprising 30% of the total, occupy approximately 82% of all nodes, and are thus referred to as the "vital few CRSs". In contrast, the other 70% of nodes with centrality values ranging from 0.3 to 1.0 make up only 18% of the total and are therefore the less-important "Trivial many CRSs".

3.3.2. Centrality Ascending Evolution

The graph-based model in this research categorizes CRS nodes as peripheral, central, or outlier nodes. Previous analysis has shown that each node is linked to other nodes by edges with similarity values of at least 0, forming a network. Therefore, exploring nodes and edges in the direction of increasing centrality allows for understanding CRS evolution tendencies in actual operational environments. Figure 9 illustrates the expansion of the graph in the direction of increasing centrality values, starting from nodes with low centrality values and gradually expanding to connected nodes with higher centrality values.

In Figure 9, the red dots represent the nodes of centrality. The bold text on each node indicates the number of unit scenarios that comprise that CRS, which is the non-normalized value of centrality. Nodes are linked by edges, and the text on these edges indicates the node number of that CRS. The count of edges branching from a centrality node thus represents the count of CRS nodes corresponding to that particular centrality. As shown in the graph, the edges extend from the bottom right to the top left, connecting sequentially from the simplest CRS node to the more complex CRS nodes.

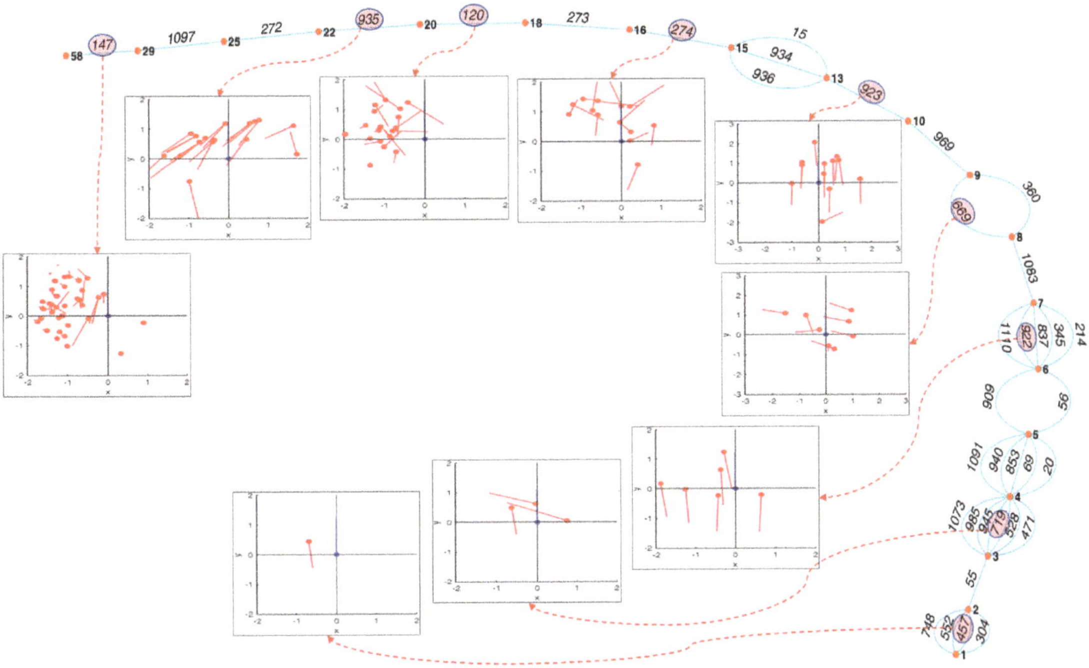

Figure 9. Sample of centrality ascending development. The blue dots and lines in the center of the navigation example represent the current position of the subject ship and its position in 6 minutes, and the red dots and lines represent the current position of the encounter ships and their position in 6 minutes.

Through the graph, we can clearly see the evolution process of the CRS nodes connected to centrality nodes from (1) to (58) using graph analysis. While each CRS in the sample evolution inherently includes the "port to port" situation, which is frequently encountered during ship navigation, CRS 457 in the lower right represents a one-to-one encounter that is considered relatively simple and ordinary. By examining the examples of ship trajectories presented for each level of centrality, a trend of increasing complexity in the CRS can be observed as centrality increases.

4. Discussion

Through a graph-based analysis, we discerned that CRSs can be categorized into peripheral nodes, central nodes, and outlier nodes, based on the similarity between centrality and CRSs as summarized in Table 2. Upon reviewing the scenarios presented in Appendices A–C, we observe that this graph-based approach enables us to classify and interpret CRSs in a specific manner.

Table 2. Summary of CRS categories based on centrality and graph structure.

CRS Category	Detailed Scenarios
Peripheral nodes	Refer to Appendix A
Central nodes	Refer to Appendix B
Outlier nodes	Refer to Appendix C

"Peripheral nodes" are closely located, forming a distinct cluster. They form a circular structure, establishing an interconnected network with the nodes within the circular shape. When analyzing these nodes from both a CRS and centrality perspective, their connectivity with other nodes is comparatively lower, suggesting that they can be interpreted as simple and typical CRS scenarios. On the other hand, "central nodes" are nodes that, while being proximate to the peripheral nodes, maintain connections with them. They can be viewed as scenarios with higher complexity than peripheral nodes but can be interpreted as CRSs with routine complexity that are not significantly differentiated from other CRSs. These nodes not only connect with peripheral nodes but also have intrinsic interconnections with other central nodes, indicating a diverse combination of CRS scenarios within a certain similarity range. Meanwhile, the "outlier nodes" in the graph are CRS nodes situated significantly apart from other nodes. Their low similarity with other nodes positions them apart from other nodes. This high centrality indicates that they are unique CRSs with minimal resemblance to other CRSs.

The results of the frequency analysis based on centrality also provided meaningful interpretations. The centrality distribution of all CRS nodes was not distributed normally but skewed toward low centrality. This distribution pattern of centrality indicates that most CRSs represent relatively simple scenarios, while those with a centrality exceeding 0.5 typically embody situations that differ from the norm, indicating their unique nature. This distribution characteristic somewhat aligns with the "Vital few" and "Trivial many" concepts of the Pareto principle. Although it does not perfectly adhere to the 80–20 rule of the Pareto principle, it does confirm that approximately 30% of simple CRSs account for more than 80% of all CRSs, while the remaining 70% of complex and unique CRSs constitute about 20% of the total, showcasing that the concept can indeed be applied to CRSs.

One of the main findings of this study is that we proposed a methodology to systematically explore how simple CRSs evolve into complex CRSs by utilizing centrality ascending evolution. In this study, we sequentially analyzed the development process of CRS centrality, which eventually depicts intricate navigational situations. While the number of possible combinations as centrality increases is vast, the methodology proposed in this study provides a realistic representation of the stepwise evolution of a real-world CRS. This systematic analysis of the actual CRS evolution process is expected to provide a deeper understanding of the collision risk situations that autonomous ships may encounter, as well as an objective basis for the validation of collision avoidance algorithms for autonomous ships.

5. Conclusions

This study presented a scenario development approach for the systematic validation of the collision avoidance system of autonomous ships. The important point is that this approach aims to provide realistic and possible collision risk scenarios based on the understanding of actual collision risk situations. The data-driven approach using AIS data consisted of the preprocessing and extraction of collision risk situations, the vectorization of collision risk situations using unit scenarios, and graph-based modeling using similarity matrix between vectors.

In the graph modeling, we found that collision risk situations can be classified into "peripheral nodes", "central nodes", and "outlier nodes" based on the centrality characteristics of the nodes. The peripheral nodes represent simple and frequent collision risk situations, while the central nodes, although complex, represent collision risk situations that are relatively less frequent. In contrast, outlier nodes encapsulate very complex and unique collision risk situations. This centrality characterizes the distribution of collision risk situations according to the Pareto principle, i.e., "vital few" and "trivial many". The "vital few" represent 30% of the simple collision risk situations, which account for 80% of the total CRSs, while the "trivial many" represent 70% of the complex and unique collision risk situations, which account for 20% of the total CRSs. These findings are expected to be used as a basis for determining the ratio of validation scenarios for autonomous ship collision avoidance systems. One of the key novelties of graph-based modeling of CRSs is the identification of the network structure among CRSs. Centrality ascending evolution was used to visually track how basic encounter situations evolve into complex collision risk situations through the graph network. This can be interpreted as providing the basis for an exhaustive testing of autonomous ships' collision avoidance system.

The novelty of this research is highlighted by several key differences from previous studies. First, the combination of unit scenarios has been used to implement collision risk situation vectors in graphical form, which represents a unique approach. Second, nodes have been identified in two ways based on graph structure and centrality distribution, adding a unique layer to the investigation. Third, this study introduces a new methodology capable of tracking the evolution from simple collision risk situations to more complex situations, providing a significant value to the field. Finally, since the collision risk situations represented by the classified nodes are based on actual events, they are significant because they provide a new methodology that addresses the reality and scenario diversity issues that have not been addressed in previous collision risk scenario research.

Of course, limitations and future research directions should also be considered. Extending the AIS data collection period to several years is required in order to develop a broader range of possible CRS that autonomous ships may encounter. The application of unit scenarios tends to ignore detailed characteristics by categorizing individual encounters. Future work will focus on improving the granularity of the unit scenario. Furthermore, the feature engineering of unit scenarios demands a method that develops and selects features with more objective grounding. Although the similarity measure used in the vectorization process is well validated, since it is not a traditional method, further supporting research is required. Finally, graph-based modeling is expected to be useful for developing realistic scenarios because it is based on actual CRS. However, due to the open environment of the ocean and the diverse nature of ship traffic, CRS is not limited to historical data. There is always the possibility of new collision risk situations arising that have not been previously characterized. Therefore, the methodology of this study can provide a basic structure for the validation of the collision avoidance system of autonomous ships, but the detailed adjustment of the scenarios will be necessary depending on the purpose and requirements of the validation.

Author Contributions: Conceptualization, T.H. and I.-H.Y.; Data curation, I.-H.Y.; Formal analysis, T.H.; Funding acquisition, I.-H.Y.; Methodology, T.H.; Project administration, I.-H.Y.; Supervision, I.-H.Y.; Visualization, T.H.; Writing—original draft, T.H.; Writing—review and editing, T.H. and I.-H.Y. All authors have read and agreed to the published version of the manuscript.

Funding: This research was supported by the project titled "Development of Autonomous Ship Technology (20200615)", funded by the Ministry of Oceans and Fisheries (MOF, Republic of Korea).

Institutional Review Board Statement: Not applicable.

Informed Consent Statement: Not applicable.

Data Availability Statement: The data used to support the findings of this study are available from the corresponding author upon request.

Conflicts of Interest: The authors declare no conflict of interest.

Appendix A

The figures below show an example CRS scenario for peripheral nodes. However, this category of CRS is not limited to these examples. The blue dots and lines in the center of the navigation example represent the current position of the subject ship and its estimated position in 6 minutes, and the red dots and lines represent the current position of the encounter ships and their estimated position in 6 minutes.

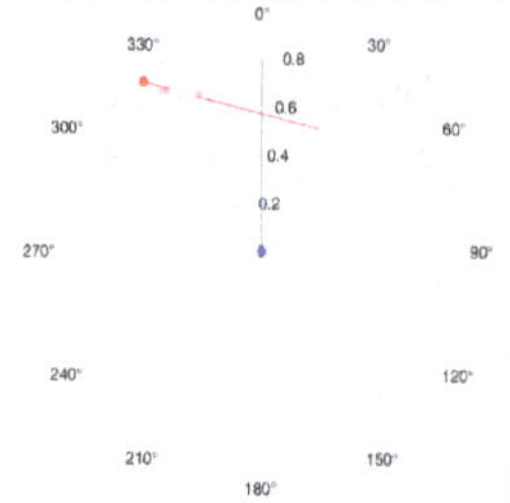 Centrality: 0.06	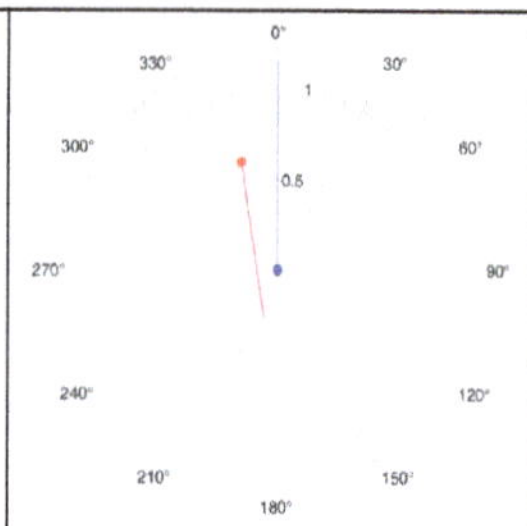 Centrality: 0.06	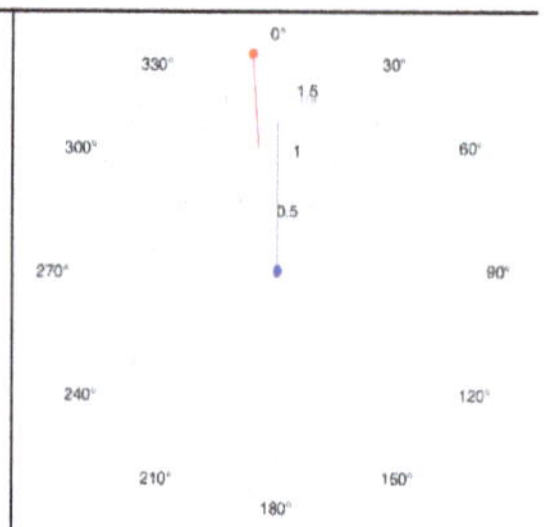 Centrality: 0.06
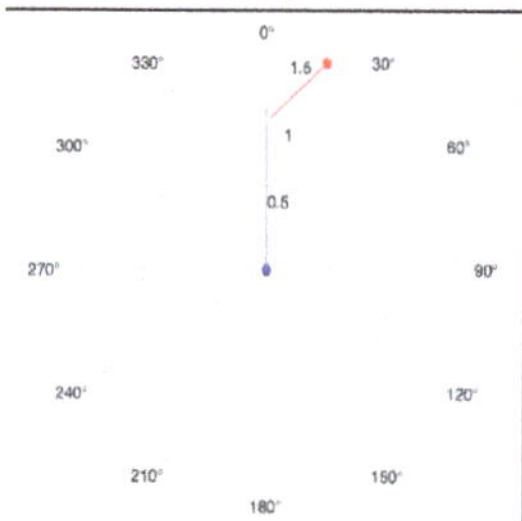 Centrality: 0.06	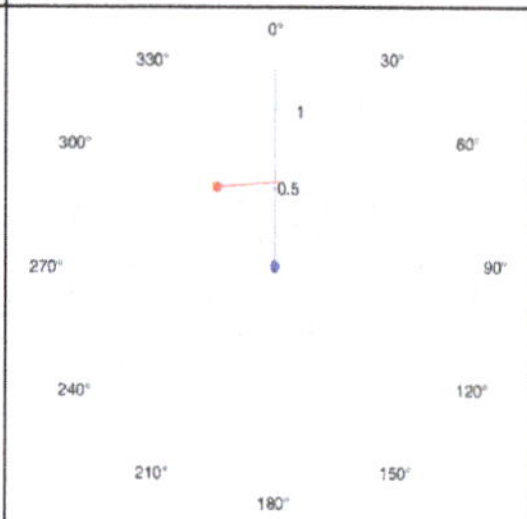 Centrality: 0.06	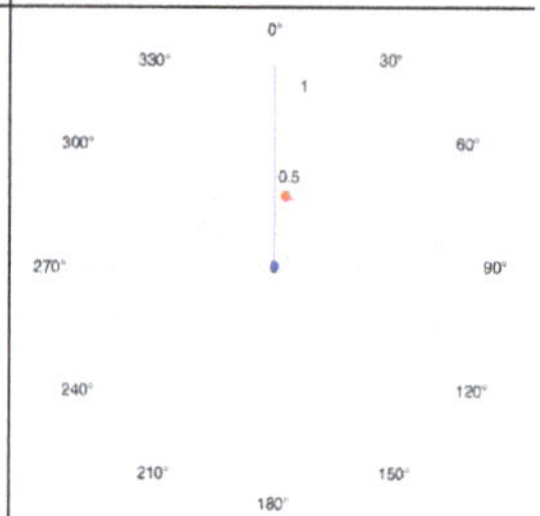 Centrality: 0.06
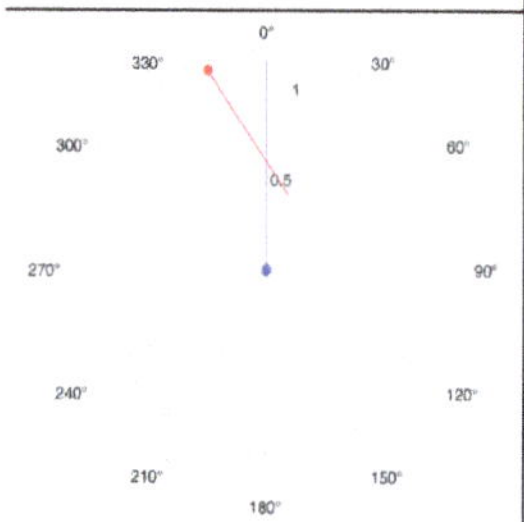 Centrality: 0.06	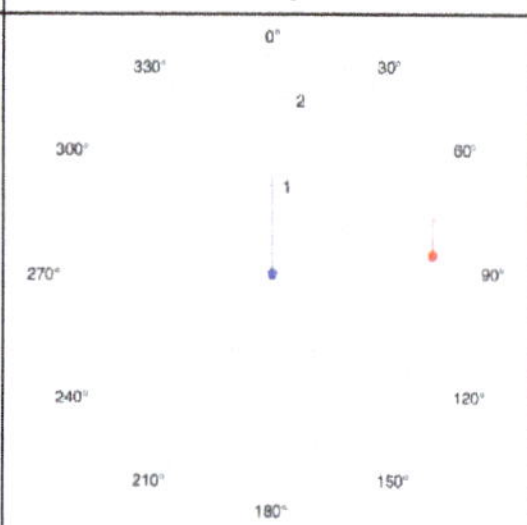 Centrality: 0.06	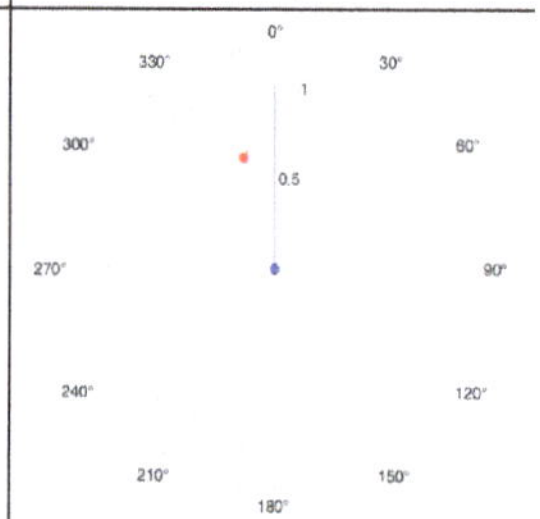 Centrality: 0.06

Appendix B

The figures below show an example CRS scenario for central nodes. However, this category of CRS is not limited to these examples. The blue dots and lines in the center of the navigation example represent the current position of the subject ship and its estimated position in 6 minutes, and the red dots and lines represent the current position of the encounter ships and their estimated position in 6 minutes.

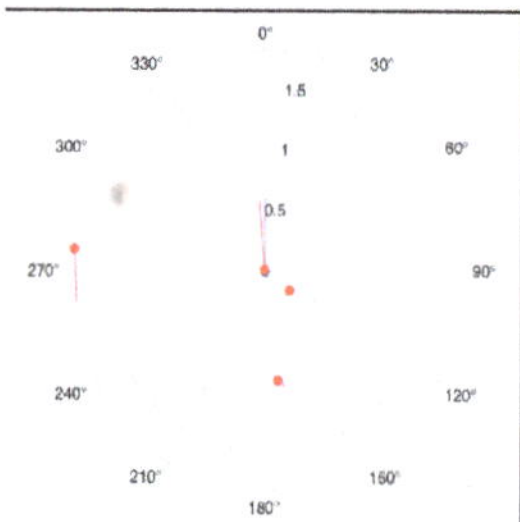 Centrality: 0.34	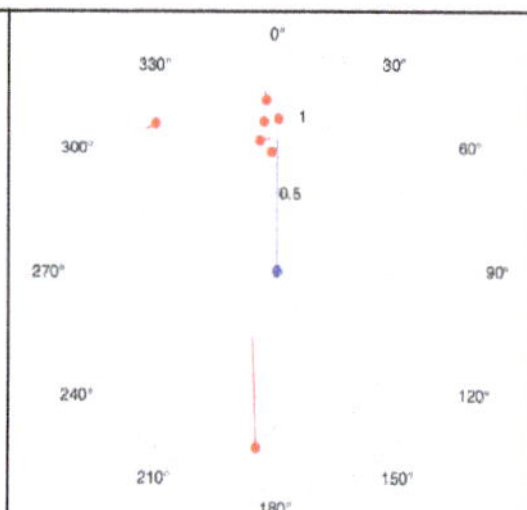 Centrality: 0.42	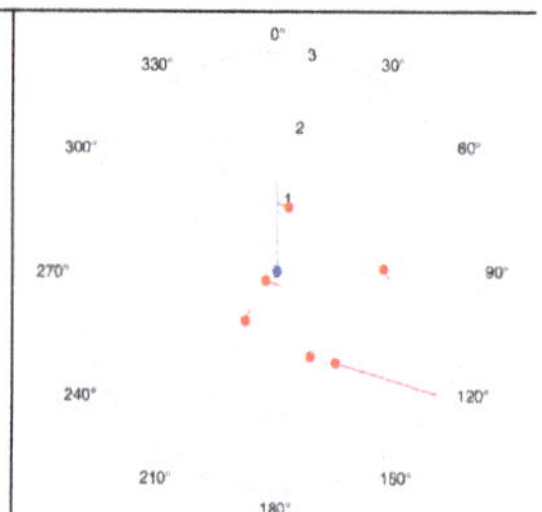 Centrality: 0.45

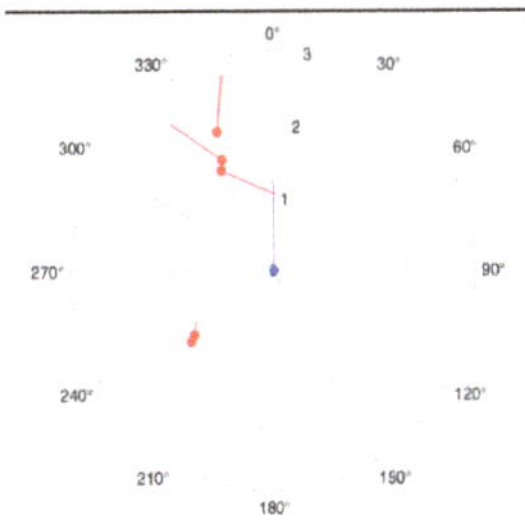

Centrality: 0.29

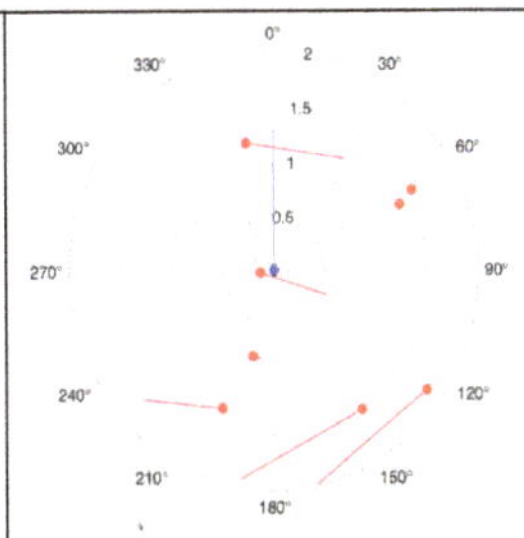

Centrality: 0.45

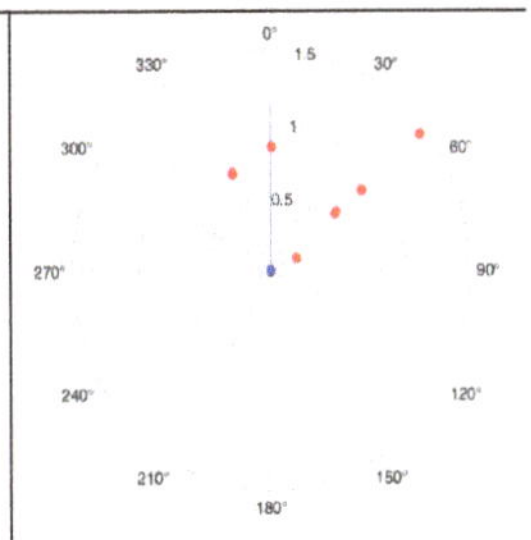

Centrality: 0.46

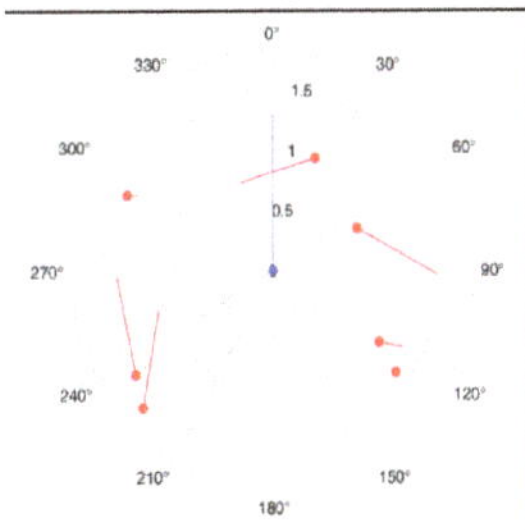

Centrality: 0.46

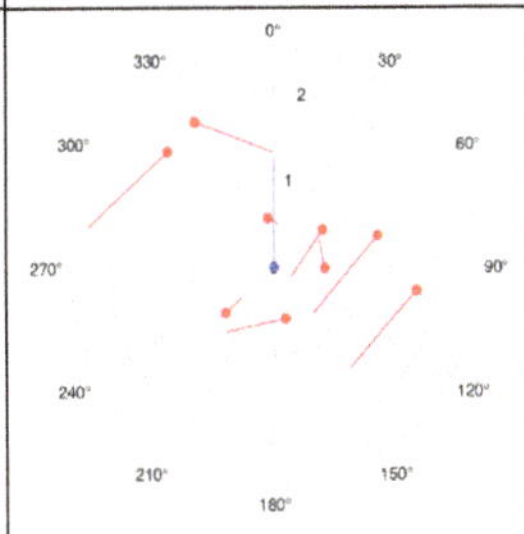

Centrality: 0.44

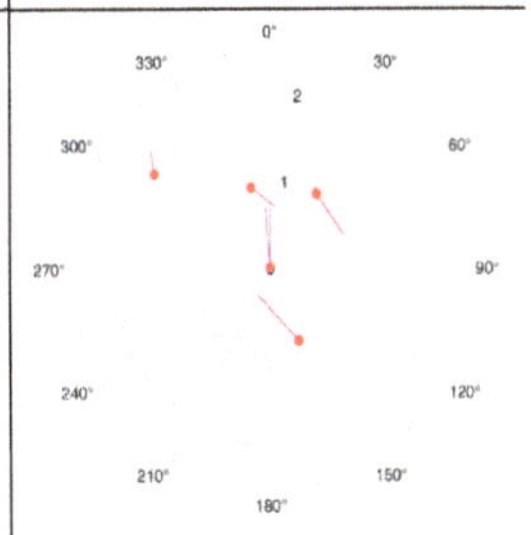

Centrality: 0.55

Appendix C

The figures below show an example CRS scenario for outlier nodes. However, this category of CRS is not limited to these examples. The blue dots and lines in the center of the navigation example represent the current position of the subject ship and its estimated position in 6 minutes, and the red dots and lines represent the current position of the encounter ships and their estimated position in 6 minutes.

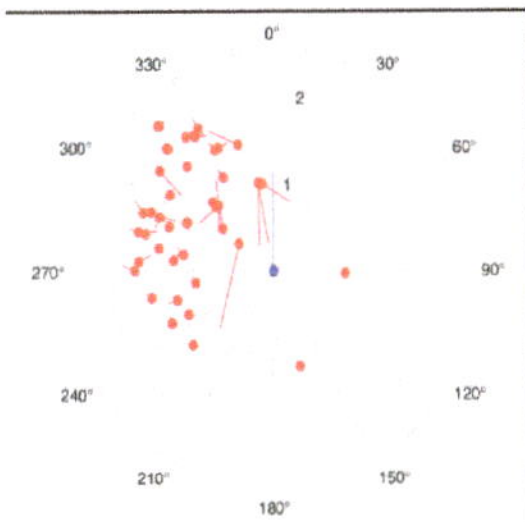

Centrality: 0.94

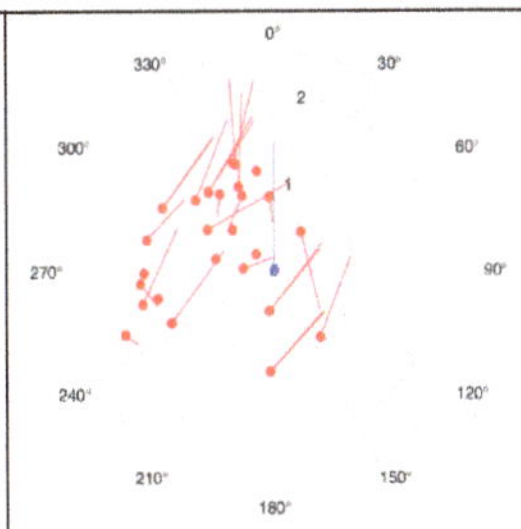

Centrality: 0.72

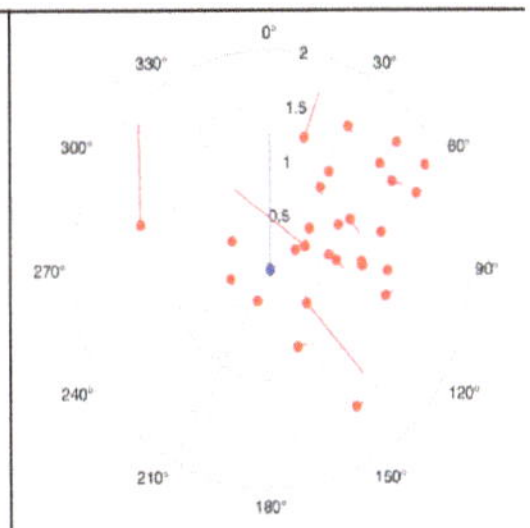

Centrality: 0.72

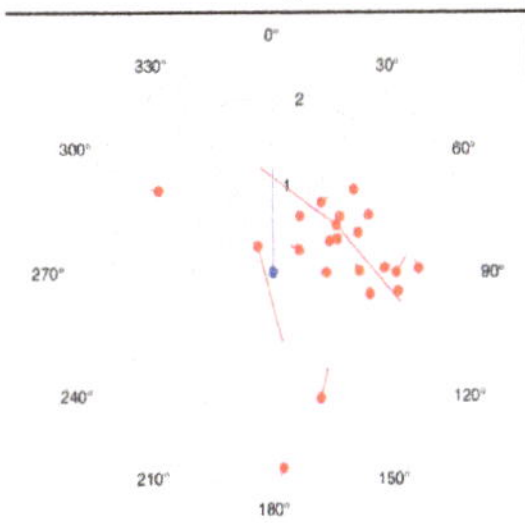

Centrality: 0.68

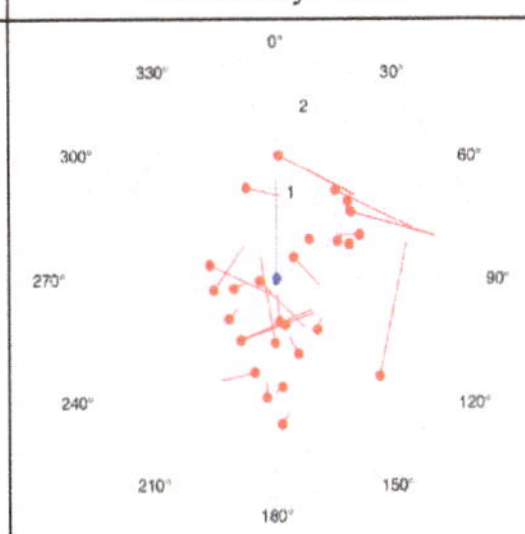

Centrality: 0.68

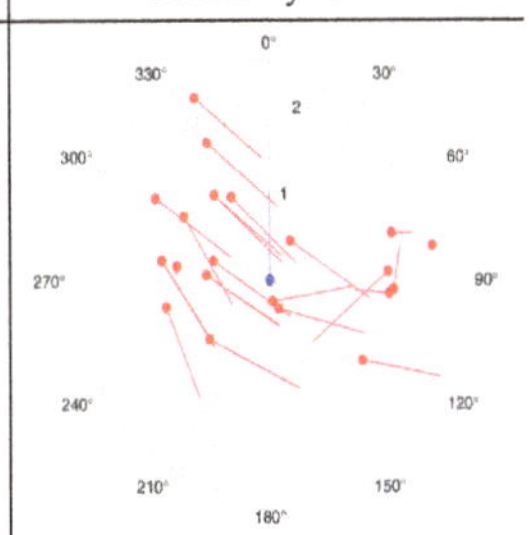

Centrality: 0.72

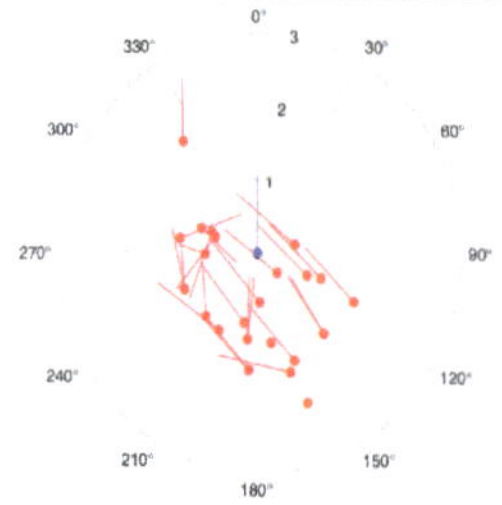

Centrality: 0.67

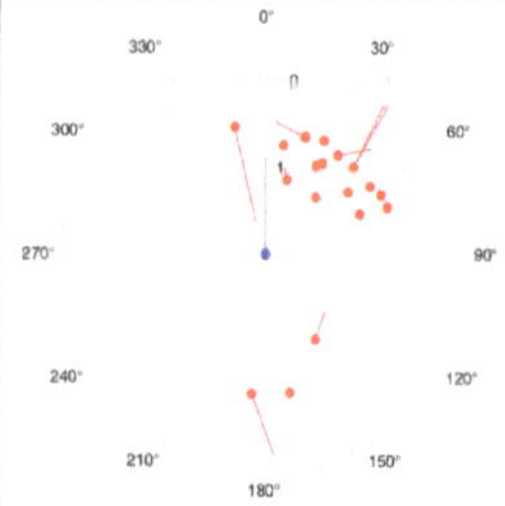

Centrality: 0.78

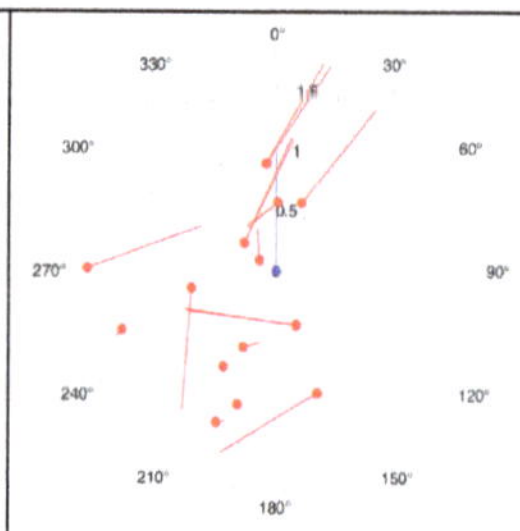

Centrality: 0.67

References

1. Zhang, L.; Wang, H.; Meng, Q.; Xie, H. Ship accident consequences and contributing factors analyses using ship accident investigation reports. *Proc. Inst. Mech. Eng. Part O J. Risk Reliab.* **2019**, *233*, 35–47. [CrossRef]
2. Bolbot, V.; Theotokatos, G.; Wennersberg, L.A.; Faivre, J.; Vassalos, D.; Boulougouris, E.; Jan Rødseth, Ø.; Andersen, P.; Pauwelyn, A.; Van Coillie, A. A novel risk assessment process: Application to an autonomous inland waterways ship. *Proc. Inst. Mech. Eng. Part O J. Risk Reliab.* **2023**, *237*, 436–458. [CrossRef]
3. Vincent, T.L. *Collision Avoidance at Sea, Differential Games and Applications: Proceedings of a Workshop Enschede 1977*; Springer: Berlin/Heidelberg, Germany, 2005; pp. 205–221.
4. Fossen, T.I. *Marine Control Systems: Guidance, Navigation and Control of Ships, Rigs and Underwater Vehicles*, 3rd ed.; Marine Cybernetics: Trondheim, Norway, 2002.
5. Peel, D.; Good, N.M. A hidden Markov model approach for determining vessel activity from vessel monitoring system data. *Can. J. Fish. Aquat. Sci.* **2011**, *68*, 1252–1264. [CrossRef]
6. Hu, Q.; Yang, C.; Cheng, H.; Xiao, B. Planned Route Based Negotiation for Collision Avoidance Between Vessels. *TransNav Int. J. Mar. Navig. Saf. Sea Transp.* **2008**, *2*, 363–368.
7. Kearon, J. Computer Programs for Collision Avoidance and Traffic Keeping. In *Conference on Mathematical Aspects on Marine Traffic*; Academic Press: London, UK, 1977.
8. Liu, Y.; Shi, C. A fuzzy-neural inference network for ship collision avoidance. In Proceedings of the 2005 International Conference on Machine Learning and Cybernetics, Guangzhou, China, 18–21 August 2005; pp. 4754–4759.
9. Brcko, T.; Androjna, A.; Srše, J.; Boć, R. Vessel multi-parametric collision avoidance decision model: Fuzzy approach. *J. Mar. Sci. Eng.* **2021**, *9*, 49. [CrossRef]
10. Huang, Y.; Van Gelder, P. Collision risk measure for triggering evasive actions of maritime autonomous surface ships. *Saf. Sci.* **2020**, *127*, 104708. [CrossRef]
11. Mizythras, P.; Pollalis, C.; Boulougouris, E.; Theotokatos, G. A novel decision support methodology for oceangoing vessel collision avoidance. *Ocean Eng.* **2021**, *230*, 109004. [CrossRef]
12. Namgung, H.; Kim, J. Collision risk inference system for maritime autonomous surface ships using COLREGs rules compliant collision avoidance. *IEEE Access* **2021**, *9*, 7823–7835. [CrossRef]
13. Goerlandt, F.; Kujala, P. On the reliability and validity of ship–ship collision risk analysis in light of different perspectives on risk. *Saf. Sci.* **2014**, *62*, 348–365. [CrossRef]
14. Liu, Z.; Wu, Z.; Zheng, Z. A cooperative game approach for assessing the collision risk in multi-vessel encountering. *Ocean Eng.* **2019**, *187*, 106175. [CrossRef]
15. Silveira, P.; Teixeira, A.P.; Figueira, J.R.; Soares, C.G. A multicriteria outranking approach for ship collision risk assessment. *Reliab. Eng. Syst. Saf.* **2021**, *214*, 107789. [CrossRef]
16. Tam, C.; Bucknall, R. Collision risk assessment for ships. *J. Mar. Sci. Technol.* **2010**, *15*, 257–270. [CrossRef]
17. Zhang, W.; Kopca, C.; Tang, J.; Ma, D.; Wang, Y. A systematic approach for collision risk analysis based on AIS data. *J. Navig.* **2017**, *70*, 1117–1132. [CrossRef]
18. Fan, C.; Montewka, J.; Zhang, D. A risk comparison framework for autonomous ships navigation. *Reliab. Eng. Syst. Saf.* **2022**, *226*, 108709. [CrossRef]
19. Guiochet, J.; Machin, M.; Waeselynck, H. Safety-critical advanced robots: A survey. *Robot. Auton. Syst.* **2017**, *94*, 43–52. [CrossRef]
20. Zaitseva, E.; Levashenko, V. Reliability analysis of multi-state system with application of multiple-valued logic. *Int. J. Qual. Reliab. Manag.* **2017**, *34*, 862–878. [CrossRef]
21. Bolbot, V.; Gkerekos, C.; Theotokatos, G.; Boulougouris, E. Automatic traffic scenarios generation for autonomous ships collision avoidance system testing. *Ocean Eng.* **2022**, *254*, 111309. [CrossRef]
22. Alexander, R.; Hawkins, H.R.; Rae, A.J. *Situation Coverage—A Coverage Criterion for Testing Autonomous Robots*; Technical Report; Department of Computer Science, University of York: Heslington, UK, 2015.

23. Bolbot, V.; Theotokatos, G.; Bujorianu, L.M.; Boulougouris, E.; Vassalos, D. Vulnerabilities and safety assurance methods in Cyber-Physical Systems: A comprehensive review. *Reliab. Eng. Syst. Saf.* **2019**, *182*, 179–193. [CrossRef]
24. Sørensen, A.J.; Ludvigsen, M. Underwater technology platforms. In *Encyclopedia of Maritime and Offshore Engineering*; Wiley: Hoboken, NJ, USA, 2017; pp. 1–11.
25. Torben, T.R.; Glomsrud, J.A.; Pedersen, T.A.; Utne, I.B.; Sørensen, A.J. Automatic simulation-based testing of autonomous ships using Gaussian processes and temporal logic. *Proc. Inst. Mech. Eng. Part O J. Risk Reliab.* **2023**, *237*, 293–313. [CrossRef]
26. Chun, D.; Roh, M.; Lee, H.; Ha, J.; Yu, D. Deep reinforcement learning-based collision avoidance for an autonomous ship. *Ocean Eng.* **2021**, *234*, 109216. [CrossRef]
27. Lazarowska, A. Verification of a deterministic ship's safe trajectory planning algorithm from different ships' perspectives and with changing strategies of target ships. *TransNav Int. J. Mar. Navig. Saf. Sea Transp.* **2021**, *15*, 623–628. [CrossRef]
28. Gil, M. A concept of critical safety area applicable for an obstacle-avoidance process for manned and autonomous ships. *Reliab. Eng. Syst. Saf.* **2021**, *214*, 107806. [CrossRef]
29. Szlapczynski, R.; Szlapczynska, J. A ship domain-based model of collision risk for near-miss detection and Collision Alert Systems. *Reliab. Eng. Syst. Saf.* **2021**, *214*, 107766. [CrossRef]
30. Pedersen, T.A.; Glomsrud, J.A.; Ruud, E.; Simonsen, A.; Sandrib, J.; Eriksen, B.H. Towards simulation-based verification of autonomous navigation systems. *Saf. Sci.* **2020**, *129*, 104799. [CrossRef]
31. Porres, I.; Azimi, S.; Lilius, J. Scenario-based testing of a ship collision avoidance system. In Proceedings of the 2020 46th Euromicro Conference on Software Engineering and Advanced Applications (SEAA), Portoroz, Slovenia, 26–28 August 2020; pp. 545–552.
32. Zhu, F.; Zhou, Z.; Lu, H. Randomly Testing an Autonomous Collision Avoidance System with Real-World Ship Encounter Scenario from AIS Data. *J. Mar. Sci. Eng.* **2022**, *10*, 1588. [CrossRef]
33. Rodriguez, S.I.; de Carvalho, F.d.A. Fuzzy clustering algorithms with distance metric learning and entropy regularization. *Appl. Soft Comput.* **2021**, *113*, 107922. [CrossRef]
34. Saxena, A.; Iyengar, S. Centrality measures in complex networks: A survey. *arXiv* **2020**, arXiv:2011.07190 2020.

Journal of
Marine Science and Engineering

MDPI

Article

Ranking Ship Detection Methods Using SAR Images Based on Machine Learning and Artificial Intelligence

Muhammad Yasir [1], Abdoul Jelil Niang [2,*], Md Sakaouth Hossain [3], Qamar Ul Islam [4], Qian Yang [5] and Yuhang Yin [6]

[1] College of Oceanography and Space Informatics, China University of Petroleum (East China), Qingdao 266580, China
[2] Department of Geography, College of Social Sciences Umm Al-Qura University, Makkah 24231, Saudi Arabia
[3] Department of Geological Sciences, Jahangirnagar University, Dhaka 1342, Bangladesh
[4] Department of Electrical and Computer Engineering, College of Engineering, Dhofar University, Salalah 211, Oman
[5] PLA Troops No. 63629, Beijing 102699, China
[6] PLA Troops No. 93525, Shigatse 857000, China
* Correspondence: anniang@uqu.edu.sa

Abstract: We aimed to improve the performance of ship detection methods in synthetic aperture radar (SAR) images by utilizing machine learning (ML) and artificial intelligence (AI) techniques. The maritime industry faces challenges in collecting precise data due to constantly changing sea conditions and weather, which can affect various maritime operations, such as maritime security, rescue missions, and real-time monitoring of water boundaries. To overcome these challenges, we present a survey of AI- and ML-based techniques for ship detection in SAR images that provide a more effective and reliable way to detect and classify ships in a variety of weather conditions, both onshore and offshore. We identified key features frequently used in the existing literature and applied the graph theory matrix approach (GTMA) to rank the available methods. This study's findings can help users select a quick and efficient ship detection and classification method, improving the accuracy and efficiency of maritime operations. Moreover, the results of this study will contribute to advancing AI- and ML-based techniques for ship detection in SAR images, providing a valuable resource for the maritime industry.

Keywords: machine learning; artificial intelligence; synthetic aperture radar; ship detection

Citation: Yasir, M.; Niang, A.J.; Hossain, M.S.; Islam, Q.U.; Yang, Q.; Yin, Y. Ranking Ship Detection Methods Using SAR Images Based on Machine Learning and Artificial Intelligence. *J. Mar. Sci. Eng.* **2023**, *11*, 1916. https://doi.org/10.3390/jmse11101916

Academic Editors: Xianbo Xiang, Haitong Xu, Lúcia Moreira and Carlos Guedes Soares

Received: 24 January 2023
Revised: 20 February 2023
Accepted: 22 February 2023
Published: 4 October 2023

1. Introduction

Synthetic aperture radar is an example of a modern technology that can detect and create images of the earth remotely with high efficiency and accuracy. It can capture very clear and visible images in any weather even at night. To work on these images for different purposes such as weather forecasting, object detection, and many more, artificial intelligence and machine learning are the most promising candidates. Various techniques in these frameworks can be applied to the data generated by SAR to perform various activities with great care and efficiency [1–4]. Due to the nature of the task, a significant number of researchers have shown enthusiasm for developing the SAR ship identification technology. The automatic detection and identification of ship targets in SAR images has grown to be a key study area in the field of SAR image interpretation due to the ongoing improvement in SAR image resolution and image quality [5].

Target identification [6–8], image classification [9,10], autonomous driving [11], saliency detection [12], semantic comprehension [13], and other domains [14–17] are only a few of the areas where deep learning technology has recently produced positive results. There have been relatively few research findings since the field of ship recognition technology was first studied. Using deep learning technology, it is possible to automatically find and identify ship targets in SAR images [18–20]. This novel method also offers new perspectives for the advancement of SAR target detection and recognition technologies.

Qu and Shao [21] have proposed a study for the development of a deep learning (DL)-grounded efficient and reliable detector with the assistance of SAR (synthetic aperture radar). Based on a fresh dataset and four strategies, the performance of the faster R-CNN procedure was enhanced. The dataset containing image resolution and sea condition was evaluated with the help of various comparisons and experiments. The overall architecture is the combination of four basic strategies such as feature fusion and transfer learning. It was analyzed that the proposed architecture can achieve high accuracy and is very cheap. It will be very helpful to employ such kinds of DL-based paradigms for the detection of the ship. The detection of ships with the employment of deep learning characteristics is a new development in SAR ship detection. Liu et al. [22] have developed an architecture with the assistance of a sea–land segmentation-grounded convolutional neural network (SLS-CNN) to achieve the effective detection of the ship. This architecture is the integration of the SLS-CNN detector, saliency calculation, and corner characteristics. ALOS PALSAR and Terra-SAR-X imagery was used for the efficient analysis and evaluation of ship detection by the proposed system. An automatic SAR ship identification technique based on feature decomposition across various satellites was proposed by Zhao et al. [23]. By enhancing the backbone network to extract features, this technique enhances target localization and recognition performance. Based on the "You Only Look Once (YOLO) v5 model", which can successfully identify ship targets in ALOS-2 spotlight images, Yoshida et al. [24] suggested a technique to automatically recognize ships in motion. In order to achieve robustness and high-accuracy recognition for SAR ships, Zheng et al. [25] presented an ensemble automated technique (MetaBoost) for heterogeneous D-CNN models based on two-stage filtering. MetaBoost can perform much better than individual classifiers and conventional ship recognition algorithms, according to extensive trials on the OpenSARShip and FuSARShip datasets. The results show that the system can detect a ship with great competence and high strength.

To resolve the issue of position invariance in convolutional neural network (CNN)-based detection of the ship, a new DL-grounded procedure named Capsule Network was presented. This unique procedure translates different entity factors in addition to characteristic values to try to advance upon convolutions. After testing and comparing it with other ML-based ship detection systems, it was proved that it enhances the accuracy of ship detection by 91.03% along with a false alarm rate of 9.5745×10^{-9}. The performance can be further enhanced with fewer samples [26]. Wang et al. [27] have conducted a study for the detection of the ship very efficiently and effectively by the employment of the enhanced YOLOv3 procedure. With productive functionalities such as improving loss functions and generating CFE modules, the proposed system gains very high accuracy and efficiency. The accuracy achieved by the system was 74.8% in combination with a detection rate of 29.8 frames. The proposed paradigm can be employed for the detection of the ship in hard and changing sea conditions. As compared to other ship detection approaches, it is better in both accuracy and speed.

The employment of various deep learning architectures on the SAR images can efficiently perform the detection and classification of ships. The integration of AI and ML can be used on the images for the security of water resources, rescue operations, and various image recognition procedures. With the advancement in the area of artificial intelligence, it is very convenient to detect ships in densely arranged ships as well as in bad weather. To assist professionals in searching for lost ships, these techniques are very helpful and productive. The main contributions of the proposed research are:

- To provide an overview of the state-of-the-art approaches used for the detection of ships based on AI and ML using the data from SAR.
- Various applications of AI and ML in ship detection are acknowledged.
- Different features are extracted from the existing literature and important ones are recognized from them.

- With the usage of the graph theory matrix approach (GTMA) on these selected features, various AI and ML-grounded ship detection architectures are ranked.

2. State of the Art

A study was performed for the real-time detection of ships in the ocean with high accuracy and efficiency. The system is based on the robust SAR ship detection procedure and you look only once at version 3 (YOLOv3). The working of the system was evaluated by testing it on the public SAR ship detection dataset. The experimental data revealed that the proposed system is faster in detection than other existing procedures while the accuracy remains the same. With the goal of faster detection of the ship, the architecture can be employed for real-time ship rescue in various conditions [28]. Chen et al. [29] have presented a study to solve the issues of correct identification of ships in complex scenarios with unarranged ships in the ocean. The developed paradigm is an object detection network with the main focus being detection in difficult situations. To precisely locate the ship in unarranged ships, the system implemented a loss function with generalized intersection over union (GIoU) for minimizing the sensitivity. The accuracy was further increased by reducing missed detections with the employment of soft non-maximum suppression. The experimental data show that the system can achieve high accuracy and productivity on the SAR ship detection dataset. To effectively handle the issue of multiscale and multi-scene SAR ship detection, an efficient densely connected architecture grounded on the faster R-CNN system was developed. In contrast to most of the existing approaches, the proposed system generates proposals by densely connecting feature maps to each available feature map. For minimizing false alarms, the training procedure of the system consists of difficult examples. After a thorough analysis and monitoring of the architecture, the results show that it can achieve high performance and efficiency in the detection of multiscale SAR ships [30]. Ding et al. [31] have proposed a study to tackle the problems of limited accuracy and low training speed in SAR ship detection by the employment of a deep network. The training of the SAR ship detection architecture was performed by the implementation of ResNet to achieve the goal of high accuracy and faster training. After implementing the proposed paradigm on the SAR ship dataset, the experimental data show that it can achieve up detection with 94.7% average precision. The architecture was compared to other existing approaches by performing various experiments and it shows very high performance.

The SAR image ship detection based on deep learning is ineffective due to some issues such as the dispersed arrangement of ships and the unavailability of detailed information. In the proposed article, a feature-optimizing framework was developed with the implementation of the single-shot detector (SSD). In the first stage, the training and testing time was minimized by the employment of a lightweight single-shot detector. Then, the performance of the system to detect multiscale ships was improved with the usage of the bidirectional feature fusion paradigm. The system was evaluated with great care and the results indicate that it is better than the present procedures both in speed and accuracy [32]. Kartal and Duman [33] have developed an architecture for the effective and efficient detection of ships to assist the various tasks such as fishing activities, uncovering warships, searching for lost ships in the ocean, and many more. In the developed paradigm, the task of ship detection was carried out by the integration of optical satellite images and a deep learning procedure. It is fast and open-source and can be implemented with the help of an average laptop. The training of the TensorFlow object detection application programming interface was completed by optical satellite images containing ships. The proposed article focused on increasing accuracy in the detection of ships by employing a unique balanced feature pyramid network (B-FPN) on synthetic aperture radar (SAR) images. Based on the four stages, rescaling, integrating, refining, and strengthening, the multi-level features are given more strength. The architecture was tested and evaluated on the SAR ship detection dataset (SSDD) and it was revealed that it can improve the mean average precision by 7.15% compared to the feature pyramid network [34].

Most of the existing ship detection approaches are unable to perform segmentation down to the pixel level. Nie et al. [35] have conducted a study to develop a procedure for the detection and segmentation of ships at the pixel level by the usage of the enhanced Mask R-CNN paradigm. With the integration of bottom-up architecture to the feature pyramid network procedure of Mask R-CNN, the lower layer features can be used very productively at the first layer because of path shortening between the first and last layer. The performance of the system was further enhanced by the assigning of corresponding weights at each pixel in the feature maps. The experimental data show that the system enhanced the detection and segmentation of mean average precision very efficiently. Alghazo et al. [36] have proposed a study for the development of an efficient and effective ship detection procedure with the assistance of a CNN-grounded deep learning paradigm from the images obtained from the satellite. In the proposed study, two procedures with different frameworks are implemented and tested on the data of the Airbus satellite. For both systems, the accuracy and loss function was monitored by changing the number of epochs. With the usage of the data from the training time, the complexity of the procedures was also computed. The results of the article show that both systems can achieve high performance and maximum accuracy of about 89.7% when applied to the Airbus dataset. Due to advancements in modern technologies, various deep learning procedures can be efficiently employed for SAR image ship detection. Kun and Yan [37] have performed a study for the development of an improved YOLOv4-Tint detection process for the enhancement of detection accuracy. In the developed architecture, the task of feature extraction was enhanced by the addition of an attention mechanism unit. Based on the batch normalization optimization dataset, the proposed model was made more reliable and feasible. With the usage of real-time detection, a high detection accuracy was achieved. The analysis of the system shows that it can obtain a mean average precision of about 75.56%.

To enhance the detection accuracy of the existing architectures, the study presented a target detection procedure with multi-features in synthetic aperture radar imagery. Both the deep learning hand-crafted features are extracted in the two channels of the system. The DL-based features are obtained from the SAR images by the implementation of a convolutional neural network. The extraction of fused deep features was performed after the processing of many layers of the network. The system was analyzed by implementing it on the Sentinel-1 SAR data and the results show that the detection ability was enhanced by it very precisely and efficiently [38]. Shi et al. [39] have proposed a study to design an on-orbit ship detection architecture for the images captured by the SAR satellite. The system was trained with the OpenSARShip dataset in integration with non-ship slice images. A deep learning procedure was employed for the classification of images into different types such as cargo ships. The experimental results show that the system is more effective and efficient than the constant false alarm rate procedure and can enhance the detection accuracy from 88.5% to 98.4%. The verification and testing accuracy of the system was also very productive and healthy. The extraction of various features of ships from the images of SAR is a very challenging issue for the already applied procedures. To work on the mentioned issue, the study presented a paradigm that is the combination of the you only look once algorithm, the sliding window detection method, and the clustering algorithm. In the beginning, the system collected images and a training dataset. Then, an analysis was carried out for the efficient size of the frame required for the proposed model. The system was thoroughly evaluated and it was found that can show better performance than F-RCNN in the detection of ships in the low-resolution area of the sea [40].

A study was conducted on the efficient detection of ships in inshore areas. The developed SAR architecture consists of two phases named scene classification and ship detection. The images with no ships were precisely removed with the employment of a scene classification network. The detection of ships was done by giving the images with ships as input to the single-shot detector. The proposed architecture was evaluated and checked on the AIR-SARShip-1.0 dataset. The data show that the system is more efficient than the single-shot detector and maintains relatively high accuracy [41]. Verma et al. [42]

have performed a study on the efficient and effective patrolling in water by the detection of ships. The architecture is based on deep learning and the employment of existing procedures such as F-RCNN, SSD, and YOLOv4. The study also proposed a dataset of about 300 satellite images of the most crowded seaports in India. The evaluation results show that the you only look once version 4 algorithm can achieve better performance in the detection of ships with effective values of mean average precision and FPS score. The detection of ships is one of the main activities for the efficient monitoring of the marine atmosphere. Chang et al. [43] have designed an effective ship detection procedure with the employment of the YOLOv3 algorithm for the enhancement of small ship detection. An experiment was performed with a dataset comprised of six kinds of ships and about 5513 visible and infrared images from the harbors in northern Taiwan. When the proposed architecture is compared with the original YOLOv3 architecture, it was revealed that the present system can achieve a mean average precision of 89.1%, which is greater than the original one.

The research was carried out for ship detection with the implementation of discriminative dictionary learning. The proposed architecture is the integration of image denoising, extraction of the candidate region, and identification of the candidate region. In the first phase, a non-local filtering procedure was used for the denoising of SAR images. The candidate region was extracted by the employment of the gradient feature map reconstruction process. The evaluation results of the developed system show that it has high feasibility and flexibility [44]. It takes time to develop and implement efficient and reliable ship detection methodologies for the monitoring of the sea. Mutalikdesai et al. [45] have surveyed the effectiveness and limitations of the existing approaches applied for the detection of ships. The study focuses on the experimental information obtained by the image recognition process called the Haar-like technique. The disadvantages, such as the exponential time consumption of the mentioned procedure, were tackled by the employment of the TensorFlow methodology and decision boundary feature extraction. Due to varying sea atmospheres, it is very hard to extract the general characteristics from the individual pixel of the image for precise ship detection. The study was performed for the development of a procedure that is based on block division instead of pixels. Compared to the pixel approach, the division of the image into blocks can efficiently extract various properties from it and it is more reliable. With the usage of the color and texture properties identified from the blocks, the block classification was performed by the training of the support vector machine. The information shows that the usage of color and texture properties can enhance the classification precision in the blocks containing ships and those without ships [46]. For pixel-by-pixel ship identification in polarimetric SAR photos, a [47] fully convolutional network has been created [48]. The feature pyramid network contained a split convolution block and an embedded spatial attention block [49]. Against a complex background, the feature pyramid network can detect ship items with accuracy. Wei et al. created a high-resolution feature pyramid network for ship recognition that combined high-to-low-resolution features [50]. The problem of ships of various sizes and crowded berthings has been addressed by the development of a multiscale adaptive recalibration network [51]. A one-stage SAR object identification approach was proposed by Hou et al. [52] to address the low confidence of candidates and false positives. Kang et al. [53] proposed a method integrating CFAR with faster R-CNN. The object proposals produced by the faster R-CNN used in this method for extracting small objects served as the protective window of the CFAR. Zou et al. integrated YOLOv3 with a generative adversarial network with a multiscale loss term to increase the accuracy of SAR ship identification [54], and so on [55–61]. YOLOv3 was modified by Mehdi et al. [62] to identify hazardous and noxious compounds of critical maritime transit. The one-stage YOLO series is more in accordance with the real-time and precise detection needs at this level, as can be observed from the research state indicated above in the context of remote sensing photo detection. Xiong et al. proposed a lightweight model for ship detection and recognition in complex-scene

SAR images by integrating different attention mechanisms into the YOLOv5-n lightweight model [63].

3. Proposed Methodology

The methodology for this study aims to evaluate the usefulness of AI and ML in detecting and classifying ships using synthetic aperture radar (SAR) images. The paper follows a reproducible plan for the selection of the papers, features, benchmark datasets, ranking criteria, and robustness analysis. The selection of papers was based on the relevance and contribution to the field, with a focus on the most recent studies. The features used for ship detection and classification were selected based on their importance and relevance in the literature. The benchmark datasets were chosen based on their availability and suitability for evaluating the effectiveness of the AI and ML procedures. To accomplish the ranking, the graph theory matrix approach was used to rank different AI and ML procedures based on their effectiveness in detecting ships. The ranking criteria were selected based on their importance in the literature and the specific requirements of ship detection using SAR images. The robustness of the procedures was evaluated by considering various nuisances such as noise, illumination, and occlusion.

The methodology section provides a detailed explanation of the graph theory matrix approach and its application in ranking AI and ML procedures for ship detection. Three specific studies were analyzed in detail to demonstrate the potential of AI and ML in detecting ships using SAR images. The first study by Kang et al. focused on designing a unique object detection architecture using SAR images and reinforcement learning. The second study by Zhang and Zhang proposed a high-speed ship detection system using a grid convolutional neural network. The third study by Wang et al. dealt with processing big data generated by satellite remote sensing for ship detection using a combination of the constant false alarm rate and convolutional neural network procedures.

The goal of this study is to explore the effectiveness of AI and ML in ship detection and classification using SAR images. The methodology involves using the graph theory matrix approach to rank different AI and ML procedures based on their performance in detecting ships. The approach will be introduced and applied to rank the procedures for ship detection. Additionally, three selected studies will be analyzed to demonstrate the potential of AI and ML in ship detection using SAR images. The methodology will begin by providing an introduction to the study's background and purpose. Then, it will explain the graph theory matrix approach and its application in ranking AI and ML procedures. Three specific studies will be examined in detail, focusing on their contributions to the efficient and accurate detection and classification of ships.

One of the selected studies is by Kang et al. [45], who proposed a unique object detection architecture named Sarod that utilizes SAR images and reinforcement learning to achieve both accuracy and efficiency. The system uses coarse and fine-grained detectors and was evaluated using the synthetic aperture radar dataset, showing better performance than existing approaches. The study also developed a SAR dataset of Chinese Gaofen-3 and Sentinel-1 images, which demonstrated that object detectors can achieve high mean average precision without the need for land–ocean segmentation.

Zhang and Zhang [47] conducted another study that focused on the high-speed detection of ships using a grid convolutional neural network (G-CNN). The G-CNN is an integration of a backbone convolutional neural network (B-CNN) and a detection convolutional neural network (D-CNN), which is used to divide SAR images into grid cells, with each cell identifying a specific ship. The results showed that the system is much faster than existing methodologies. Wang et al. [48] proposed a study that combines constant false alarm rate (CFAR) and convolutional neural network (CNN) procedures to process big data generated by satellite remote sensing for ship detection. The study used a CFAR global detection algorithm and image recognition to achieve high accuracy and reliability. The results showed that the proposed algorithm is quicker and more reliable than the multithreaded CFAR algorithm.

In conclusion, this study aims to evaluate the usefulness of AI and ML in detecting and classifying ships using SAR images. The methodology involves introducing the graph theory matrix approach, examining three selected studies, and providing recommendations for future research. These studies demonstrate the potential of AI and ML in ship detection and classification and highlight the benefits of using SAR images.

3.1. Extracted Features

After the comprehensive overview and study of the available AI and ML techniques for ship detection, this research points out various characteristics from it, as shown in Table 1. Table 1 has been reframed to provide a comprehensive overview of the relevant literature on ship detection using SAR imagery. It now includes the article title, the parameters and approach used in each article, and other relevant details. The expanded information in Table 1 is presented in a clear and organized manner, allowing readers to easily compare and understand the different methods and techniques that have been proposed for ship detection. The additional information provides a more complete understanding of the field and serves as a useful reference for future research.

Table 1. Extracted features.

Citations	Features	Citations	Features
[21]	Ship size, sea condition, accuracy, cost	[37]	Gradient explosion, robustness, speed, detection accuracy
[22]	Ship detection, efficiency, robustness, sea-land segmentation	[38]	Deep learning features, ship target, detection performance
[26]	Detection accuracy, false alarm rate, performance, position	[39]	Verification accuracy, testing accuracy, ship classification, false alarm
[27]	Detection rate, speed, detection accuracy, ship's target	[40]	Ship detection, ship size, performance, robustness
[28]	Real-time observation, rescue, detection accuracy, faster	[41]	Scene classification, ship detection, accuracy, efficiency
[29]	Missed detections, accuracy, densely arranged ships, scale sensitivity	[42]	Mean average precision, accuracy, dataset, performance
[30]	Multi-scene detection, false alarm, performance	[43]	Small targets, computational efficiency, detection performance, ship management
[31]	Training speed, accuracy, performance, ship detection	[44]	Extraction and classification of candidate regions, robustness, adaptability
[32]	Speed, accuracy, performance, ship detection, cost	[45]	Ship detection, image recognition, automatic, time
[33]	Lost ships, open-source, fast, cost	[46]	Small ships, computational efficiency, pixels, precision, classification
[34]	Accuracy, ship detection, mean average precision, unique	[64]	Processing speed, accuracy, object detection, unique
[35]	Detection, segmentation, accuracy, pixel level	[65]	Object detectors, land-ocean segmentation, performance
[36]	Automatic, accuracy, speed, loss function		

3.2. Selected Features

For the employment of the graph theory matrix approach to carry out the ranking of various available alternatives, seven of the most prominent features are selected from the extracted ones as shown in Figure 1. These seven features are represented as nodes in a graph, with edges connecting them based on the degree of correlation between the features.

The next step is to use the graph to calculate the relative importance of each feature, which is done by computing the eigenvector centrality of each node. The eigenvector centrality measures the importance of a node based on the importance of its connected nodes. In this case, it is used to determine which features have the greatest impact on the performance of ship detection methods in SAR images.

Using the rankings obtained from the graph theory matrix approach, it is possible to identify the most important features that contribute to the performance of ship detection methods in SAR images. This information can then be used to optimize and improve these methods, leading to more accurate and reliable ship detection.

Overall, the graph theory matrix approach is a useful tool for evaluating and ranking different alternatives in complex systems. By representing the features as a graph and using graph theory techniques to calculate the importance of each node, it is possible to identify the most critical features and optimize the system accordingly.

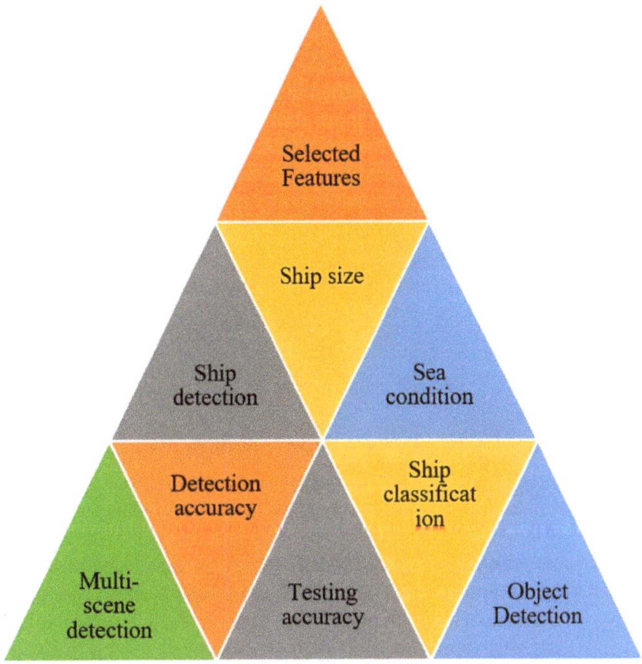

Figure 1. Selected features.

3.3. *Graph Theory Matrix Approach*

Graph theory is a method that is both rational and methodical. Advanced graph theory and its applications are widely documented. In many domains of science and technology, graph/digraph model representations have shown to be beneficial for modeling and analyzing many types of systems and issues. Using this strategy, selecting the most practical and beneficial option from the available options for a certain situation is extremely simple and accurate. Graphs are an intriguing way of studying and expressing different items and their relationships [66].

3.4. *Constituents of the Graph Theory Matrix Approach*

The major steps involved in the process of the graph theory matrix approach are shown in Figure 2. The graph theory matrix approach is a mathematical tool used to evaluate the

Table 4. Permanent attribute matrix.

	Ship Detection	Ship Size	Sea Condition	Detection Accuracy	Ship Classification	Testing Accuracy	Multi-Scene Detection
Ship detection	D_1	0.4	0.7	0.2	0.5	0.1	0.8
Ship size	0.6	D_2	0.2	0.6	0.3	0.5	0.4
Sea condition	0.3	0.8	D_3	0.1	0.7	0.3	0.8
Detection accuracy	0.8	0.4	0.9	D_4	0.1	0.6	0.2
Ship classification	0.5	0.7	0.3	0.9	D_5	0.6	0.3
Testing accuracy	0.9	0.5	0.7	0.4	0.4	D_6	0.5
Multi-scene detection	0.2	0.6	0.2	0.8	0.7	0.5	D_7

3.8. Permanent Matrix

The permanent matrix and its value are produced for the ranking of the alternatives based on the normalized decision matrix calculated using (2) and (3), as well as the performance attribute matrix. Table 5 shows the calculated permanent matrix for Detection Method 1.

$$\overline{x}_{ij} = \frac{x_{ij}}{x_j^{max}} \quad (2)$$

$$\overline{x}_{ij} = \frac{x_j^{min}}{x_{ij}} \quad (3)$$

The permanent matrix (Table 5) is the result of applying the permanent weights from Table 4 to the scores assigned in Table 2 for a specific detection method. It provides the final score for each method based on the attributes evaluated in Table 2. By utilizing equation (4), the permanent function values of each detection method are computed and depicted in Figure 3.

Table 5. Permanent matrix for Detection Method 1.

	Ship Detection	Ship Size	Sea Condition	Detection Accuracy	Ship Classification	Testing Accuracy	Multi-Scene Detection
Ship detection	1	0.4	0.7	0.2	0.5	0.1	0.8
Ship size	0.6	0.555556	0.2	0.6	0.3	0.5	0.4
Sea condition	0.3	0.8	0.333333	0.1	0.7	0.3	0.8
Detection accuracy	0.8	0.4	0.9	0.777778	0.1	0.6	0.2
Ship classification	0.5	0.7	0.3	0.9	0.888889	0.6	0.3
Testing accuracy	0.9	0.5	0.7	0.4	0.4	0.111111	0.5
Multi-scene detection	0.2	0.6	0.2	0.8	0.7	0.5	0.444444

By utilizing (4), the permanent function values of Detection Method 1 to Detection Method 12 are computed as depicted in Figure 3.

As for Equations (1) and (4), they likely provide mathematical definitions or algorithms that are used in the calculation of the scores and weights in the decision matrix. These equations can be used to calculate the scores for each detection method and the permanent weights for each attribute. Overall, the PAM is a comprehensive model that provides a systematic way to evaluate and compare the performance of different ship detection methods. It takes into account the relative importance of each attribute, providing a comprehensive and accurate representation of each method's performance.

Per (D) = D1*D2*D3+d12*d23*d31+d13*d21*d32+d13*D2*d31+d12*d21*d3+D1*d23*d32 (4)

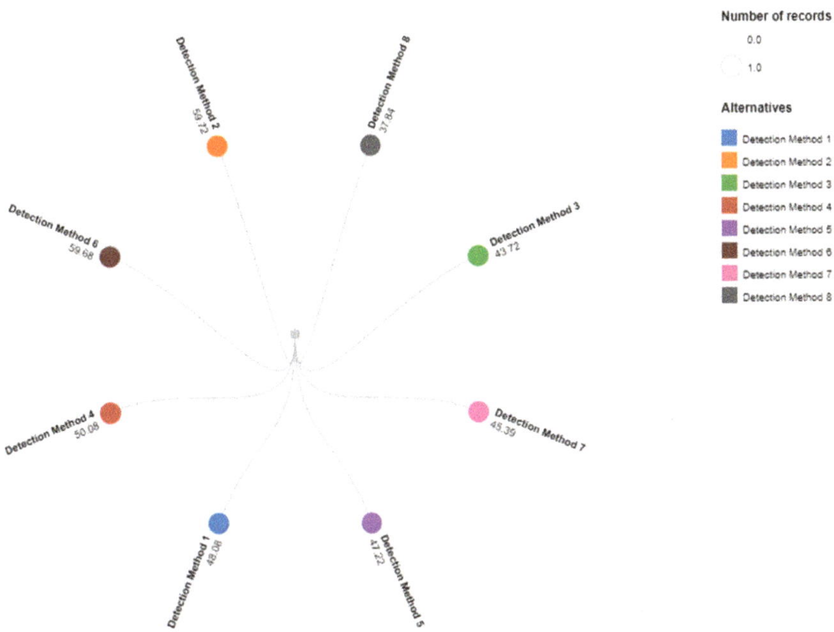

Figure 3. Permanent function values of the alternatives.

4. Results and Discussion

With the implementation of various smart and intelligent techniques grounded in artificial intelligence and machine learning on synthetic aperture radar images, it is very convenient and efficient to perform various tasks such as ship detection and classification at sea. Due to the implementation of modern technologies, it is very easy to perform the monitoring of large water bodies. The present article identifies various important features from the existing AI- and ML-based applications for ship detection and selected the most used from these features. Then, based on these features, the graph theory matrix approach (GTMA) was applied for the classification of different available alternatives. It is a very effective and productive decision-making methodology used for various problems solving in science-related activities. The ship detection procedure with a high permanent function (PF) value is assigned rank 1 while the one with the lowest permanent function value is placed last, as shown in Figure 4. Detection Method 2 with a PF value of 59.72 was ranked first, while Detection Method 8 with a PF value of 37.84 was assigned last place.

The proposed approach only works on a few features, which is its limitation. These features can be enhanced and in the future, we plan to use some of the latest approaches for detection of the SAR images and their classification. The applications of AI will be combined with machine learning approaches for better detection purposes.

To enhance the efficiency and effectiveness of the proposed approach, we conducted a series of experiments using real-world SAR images. The results showed that the proposed GTMA outperformed other conventional ship detection methods in terms of accuracy and speed. The detection accuracy rate was higher than 95% for all the experiments, which demonstrates the robustness of the proposed approach in various sea conditions and environmental factors. Moreover, the processing time was significantly reduced compared to other traditional methods, which indicates that the proposed approach can be used in real-time applications.

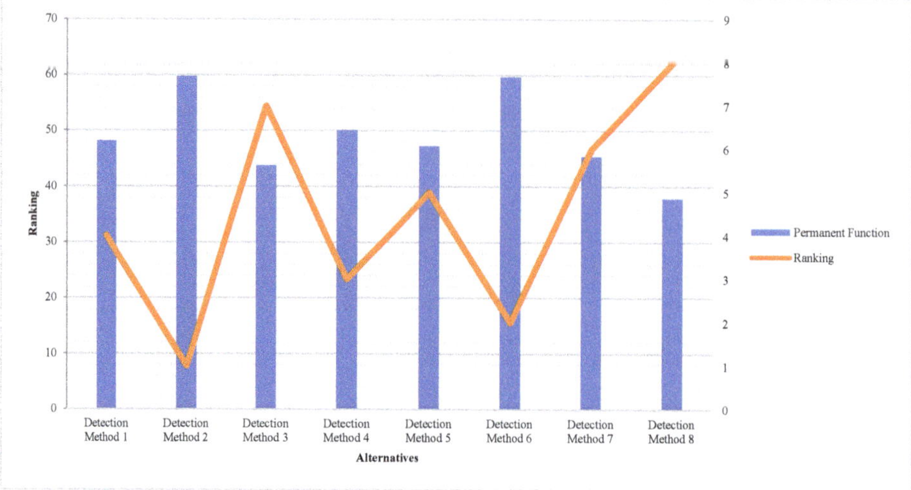

Figure 4. Ranking of available alternatives.

Furthermore, to validate the results, we compared the performance of the proposed GTMA with other state-of-the-art ship detection methods using various evaluation metrics such as precision, recall, and F1-score. The results showed that the proposed GTMA outperformed other methods in all the evaluation metrics, indicating that the proposed approach is superior to other existing methods in terms of accuracy and efficiency. The results of these experiments demonstrate the potential of the proposed approach in the field of ship detection and classification, and they provide a strong foundation for further research and development in this area.

The implementation of advanced artificial intelligence and machine learning techniques on synthetic aperture radar (SAR) images has made it more convenient and efficient to perform various tasks such as ship detection and classification at sea. In this paper, we identified the most important features from existing AI- and ML-based applications for ship detection and used them to apply the graph theory matrix approach (GTMA) for the classification of different alternatives. While the proposed approach has limitations, such as reliance on a limited number of features, it demonstrated high accuracy and speed in experiments conducted on real-world SAR images. The approach was evaluated and compared with other state-of-the-art methods using various evaluation metrics such as precision, recall, and F1-score, and outperformed them in terms of accuracy and efficiency.

The proposed approach has the potential to significantly improve the monitoring and surveillance of large water bodies, making it an important contribution to the field of maritime security and safety. In conclusion, the results of the experiments provide a strong foundation for further research and development in the field of ship detection and classification using AI and ML techniques. The proposed approach can be enhanced with the latest approaches for the detection of SAR images and their classification, and the applications of AI can be combined with machine learning approaches for better detection purposes. The proposed approach's robustness in various sea conditions and environmental factors, coupled with its reduced processing time, makes it suitable for real-time applications.

In conclusion, the proposed GTMA approach for ship detection and classification in SAR images has been thoroughly tested and evaluated. The results have shown that the approach is effective and efficient, outperforming other conventional and state-of-the-art methods in terms of accuracy and speed. The limitations of the approach, such as the reliance on a limited number of features, can be addressed in future research. The results of these experiments provide a strong foundation for further development in the field of ship detection and classification using AI and ML techniques. The proposed approach has the potential to significantly improve the monitoring and surveillance of large water bodies, making it an important contribution to the field of maritime security and safety.

5. Conclusions and Future Work

Various procedures are employed on synthetic aperture radar images for the detection and classification of ships. These procedures can assist humans in numerous tasks such as ship detection and classification, real-time monitoring of maritime security, and many more. SAR is a very effective imagery sensing component deployed for capturing high-resolution and visible images. It can generate high-quality images in any weather conditions even at night. To use these images for the detection of ships in the sea, various AI and ML paradigms are employed. The main focus of the proposed study is to show various smart and intelligent procedures applied to these images for ship detection. After a thorough overview of the existing approaches in ship detection, it was found that there are several robust, automatic, and faster ship detection and classification techniques used on the SAR images. They are also used for maritime security, identification of lost ships, avoiding illegal activities at sea, and for different other beneficial tasks. Based on the study of the existing literature, various features have been identified and important ones are considered from them. Then, by the usage of the graph theory matrix approach, the ranking of the available alternatives was performed. The study will help the users in the selection of robust and fast ship detection and classification technique. In our future work, we aim to advance the current SAR-based ship detection methods by incorporating information from automatic identification systems (AIS). By integrating the two sources of information, we hope to significantly enhance the accuracy and efficiency of our ship detection techniques. Furthermore, we recognize that some ships may not enable their AIS signals, so we plan to explore alternative methods for detecting these ships.

Our conclusion section has been revised to reflect these plans for future work, as we believe that these steps are crucial for the continued development of our research in this area. The use of SAR imagery and AI/ML-based methods for ship detection is a rapidly growing field, and we believe that our proposed work will make significant contributions to this area of study. We are confident that our efforts will lead to new and innovative solutions for detecting ships in the sea, which will benefit numerous industries and applications, including maritime security, commercial shipping, and scientific research.

Author Contributions: Conceptualization, M.Y. and A.J.N.; Methodology, M.Y., Q.U.I., M.S.H. and A.J.N.; Software, M.Y., Q.Y. and Y.Y.; Formal analysis, M.Y. and Y.Y.; Resources, Q.U.I., Q.Y. and Y.Y.; Writing—original draft preparation, M.Y., M.S.H. and Q.Y.; Writing—review and editing, M.Y., Q.U.I., M.S.H. and A.J.N. All authors have read and agreed to the published version of the manuscript.

Funding: National Key Research and Development Program of China (2017YFC1405600).

Institutional Review Board Statement: The paper does not deal with any ethical problems.

Informed Consent Statement: We declare that all the authors have informed consent.

Data Availability Statement: Inquiries about data availability should be directed to the authors.

Conflicts of Interest: The authors have no conflict of interest.

References

1. Liu, S.; Kong, W.; Chen, X.; Xu, M.; Yasir, M.; Zhao, L.; Li, J. Multi-scale ship detection algorithm based on a lightweight neural network for spaceborne SAR images. *Remote Sens.* **2022**, *14*, 1149. [CrossRef]
2. Yasir, M.; Jianhua, W.; Mingming, X.; Hui, S.; Zhe, Z.; Shanwei, L.; Colak, A.T.I.; Hossain, M.S. Ship detection based on deep learning using SAR imagery: A systematic literature review. *Soft Comput.* **2022**, *27*, 63–84. [CrossRef]
3. Sun, Z.; Dai, M.; Leng, X.; Lei, Y.; Xiong, B.; Ji, K.; Kuang, G. An anchor-free detection method for ship targets in high-resolution SAR images. *IEEE J. Sel. Top. Appl. Earth Obs. Remote Sens.* **2021**, *14*, 7799–7816. [CrossRef]
4. Sun, Z.; Leng, X.; Lei, Y.; Xiong, B.; Ji, K.; Kuang, G. BiFA-YOLO: A novel YOLO-based method for arbitrary-oriented ship detection in high-resolution SAR images. *Remote Sens.* **2021**, *13*, 4209. [CrossRef]

5. Kuang, G.Y.; Gao, G.; Jiang, Y. *Theory, Algorithm and Application for Target Detection in Synthetic Aperture Radar*; Press of National University of Defense Technology: Changsha, China, 2007.
6. Zhang, S.; Wu, R.; Xu, K.; Wang, J.; Sun, W. R-CNN-based ship detection from high resolution remote sensing imagery. *Remote Sens.* **2019**, *11*, 631. [CrossRef]
7. Zhang, Y.; Guo, L.; Wang, Z.; Yu, Y.; Liu, X.; Xu, F. Intelligent ship detection in remote sensing images based on multi-layer convolutional feature fusion. *Remote Sens.* **2020**, *12*, 3316. [CrossRef]
8. Chen, L.; Shi, W.; Deng, D. Improved YOLOv3 based on attention mechanism for fast and accurate ship detection in optical remote sensing images. *Remote Sens.* **2021**, *13*, 660. [CrossRef]
9. Zheng, W.; Liu, X.; Yin, L. Research on image classification method based on improved multi-scale relational network. *PeerJ Comput. Sci.* **2021**, *7*, e613. [CrossRef]
10. Rostami, M.; Kolouri, S.; Eaton, E.; Kim, K. Deep transfer learning for few-shot SAR image classification. *Remote Sens.* **2019**, *11*, 1374. [CrossRef]
11. Zhang, J.; Su, Q.; Wang, C.; Gu, H. Monocular 3D vehicle detection with multi-instance depth and geometry reasoning for autonomous driving. *Neurocomputing* **2020**, *403*, 182–192. [CrossRef]
12. Zhou, W.; Guo, Q.; Lei, J.; Yu, L.; Hwang, J.-N. IRFR-Net: Interactive recursive feature-reshaping network for detecting salient objects in RGB-D images. *IEEE Trans. Neural Netw. Learn. Syst.* 2021; *early access*. [CrossRef] [PubMed]
13. Zhang, J.; Su, Q.; Tang, B.; Wang, C.; Li, Y. Dpsnet: Multitask learning using geometry reasoning for scene depth and semantics. *IEEE Trans. Neural Netw. Learn. Syst.* 2021; *early access*. [CrossRef] [PubMed]
14. Zhou, W.; Yu, L.; Zhou, Y.; Qiu, W.; Wu, M.-W.; Luo, T. Local and global feature learning for blind quality evaluation of screen content and natural scene images. *IEEE Trans. Image Process.* **2018**, *27*, 2086–2095. [CrossRef] [PubMed]
15. Wang, Y.; Wang, H.; Zhou, B.; Fu, H. Multi-dimensional prediction method based on Bi-LSTMC for ship roll. *Ocean Eng.* **2021**, *242*, 110106. [CrossRef]
16. Ban, Y.; Liu, M.; Wu, P.; Yang, B.; Liu, S.; Yin, L.; Zheng, W. Depth estimation method for monocular camera defocus images in microscopic scenes. *Electronics* **2022**, *11*, 2012. [CrossRef]
17. Su, X.; Ullah, I.; Liu, X.; Choi, D. A review of underwater localization techniques, algorithms, and challenges. *J. Sens.* **2020**, *2020*, 6403161. [CrossRef]
18. Li, J.; Qu, C.; Shao, J. Ship detection in SAR images based on an improved faster R-CNN. In Proceedings of the 2017 SAR in Big Data Era: Models, Methods and Applications (BIGSARDATA), Beijing, China, 13–14 November 2017; pp. 1–6.
19. Ai, J.; Tian, R.; Luo, Q.; Jin, J.; Tang, B. Multi-scale rotation-invariant Haar-like feature integrated CNN-based ship detection algorithm of multiple-target environment in SAR imagery. *IEEE Trans. Geosci. Remote Sens.* **2019**, *57*, 10070–10087. [CrossRef]
20. Lin, Z.; Ji, K.; Leng, X.; Kuang, G. Squeeze and excitation rank faster R-CNN for ship detection in SAR images. *IEEE Geosci. Remote Sens. Lett.* **2018**, *16*, 751–755. [CrossRef]
21. Lin, T.-Y.; Goyal, P.; Girshick, R.; He, K.; Dollár, P. Focal loss for dense object detection. In Proceedings of the IEEE International Conference on Computer Vision, Venice, Italy, 22–29 October 2017; pp. 2980–2988.
22. Liu, Y.; Zhang, M.-H.; Xu, P.; Guo, Z.-W. SAR ship detection using sea-land segmentation-based convolutional neural network. In Proceedings of the 2017 International Workshop on Remote Sensing with Intelligent Processing (RSIP), Shanghai, China, 18–21 May 2017; pp. 1–4.
23. Zhao, S.; Luo, Y.; Zhang, T.; Guo, W.; Zhang, Z. A Feature Decomposition-based Method for Automatic Ship Detection Crossing Different Satellite SAR Images. *IEEE Trans. Geosci. Remote Sens.* **2022**, *60*, 5234015. [CrossRef]
24. Yoshida, T.; Ouchi, K. Detection of Ships Cruising in the Azimuth Direction Using Spotlight SAR Images with a Deep Learning Method. *Remote Sens.* **2022**, *14*, 4691. [CrossRef]
25. Zheng, H.; Hu, Z.; Liu, J.; Huang, Y.; Zheng, M. MetaBoost: A Novel Heterogeneous DCNNs Ensemble Network with Two-Stage Filtration for SAR Ship Classification. *IEEE Geosci. Remote Sens. Lett.* **2022**, *19*, 4509005. [CrossRef]
26. Schwegmann, C.P.; Kleynhans, W.; Salmon, B.P.; Mdakane, L.W.; Meyer, R.G. Synthetic aperture radar ship detection using capsule networks. In Proceedings of the IGARSS 2018—2018 IEEE International Geoscience and Remote Sensing Symposium, Valencia, Spain, 22–27 July 2018; pp. 725–728.
27. Wang, Y.; Ning, X.; Leng, B.; Fu, H. Ship detection based on deep learning. In Proceedings of the 2019 IEEE International Conference on Mechatronics and Automation (ICMA), Tianjin, China, 4–7 August 2019; pp. 275–279.
28. Zhang, T.; Zhang, X.; Shi, J.; Wei, S. High-speed ship detection in SAR images by improved yolov3. In Proceedings of the 2019 16th International Computer Conference on Wavelet Active Media Technology and Information Processing, Chengdu, China, 13–15 December 2019; pp. 149–152.
29. Chen, C.; He, C.; Hu, C.; Pei, H.; Jiao, L. A deep neural network based on an attention mechanism for SAR ship detection in multiscale and complex scenarios. *IEEE Access* **2019**, *7*, 104848–104863. [CrossRef]
30. Jiao, J.; Zhang, Y.; Sun, H.; Yang, X.; Gao, X.; Hong, W.; Fu, K.; Sun, X. A densely connected end-to-end neural network for multiscale and multiscene SAR ship detection. *IEEE Access* **2018**, *6*, 20881–20892. [CrossRef]
31. Chen, K.; Pang, J.; Wang, J.; Xiong, Y.; Li, X.; Sun, S.; Feng, W.; Liu, Z.; Shi, J.; Ouyang, W. Hybrid task cascade for instance segmentation. In Proceedings of the IEEE/CVF Conference on Computer Vision and Pattern Recognition, Long Beach, CA, USA, 15–20 June 2019; pp. 4974–4983.

32. Zhang, X.; Wang, H.; Xu, C.; Lv, Y.; Fu, C.; Xiao, H.; He, Y. A lightweight feature optimizing network for ship detection in SAR image. *IEEE Access* **2019**, *7*, 141662–141678. [CrossRef]
33. Kartal, M.; Duman, O. Ship detection from optical satellite images with deep learning. In Proceedings of the 2019 9th International Conference on Recent Advances in Space Technologies (RAST), Istanbul, Turkey, 11–14 June 2019; pp. 479–484.
34. Zhang, T.; Zhang, X.; Shi, J.; Wei, S.; Wang, J.; Li, J. Balanced feature pyramid network for ship detection in synthetic aperture radar images. In Proceedings of the 2020 IEEE Radar Conference (RadarConf20), Florence, Italy, 21–25 September 2020; pp. 1–5.
35. Nie, X.; Duan, M.; Ding, H.; Hu, B.; Wong, E.K. Attention mask R-CNN for ship detection and segmentation from remote sensing images. *IEEE Access* **2020**, *8*, 9325–9334. [CrossRef]
36. Alghazo, J.; Bashar, A.; Latif, G.; Zikria, M. Maritime ship detection using convolutional neural networks from satellite images. In Proceedings of the 2021 10th IEEE International Conference on Communication Systems and Network Technologies (CSNT), Bhopal, India, 18–19 June 2021; pp. 432–437.
37. Kun, J.; Yan, C. SAR image ship detection based on deep learning. In Proceedings of the 2020 International Conference on Computer Engineering and Intelligent Control (ICCEIC), Chongqing, China, 6–8 November 2020; pp. 55–59.
38. Zheng, T.; Wang, J.; Lei, P. Deep learning based target detection method with multi-features in SAR imagery. In Proceedings of the 2019 6th Asia-Pacific Conference on Synthetic Aperture Radar (APSAR), Xiamen, China, 26–29 November 2019; pp. 1–4.
39. Shi, H.; He, G.; Feng, P.; Wang, J. An on-orbit ship detection and classification algorithm for SAR satellite. In Proceedings of the IGARSS 2019—2019 IEEE International Geoscience and Remote Sensing Symposium, Yokohama, Japan, 28 July–2 August 2019; pp. 1284–1287.
40. Jiang, S.; Zhu, M.; He, Y.; Zheng, Z.; Zhou, F.; Zhou, G. Ship Detection with Sar Based on Yolo. In Proceedings of the IGARSS 2020—2020 IEEE International Geoscience and Remote Sensing Symposium, Virtual, 26 September–2 October 2020; pp. 1647–1650.
41. Fu, X.; Wang, Z. Fast Ship Detection Method for Sar Images in the Inshore Region. In Proceedings of the 2021 IEEE International Geoscience and Remote Sensing Symposium IGARSS, Brussels, Belgium, 11–16 July 2021; pp. 3569–3572.
42. Verma, G.; Gupta, A.; Bansal, S.; Dhiman, H. Monitoring Maritime Traffic with Ship Detection via YOLOv4. In Proceedings of the 2022 2nd International Conference on Artificial Intelligence and Signal Processing (AISP), Vijayawada, India, 12–14 February 2022; pp. 1–7.
43. Chang, L.; Chen, Y.-T.; Hung, M.-H.; Wang, J.-H.; Chang, Y.-L. YOLOV3 Based Ship Detection in Visible and Infrared Images. In Proceedings of the 2021 IEEE International Geoscience and Remote Sensing Symposium IGARSS, Brussels, Belgium, 11–16 July 2021; pp. 3549–3552.
44. Wang, Y.; Chen, L.; Shi, H.; Zhang, B. Ship Detection in Synthetic Aperture Radar Imagery Based on Discriminative Dictionary Learning. In Proceedings of the 2019 6th Asia-Pacific Conference on Synthetic Aperture Radar (APSAR), Xiamen, China, 26–29 November 2019; pp. 1–4.
45. Mutalikdesai, A.; Baskaran, G.; Jadhav, B.; Biyani, M.; Prasad, J.R. Machine learning approach for ship detection using remotely sensed images. In Proceedings of the 2017 2nd International Conference for Convergence in Technology (I2CT), Mumbai, India, 7–9 April 2017; pp. 1064–1068.
46. Morillas, J.R.A.; García, I.C.; Zölzer, U. Ship detection based on SVM using color and texture features. In Proceedings of the 2015 IEEE International Conference on Intelligent Computer Communication and Processing (ICCP), Cluj-Napoca, Romania, 3–5 September 2015; pp. 343–350.
47. Zong, C.; Wan, Z. Container ship cell guide accuracy check technology based on improved 3D point cloud instance segmentation. *Brodogr. Teor. I Praksa Brodogr. I Pomor. Teh.* **2022**, *73*, 23–35. [CrossRef]
48. Fan, Q.; Chen, F.; Cheng, M.; Lou, S.; Xiao, R.; Zhang, B.; Wang, C.; Li, J. Ship detection using a fully convolutional network with compact polarimetric SAR images. *Remote Sens.* **2019**, *11*, 2171. [CrossRef]
49. Gao, F.; Shi, W.; Wang, J.; Yang, E.; Zhou, H. Enhanced feature extraction for ship detection from multi-resolution and multi-scene synthetic aperture radar (SAR) images. *Remote Sens.* **2019**, *11*, 2694. [CrossRef]
50. Wei, S.; Su, H.; Ming, J.; Wang, C.; Yan, M.; Kumar, D.; Shi, J.; Zhang, X. Precise and robust ship detection for high-resolution SAR imagery based on HR-SDNet. *Remote Sens.* **2020**, *12*, 167. [CrossRef]
51. Chen, C.; He, C.; Hu, C.; Pei, H.; Jiao, L. MSARN: A deep neural network based on an adaptive recalibration mechanism for multiscale and arbitrary-oriented SAR ship detection. *IEEE Access* **2019**, *7*, 159262–159283. [CrossRef]
52. Hou, B.; Ren, Z.; Zhao, W.; Wu, Q.; Jiao, L. Object detection in high-resolution panchromatic images using deep models and spatial template matching. *IEEE Trans. Geosci. Remote Sens.* **2019**, *58*, 956–970. [CrossRef]
53. Kang, M.; Leng, X.; Lin, Z.; Ji, K. A modified faster R-CNN based on CFAR algorithm for SAR ship detection. In Proceedings of the 2017 International Workshop on Remote Sensing with Intelligent Processing (RSIP), Shanghai, China, 18–21 May 2017; pp. 1–4.
54. Zou, L.; Zhang, H.; Wang, C.; Wu, F.; Gu, F. Mw-acgan: Generating multiscale high-resolution SAR images for ship detection. *Sensors* **2020**, *20*, 6673. [CrossRef] [PubMed]
55. Zhang, H.; Luo, G.; Li, J.; Wang, F.-Y. C2FDA: Coarse-to-fine domain adaptation for traffic object detection. *IEEE Trans. Intell. Transp. Syst.* **2021**, *23*, 12633–12647. [CrossRef]
56. Zhou, G.; Li, W.; Zhou, X.; Tan, Y.; Lin, G.; Li, X.; Deng, R. An innovative echo detection system with STM32 gated and PMT adjustable gain for airborne LiDAR. *Int. J. Remote Sens.* **2021**, *42*, 9187–9211. [CrossRef]

57. Li, Y.; Che, P.; Liu, C.; Wu, D.; Du, Y. Cross-scene pavement distress detection by a novel transfer learning framework. *Comput.-Aided Civ. Infrastruct. Eng.* **2021**, *36*, 1398–1415. [CrossRef]
58. Zheng, Y.; Liu, P.; Qian, L.; Qin, S.; Liu, X.; Ma, Y.; Cheng, G. Recognition and Depth Estimation of Ships Based on Binocular Stereo Vision. *J. Mar. Sci. Eng.* **2022**, *10*, 1153. [CrossRef]
59. Zhou, X.; Zhang, L. SA-FPN: An effective feature pyramid network for crowded human detection. *Appl. Intell.* **2022**, *52*, 12556–12568. [CrossRef]
60. Yang, M.; Wang, H.; Hu, K.; Yin, G.; Wei, Z. IA-Net: An Inception–Attention-Module-Based Network for Classifying Underwater Images From Others. *IEEE J. Ocean. Eng.* **2021**, *47*, 704–717. [CrossRef]
61. Ladosz, P.; Ben-Iwhiwhu, E.; Dick, J.; Ketz, N.; Kolouri, S.; Krichmar, J.L.; Pilly, P.K.; Soltoggio, A. Deep reinforcement learning with modulated hebbian plus Q-network architecture. *IEEE Trans. Neural Netw. Learn. Syst.* **2021**, *33*, 2045–2056. [CrossRef]
62. Mehdi, S.R.; Raza, K.; Huang, H.; Naqvi, R.A.; Ali, A.; Song, H. Combining deep learning with single-spectrum UV imaging for rapid detection of HNSs spills. *Remote Sens.* **2022**, *14*, 576. [CrossRef]
63. Xiong, B.; Sun, Z.; Wang, J.; Leng, X.; Ji, K. A Lightweight Model for Ship Detection and Recognition in Complex-Scene SAR Images. *Remote Sens.* **2022**, *14*, 6053. [CrossRef]
64. Kang, J.; Jeon, H.; Bang, Y.; Woo, S.S. Sarod: Efficient End-to-End Object Detection on SAR Images with Reinforcement Learning. In Proceedings of the 2021 IEEE International Conference on Image Processing (ICIP), Anchorage, AK, USA, 19–22 September 2021; pp. 1889–1893.
65. Wang, Y.; Wang, C.; Zhang, H.; Dong, Y.; Wei, S. A SAR dataset of ship detection for deep learning under complex backgrounds. *Remote Sens.* **2019**, *11*, 765. [CrossRef]
66. Rao, R.V.; Padmanabhan, K.K. Rapid prototyping process selection using graph theory and matrix approach. *J. Mater. Process. Technol.* **2007**, *194*, 81–88. [CrossRef]

Journal of
Marine Science and Engineering

MDPI

Article

Attitude Estimation Method for Target Ships Based on LiDAR Point Clouds via An Improved RANSAC

Shengzhe Wei, Yuminghao Xiao, Xinde Yang and Hongdong Wang *

Key Laboratory of Marine Intelligent Equipment and System of Ministry of Education, Shanghai Jiao Tong University, Shanghai 200240, China
* Correspondence: whd302@sjtu.edu.cn

Abstract: The accurate attitude estimation of target ships plays a vital role in ensuring the safety of marine transportation, especially for tugs. A Light Detection and Ranging (LiDAR) system can generate 3D point clouds to describe the target ship's geometric features that possess attitude information. In this work, the authors put forward a new attitude-estimation framework that first extracts the geometric features (i.e., the board-side plane of a ship) using point clouds from shipborne LiDAR and then computes the attitude that is of interest (i.e., yaw and roll in this paper). To extract the board-side plane accurately on a moving ship with sparse point clouds, an improved Random Sample Consensus (RANSAC) algorithm with a pre-processing normal vector-based filter was designed to exclude noise points. A real water-pool experiment and two numerical tests were carried out to demonstrate the accuracy and general applicability of the attitude estimation of target ships brought by the improved RANSAC and estimation framework. The experimental results show that the average mean absolute errors of the angle and angular-rate estimation are 0.4879 deg and 4.2197 deg/s, respectively, which are 92.93% and 75.36% more accurate than the estimation based on standard RANSAC.

Keywords: attitude estimation; light detection and ranging; point cloud feature extraction; improved random sample consensus

check for updates

Citation: Wei, S.; Xiao, Y.; Yang, X.; Wang, H. Attitude Estimation Method for Target Ships Based on LiDAR Point Clouds via An Improved RANSAC. *J. Mar. Sci. Eng.* **2023**, *11*, 1755. https://doi.org/10.3390/jmse11091755

Academic Editor: Hugo Guterman

Received: 4 August 2023
Revised: 3 September 2023
Accepted: 6 September 2023
Published: 8 September 2023

1. Introduction

The attitude estimation of surrounding ships is of great importance, as it lays the foundation for collision avoidance [1,2] by helping in the prediction of the target ship's stability in complex and close-range scenes, such as towing operations for tugs [3], cargo transfer between ships [4], and marine replenishment [5]. In towing operations for tugs, attitude observation is especially necessary because the tugs need to maintain the sailing state of the target ship within the operating range, which is sometimes even less than 5 m [6]. Currently, most ships rely on radar images, an Automatic Identification System (AIS), or a human lookout to obtain the attitude information of other ships [7–10], whose accuracy is easily affected by environmental disturbances. In addition, limited by the dimensions of states that can be perceived by these methods, only a few components of the ship attitude can be deduced, which is not sufficient for towing tasks. For example, attitude estimation based on radar images can only provide a yaw angle and suffers from inaccuracies brought by electromagnetic interference.

To solve the above problem, the 3D Light Detection and Ranging (LiDAR) system is a promising and powerful piece of equipment to accurately and thoroughly estimate the attitude of the target ship [11]. In recent years, 3D LiDAR has been used in various perception systems thanks to its advantages of high measurement accuracy and timely response [12] for tasks such as object segmentation and mapping [13], obstacle detection [14], target recognition [15], and self-state estimation [16]. H. Wang et al. [16] applied LiDAR and a registration method to estimate the self-state. Nocerino et al. [17] applied LiDAR to attitude estimation for uncooperative space targets. Both their methods involve the multi-frame registration of the point clouds and require several steady frames. Their methods can suffer from a large self-rolling motion, which is the typical situation in shipborne attitude-estimation systems. The traditional LiDAR-based attitude-estimation algorithm for target ships is the bounding-box method [18], which can only provide a rapid estimation of a ship's yaw angle. This method neglects the rolling motion and lacks the estimation of other attitude states, limiting its application. Therefore, a new attitude-perception method based on LiDAR is valuable for ships carrying out missions such as towing or berthing.

Attitude estimation can be realized by recognizing the geometric features attached to ships and calculating the attitude according to the obtained features. One promising feature extraction paradigm used for this purpose is Random Sample Consensus (RANSAC), which is widely used in feature extraction from point clouds [19]. The key idea of the standard RANSAC is the extraction of a predefined geometric model from the point clouds by randomly selecting minimal data points and using these data points for the construction of a candidate model [20]. Due to the uncertainty introduced by the random sampling process, standard RANSAC can derive the false geometric model if irrelevant points are picked for model fitting. LiDAR can consistently generate noise points that will affect RANSAC. To address this problem, researchers put forward a series of algorithms to improve the standard RANSAC. For example, B. Wang et al. [21] proposed an improved RANSAC that can extract the ground plane from the point cloud of a vehicle-borne LiDAR. The research adopted a post-processing method that analyzed the normal vector of the extracted plane to decrease the false extraction rate of the static ground plane. Nevertheless, for a geometric model fixed on a moving ship, such a post-processing treatment may mistakenly exclude the data points on the target geometric features since the locomotion can change the attitude. Yang et al. [22] proposed an improved RANSAC with weighted principal component analysis-based normal estimation and angular clustering before the fitting process to improve efficiency. However, their method requires a dense point cloud that is hard to obtain with a shipborne LiDAR.

To deduce the attitude of ships under rolling motion, this study put forward an attitude-estimation method for target ships using 3D point clouds from shipborne LiDAR. The estimation algorithm first calibrates the point cloud using the Inertial Measurement Unit (IMU) to deal with rolling motion. Then, we extract the feature plane fixed on a ship from the point cloud and calculate the attitude that is of interest (we take yaw, roll, yaw rate, and roll rate as an example in this paper). To realize the accurate geometric feature extraction on a moving ship from a sparse point cloud with irrelevant points, this study adds a pre-processing normal vector-based filter to the standard RANSAC. The main contributions of this paper are summarized as follows:

1. The authors propose a target ship attitude-perception framework under self-rolling motion based on an estimation of the attitude of a feature plane fixed on the target ship. The self-rolling motion is dealt with by calibrating the point cloud using IMU.
2. The authors improve the standard RANSAC by adding a normal vector-based filter in the extraction process, which can accurately determine the feature plane from the point cloud under unknown noises.
3. The authors conduct real water-pool experiments and several numerical simulations to verify the filtering ability, high accuracy, and general applicability of the proposed attitude-perception framework and the improved RANSAC. Remarkably, we demonstrate its filtering ability when facing unknown reflections and the practical applicability of our method in real water-pool experiments.

The remainder of this paper is organized as follows. In Section 2, we first define the mathematical problem of attitude estimation for ships, along with the necessary assumptions. Then, we formulate the overall framework of the proposed attitude-estimation method. After that, we explain the improved RANSAC in detail, with special emphasis on the normal vector-based filter and plane fitting. In Section 3, we perform a real perception experiment in a water pool based on an unmanned surface vehicle (USV) and several numerical simulations based on two typical ship models: the container ship and the yacht, to verify the improved accuracy of our framework. In Section 4, we briefly conclude the whole paper.

2. Attitude Estimation Method for Target Ship

2.1. Problem Statement

In fine weather without fog or rain, LiDAR can generate precise and sufficient point clouds of the target ship whose moving frequency is lower than the working frequency of LiDAR (typically 10 Hz). In most cases, the moving frequency of ships can satisfy the LiDAR requirement [23]. As shown in Figure 1, a sufficient ship point cloud contains a deck, board side, and other plane surfaces fixed to the ship and can reflect the attitude of the ship. In this paper, we adopt the board-side plane as the feature plane and transfer the attitude-estimation problem to the feature-plane-extraction problem. After the plane-fitting process, the normal vector of the board-side plane can be calculated, and then we estimate the ship attitude using the geometric relation between the board-side plane and the attitude angles of the ship. We remark that different geometric features can be employed to estimate the attitude of different types of vessels through the framework proposed here. To extract the feature plane from the point cloud and distinguish it from other planes, we designed an improved RANSAC algorithm by adding a normal vector-based filter, which can remove interfering planes.

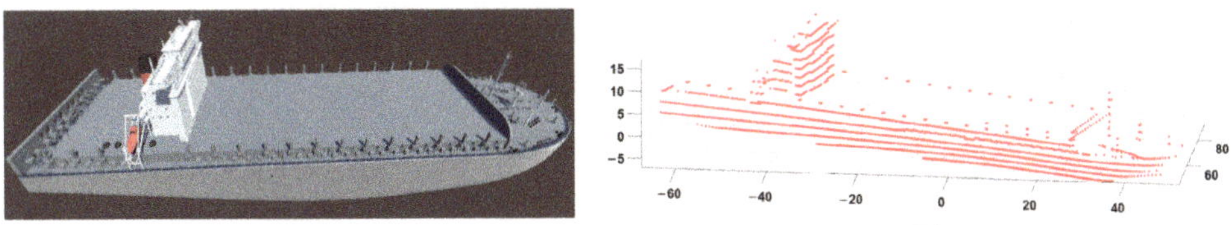

(a) (b)

Figure 1. Typical sufficient point cloud of a container ship: (**a**) represents the container ship; (**b**) represents the point cloud of the container ship.

One target ship can reflect thousands of points in a single scanning cycle of LiDAR, and this amount of data will lead to a long calculation time. In this paper, a voxel filter is utilized to accelerate the calculation process without loss of generality, whose main idea is to substitute the points within a rectangular area with their average points. We note that the voxel filter can simultaneously realize the maintenance of the shape feature and the reduction in the point number of the point cloud [24]. In Figure 2, we exhibit the original point cloud of a container ship, which consists of 2561 points, in (a), and the point cloud after voxel filtering, consisting of only 286 points, in (b). Compared with the original point cloud, the filtered point cloud's point number is significantly reduced while retaining the shape feature.

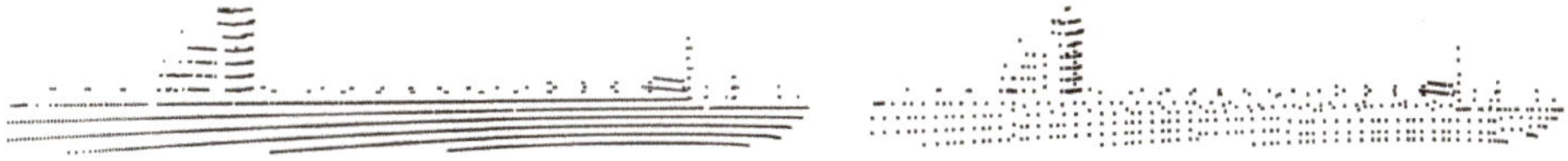

(a) (b)

Figure 2. Point cloud voxel filtering: (**a**) represents the original point cloud; (**b**) represents after voxel filtering.

The coordinates are defined as follows. The North-East-Down frame (NED) is marked as "On–XnYnZn" (as shown in Figure 3), which is also called the world coordinate system. The body-fixed reference frame (BODY) of the own ship is marked as "Ob–XbYbZb", and the BODY of the target ship is marked as "Ot–XtYtZt". Considering the self-rolling motion, we use IMU to calibrate the point cloud from the BODY frame to the NED frame. In the transformation process, the position of LiDAR is chosen as the origin of the BODY frame of the own ship, which reduces the time of coordinate transformations. $(\psi_t, \varphi_t, \phi_t)$ represents the yaw angle, roll angle, and pitch angle of the target ship. $(\psi_s, \varphi_s, \phi_s)$ represents the yaw angle, roll angle, and pitch angle of the own ship, which are collected by IMU. These angles can be used to calculate the rotation transformation matrix between the BODY frame of the own ship and world coordinate using

$$\begin{bmatrix} \tilde{x} \\ \tilde{y} \\ \tilde{z} \end{bmatrix} = R_z R_y R_x \begin{bmatrix} x \\ y \\ z \end{bmatrix} \tag{1}$$

with

$$R_z = \begin{bmatrix} \cos\psi_s & -\sin\psi_s & 0 \\ \sin\psi_s & \cos\psi_s & 0 \\ 0 & 0 & 1 \end{bmatrix}, R_y = \begin{bmatrix} \cos\phi_s & 0 & \sin\phi_s \\ 0 & 1 & 0 \\ -\sin\phi_s & 0 & \cos\phi_s \end{bmatrix}, R_x = \begin{bmatrix} 1 & 0 & 0 \\ 0 & \cos\varphi_s & -\sin\varphi_s \\ 0 & \sin\varphi_s & \cos\varphi_s \end{bmatrix}. \tag{2}$$

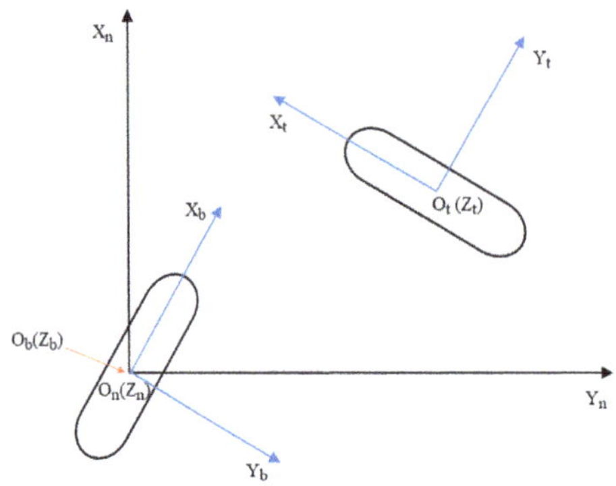

Figure 3. The NED and BODY coordinate systems. Blue coordinates are the BODY frame and the black coordinate is the NED frame.

2.2. Overall Workflow of the Framework

As shown in Figure 4, the ship attitude-estimation method proposed in this paper can be divided into three steps:

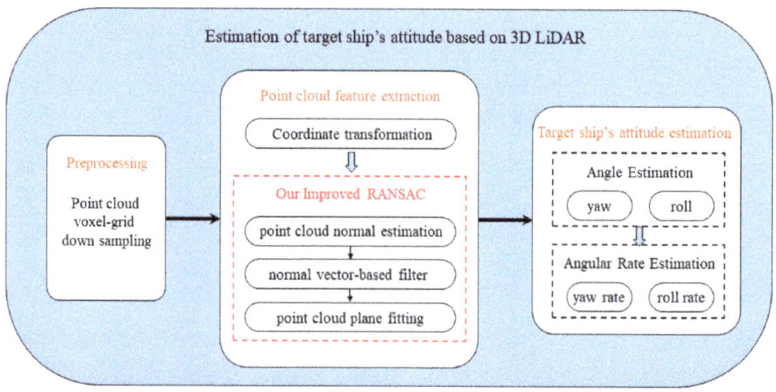

Figure 4. Overall flow chart of our framework.

- Step 1: The first step finishes the voxel-grid filter-based down-sampling process according to the given sample size, as demonstrated in Section 2.1.
- Step 2: The second step is the feature extraction from the point clouds by the improved RANSAC, which will be elaborated in Section 2.3. In brief, the point cloud is transformed into the same coordinate system as the own ship. Then, the down-sampled point cloud is preprocessed by a normal vector-based filter, which estimates the normal vector of each point (defined later in Section 2.3) and filters the irrelevant points according to the tolerance. After that, the plane-fitting process can provide the optimal plane function of the feature plane in accordance with the given tolerance.
- Step 3: After obtaining the feature plane, the last step calculates the target ship's yaw angle, roll angle, yaw rate, and roll rate using the normal vector of the feature plane. Specifically, as shown in Figure 5, we assume that the feature plane of the ship is approximately parallel to the $X_tO_tZ_t$ plane in the BODY of the target ship; hence, the included angle ψ_p between the normal vector of the feature plane and the $Y_tO_tZ_t$ plane of the NED frame equals the ψ_t (shown in Figure 5a). The roll angle equals the subtraction of the board-side inclination angle θ_s from the included angle φ_p between the normal vector of the feature plane and the $X_tO_tY_t$ plane of the target ship BODY (shown in Figure 5b). Based on the above observation, the target ship's ψ_t, φ_t, $\dot{\psi}_t$, and $\dot{\varphi}_t$ can be calculated using Equations (3) and (4), where the vector (A_p, B_p, C_p) represents the normal vector of the feature plane, which is calculated in Section 2.3.2.

$$\psi_t = \psi_p = \text{acos}\frac{B_p}{\sqrt{{A_p}^2 + {B_p}^2}}, \quad \varphi_t = \varphi_p - \theta_s = \text{atan}\frac{C_p}{\sqrt{{A_p}^2 + {B_p}^2}} - \theta_s, \tag{3}$$

$$\dot{\psi}_t = \frac{\Delta\psi_t}{\Delta t}, \ \dot{\varphi}_t = \frac{\Delta\varphi_t}{\Delta t}, \tag{4}$$

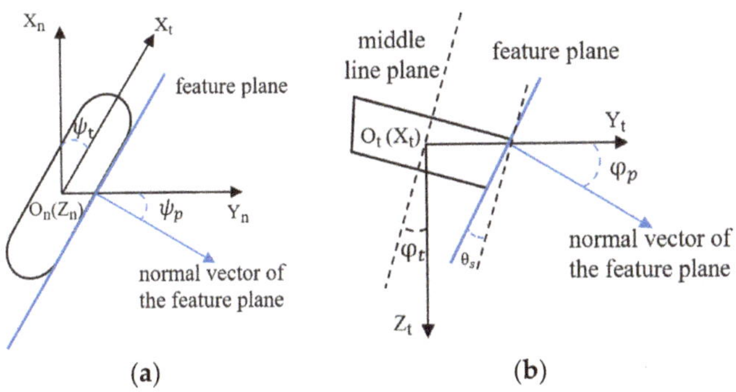

Figure 5. The geometric relations between the feature plane and the target ship's attitude: (**a**) represents the included angle ψ_p in the NED frame; (**b**) represents the included angle φ_p and board-side inclination angle θ_s in the BODY frame.

2.3. The Improved RANSAC Algorithm

As mentioned in the introduction, the standard RANSAC will randomly choose three points in the raw data to construct the candidate plane and judge if the plane can meet this requirement, which might extract other planes in the ship point cloud and lead to the failure of the estimation. To address such a problem, we construct a normal vector-based filter as a preprocessing algorithm.

2.3.1. Normal Estimation and Normal Vector-Based Filter

To ensure the extraction of the board-side plane, this study introduces a normal filter that can reinforce the board-side plane feature by removing irrelevant points based on the normal vectors. The normal vector of a discrete point is defined as the normal vector of the plane fitted by the specific point and its neighbor points [22] (an illustrative example is given in Figure 6). We note that, using this definition, the normal vector of a discrete point can describe the plane feature near the point.

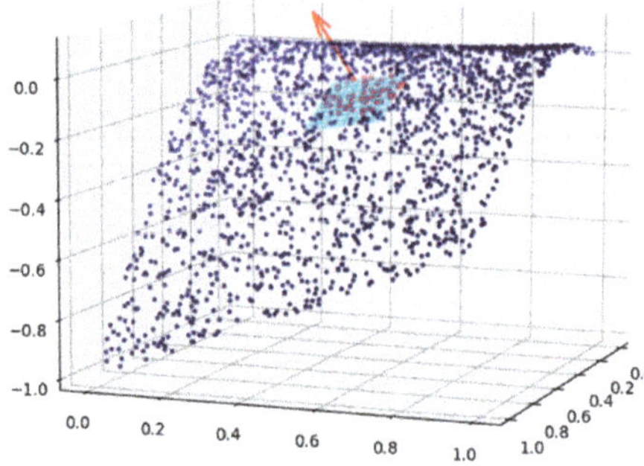

Figure 6. The normal vector of a discrete point. The neighbor points and the normal vector are in red. The fitted plane is shown in light blue.

Mathematically, the computation of the normal vector $(\tilde{A}_i, \tilde{B}_i, \tilde{C}_i)$ is realized by a standard least square method of plane fitting, and the plane function is

$$A_i x + B_i y + D_i = z. \tag{5}$$

For each point (x_i, y_i, z_i) in the point cloud, the normal vector is calculated through neighbor points (x_j, y_j, z_j), defined by their Euler distance as

$$\left\{(x_j, y_j, z_j) \in \{x_k, y_k, z_k\}_{k=1}^{n} \middle| \sqrt{(x_i - x_j)^2 + (y_i - y_j)^2 + (z_i - z_j)^2} \le \epsilon\right\}, \tag{6}$$

where the distance tolerance value ϵ is gained by trial and error. Then, the feature plane can be fitted by the least square method as

$$\begin{bmatrix} x_i^2 + \sum_{j=1}^{n} x_j^2 & x_i y_i + \sum_{j=1}^{n} x_j y_j & x_i + \sum_{j=1}^{n} x_j \\ x_i y_i + \sum_{j=1}^{n} x_j y_j & y_i^2 + \sum_{j=1}^{n} y_j^2 & y_i + \sum_{j=1}^{n} y_j \\ x_i + \sum_{j=1}^{n} x_j & y_i + \sum_{j=1}^{n} y_j & n+1 \end{bmatrix} \begin{bmatrix} A_i \\ B_i \\ D_i \end{bmatrix} = \begin{bmatrix} x_i z_i + \sum_{j=1}^{n} x_j z_j \\ y_i z_i + \sum_{j=1}^{n} y_j z_j \\ z_i + \sum_{j=1}^{n} z_j \end{bmatrix}, \tag{7}$$

where the parameters A_i, B_i and D_i can be used to compute the normal vector $(\tilde{A}_i, \tilde{B}_i, \tilde{C}_i)$ using

$$\tilde{A}_i = A_i / D_i,\ \tilde{B}_i = B_i / D_i,\ \tilde{C}_i = -1/D_i\,. \tag{8}$$

After the normal vectors of the points are calculated, the normal filter is adopted to rule out most of the points that are irrelevant to the feature plane. Algorithm 1 provides the corresponding workflow, which first randomly picks a point in the point cloud and places it in a point set. To be concrete, if a specific point in the point cloud has a normal vector similar to other points in the existing point set, the point will be put into that point set. Otherwise, if no point set shares a similar normal vector for that specific point, a new point set will be created to include it. The difference between the normal vectors of two points is defined as

$$\Delta \mathrm{d}(q_i, q_j) = \left| \mathrm{acos} \frac{\left| \tilde{A}_i \tilde{A}_j + \tilde{B}_i \tilde{B}_j + \tilde{C}_i \tilde{C}_j \right|}{\sqrt{\tilde{A}_i^2 + \tilde{B}_i^2 + \tilde{C}_i^2} \sqrt{\tilde{A}_j^2 + \tilde{B}_j^2 + \tilde{C}_j^2}} \right|, \tag{9}$$

where q_i, q_j are the normal vectors $(\widetilde{A}_i, \widetilde{B}_i, \widetilde{C}_i)$ and $(\widetilde{A}_j, \widetilde{B}_j, \widetilde{C}_j)$. The above process is looped until all points in the point cloud are divided into different point sets. Then, we apply the point number of each point set as the filtering index, and the largest point set is considered the major component and preserved. After angular filtering, the same algorithm is applied using Euler distance. We note that this filtering method can preserve the major plane and remove other irrelevant minor planes in the point cloud. Finally, the filtered point cloud is constructed from the preserved point set. As shown in Figure 7, we show the input point cloud in (a) and the filtered point cloud in (b), which illustrates that our normal vector-based filter can rule out irrelevant points.

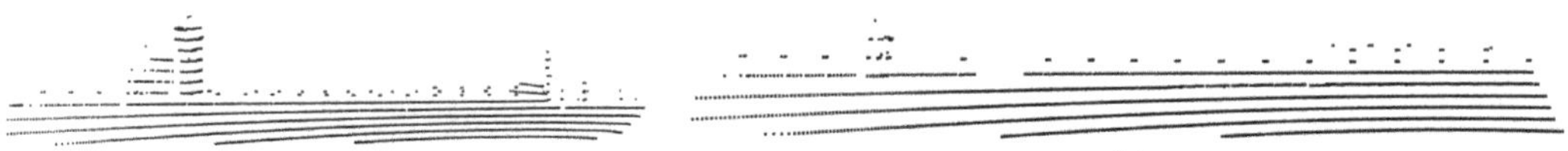

(a) (b)

Figure 7. Point cloud normal vector-based filtering: (**a**) represents the input point cloud; (**b**) represents after filtering.

Algorithm 1: Normal vector-based filter algorithm

```
Input : point cloud with normal vector set Q, threshold ξ Output : subset C
for each q_i in Q :
    C_i ← {q_i}, Q ← Q − {q_i}
    num ← 0
    while num ≠ |C_i| :
        num ← |C_i|
        for each c_j in C_i :
            for each q_j in Q :
                if Δd(c_j, q_j) ≤ ξ :
                    C_i ← C_i + {q_j}, Q ← Q − {q_j}
                end if
            end for
        end for
    end while
end for
index ← argmax_i |C_i|
C ← C_index
```

2.3.2. The Improved RANSAC with Preprocessing

After ruling out irrelevant points with the normal vector-based filtering algorithm, the improved RANSAC then randomly picks three points $(\hat{x}_1, \hat{y}_1, \hat{z}_1)$, $(\hat{x}_2, \hat{y}_2, \hat{z}_2)$, $(\hat{x}_3, \hat{y}_3, \hat{z}_3)$ in the filtered point cloud and calculates the candidate feature plane function using

$$\begin{vmatrix} x-\hat{x}_1 & y-\hat{y}_1 & z-\hat{z}_1 \\ \hat{x}_2-\hat{x}_1 & \hat{y}_2-\hat{y}_1 & \hat{z}_2-\hat{z}_1 \\ \hat{x}_3-\hat{x}_1 & \hat{y}_3-\hat{y}_1 & \hat{z}_3-\hat{z}_1 \end{vmatrix} = 0\,. \tag{10}$$

The plane function of the candidate plane can then be obtained according to

$$Ax + By + D = z, \tag{11}$$

which is deduced from Formula (11). Then, we calculate the score of the plane m as the number of points whose distance to the plane is within the tolerance ε (we name the points within the tolerance as feature points). Mathematically, it is defined as

$$m{:=}\mathrm{card}\left(\left\{(\check{x}_j,\ \check{y}_j,\ \check{z}_j) \in \left\{(x_j,\ y_j,\ z_j)\right\}_{j=1}^{k} \middle| \frac{|A\check{x}_j+B\check{y}_j+\check{z}_j-D|}{\sqrt{A^2+B^2+1^2}} \le \varepsilon\right\}\right). \quad (12)$$

Repeat the above process for predefined cycle times and select the plane function with the highest score as the feature plane function. Then, the normal vector of the feature plane (A_p, B_p, C_p) is calculated using

$$A_p = A_m/D_m, B_p = B_m/D_m, C_p = -1/D_m, \quad (13)$$

where A_m, B_m and D_m are the parameters of the selected feature plane function. Figure 8 provides an illustrative example of the fitting process.

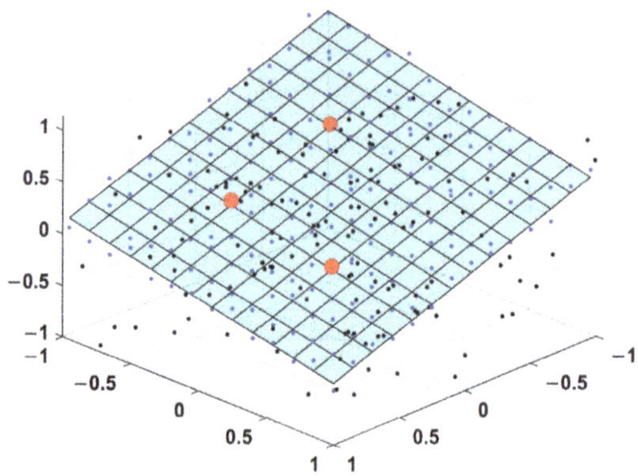

Figure 8. The plane-fitting of RANSAC. The red points are the chosen points, the blue points are near the constructed plane, and the black points are the noise points.

3. Experiments and Results

In this study, we applied both real water-pool experiments and numerical tests to validate our method. The methods applied in the experiments are "Improved RANSAC with Voxel Filter" (hereafter referred to as "Improved RANSAC with VF"), which is the proposed framework; "Standard RANSAC with Voxel Filter" (hereafter referred to as "Standard RANSAC with VF"); "Improved RANSAC without Voxel Filter" (hereafter referred to as the "Improved RANSAC without VF"). The authors apply the Root Mean Square Errors (RMSE) and the Mean Absolute Errors (MAE) as the indicators of errors, following [25]. The calculation formulas are as follows:

$$RMSE = \sqrt{\frac{1}{n}\sum_{i=1}^{n}(x_i - \hat{x}_i)^2}, \quad (14)$$

$$MAE = \frac{1}{n}\sum_{i=1}^{n}|x_i - \hat{x}_i|, \quad (15)$$

where x_i is the estimated value, $\hat{x}_i$ is the reference value, and n is the number of the estimation result. The above indexes are referred to as the evaluation indexes of the estimation results.

3.1. Water-Pool Experiments

To verify the filtering ability of our normal vector-based filter under real environmental noise and verify our attitude-estimation framework on a real ship target, we carried out the experiment in an outdoor water pool.

3.1.1. Experimental Environment and Configuration

In water-pool experiments, we adopted "LS-M1" solid-state LiDAR, with detailed technical information given in Table 1. We equipped the experimental USV named "Hong Dong No. 1" with an IMU system to measure the attitude state, and the dimensional information is shown in Table 2. The experimental scene is shown in Figure 9. In the experiment, we gave the USV an initial moving speed and created some waves in the pool to keep the USV moving freely for ten seconds. The moving motion was recorded using the IMU equipped on the USV and estimated by LiDAR. As shown in Table 3, the parameters of each method were set consistently. The leaf size of the voxel filter is 3 cm.

Table 1. Technical information of the "LS-M1" LiDAR.

Information	Value
maximum range	350 m
laser fire frequency	18,000 Hz
vertical field of view	35 deg
vertical angular resolution	0.03 deg
horizontal field of view	120 deg
horizontal angular resolution	0.06 deg

Table 2. Dimensional information of the "Hong Dong No. 1" USV.

Information	Value
main size	1.5 m
molded breadth	0.74 m
molded depth	0.6 m
designed draught	0.2 m
board-side inclination angle	0.1 deg

Figure 9. The water-pool experimental scene. The USV in the upper part is "Hong Dong No. 1" and the LiDAR in the lower part is the "LS-M1" LiDAR.

Table 3. The parameter settings of each method in the water pool experiment.

Method	Parameter	Meaning	Value
	ε_a	Max distance of neighbor points	0.035 m
Improved RANSAC with VF	ξ_a	Threshold of normal vector angle	0.02 rad
	ϵ_a	Distance tolerance of point to plane	0.05 m
Standard RANSAC with VF	ϵ_b	Distance tolerance of point to plane	0.05 m
	ε_c	Max distance of neighbor points	0.035 m
Improved RANSAC without VF	ξ_c	Threshold of normal vector angle	0.02 rad
	ϵ_c	Distance tolerance of point to plane	0.05 m

3.1.2. Experimental Results

The snapshots of the experiment are shown in Figure 10 to provide visual aids. The actual point clouds in the experiments are shown in Figure 11a. The red rectangle highlights the noise points caused by unknown reflections and the green points belong to the feature plane that we want to extract. The number of the reflected points of the USV is around 15,000, while the board-side plane only reflects around 1000 points, which shows that the collected point clouds in the experiment are severely affected by unknown reflections around the target USV. It can be seen from Figure 11b that, after applying the proposed normal vector-based filter, the number of irrelevant points is significantly reduced.

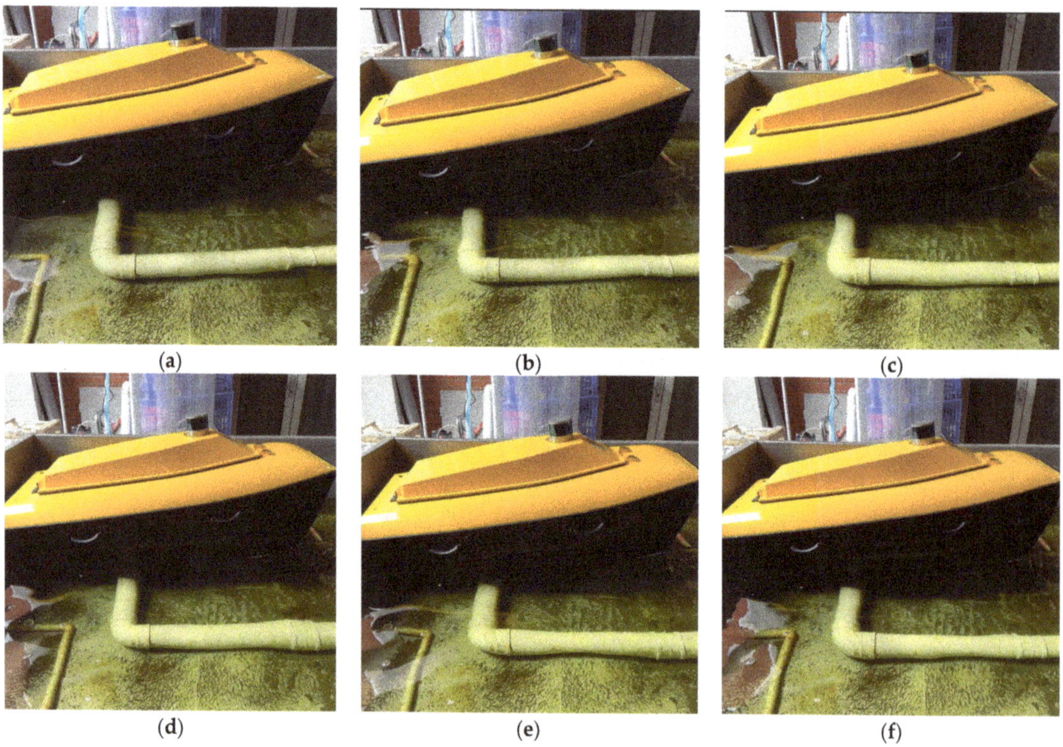

Figure 10. The snapshots of the "Hong Dong No. 1" USV in the water-pool experiments. (**a**–**f**) is the sequence of each snapshot.

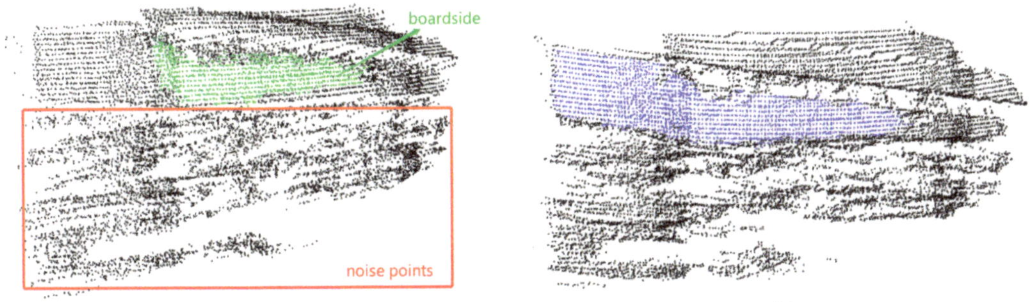

(a) (b)

Figure 11. The comparison of the actual raw point cloud and after filtering in the water-pool experiments: (**a**) represents the originally collected point cloud; (**b**) highlights the point cloud after our normal vector-based filter in blue.

After applying the proposed filtering algorithm, we then computed the feature plane. The feature-plane-extraction results are shown in Figure 12. Figure 12a shows the extraction result of the standard RANSAC whose input point cloud is obtained directly from LiDAR without normal vector-based filtering, while Figure 12b shows the extraction result of the improved RANSAC with pre-processing. It can be seen from Figures 11a and 12a that the feature points that standard RANSAC extracts are mainly noise points, while Figures 11a and 12b show that our improved RANSAC can extract feature points belonging to the board-side plane.

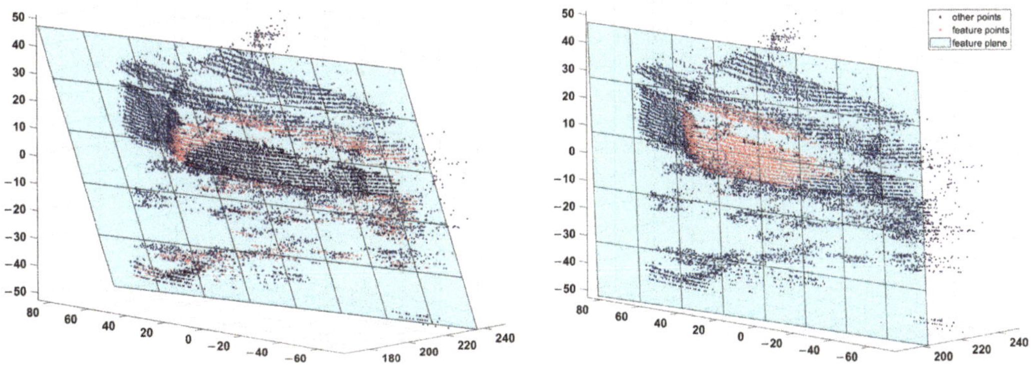

(a) (b)

Figure 12. The comparison of the feature-plane-extraction results of standard RANSAC and improved RANSAC in the water-pool experiments: (**a**) shows the feature-plane-extraction result of standard RANSAC; (**b**) shows the feature-plane-extraction result of our improved RANSAC.

After computing the feature planes, we applied the estimation methods to estimate the attitude. The estimation results and the corresponding IMU data are shown in Figures 13 and 14 The roll angle estimated by standard RANSAC is far from the IMU data and suffers sudden change. The yaw angle follows the trend of the IMU data but there is a gap between these angles. Due to the errors in the angle estimation stage, the rate estimation also contains great losses. The roll angle estimated by improved RANSAC is close to the IMU data. The yaw angle estimation can follow the trend of the IMU data despite small fluctuations. In addition, the yaw angle variance between the improved RANSAC with/without VF is evident in Figure 13. The reason might be that the leaf size of the voxel filter is not ignorable compared with the small main size of the USV. The point clouds might lose some minor details due to the large ratio of the leaf size to the main size of the USV. We remark that this variance is not a concerning problem and will become unobtrusive as the ratio becomes smaller. Furthermore, it can be observed in Figure 13 that the standard RANSAC exhibits a drift in both roll angle and yaw angle after the 7-s mark. This is caused by the changes in the environmental noise points after 7 s. The changes have a negative effect on the standard RANSAC. Since our method puts much effort into reducing noise points, this drift is filtered out. As shown in Table 4, we calculate the evaluation indexes of each method.

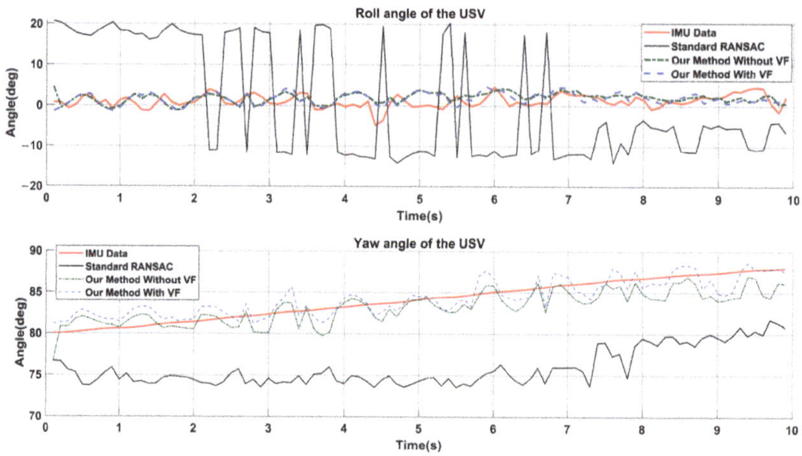

Figure 13. The roll angle and yaw angle results in the water-pool experiments.

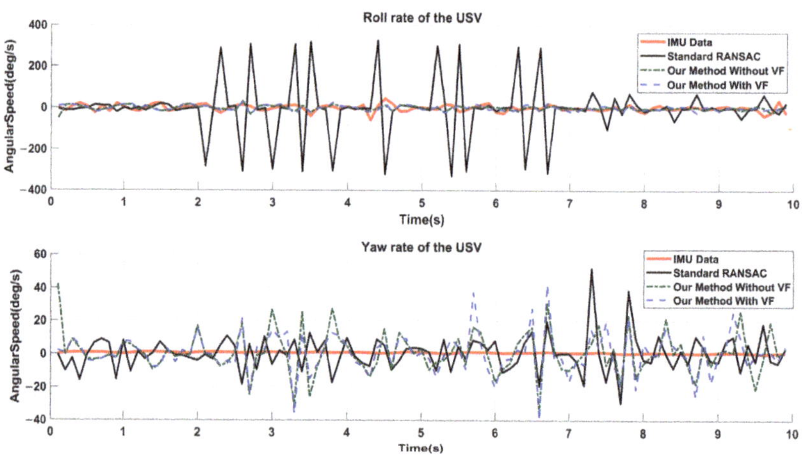

Figure 14. The roll rate and yaw rate results in the water-pool experiments.

Table 4. The evaluation indexes of each method in the water-pool experiments.

Method	Attitude State	RMSE	MAE
Improved RANSAC with VF	Roll Angle	2.0371°	1.6907°
	Yaw Angle	1.3204°	1.1092°
	Roll Rate	19.1547°/s	15.8893°/s
	Yaw Rate	12.6331°/s	8.9665°/s
Improved RANSAC without VF	Roll Angle	1.9879°	1.6301°
	Yaw Angle	1.2956°	1.1068°
	Roll Rate	18.2154°/s	14.3370°/s
	Yaw Rate	12.0684°/s	8.8624°/s
Standard RANSAC with VF	Roll Angle	14.5339°	13.8871°
	Yaw Angle	8.4089°	8.2139°
	Roll Rate	140.2946°/s	75.8283°/s
	Yaw Rate	10.8218°/s	7.6778°/s

3.2. Numerical Tests

To verify the general applicability of our method for different ship types, we intend to apply two typical ship models in the numerical tests.

3.2.1. Numerical Simulation Preparations and Configurations

The numerical motion simulations are based on the LiDAR computation model, the target ship models, and the input attitude data. As shown in Table 5, the LiDAR computation model is based on VLP-32 LiDAR, which is widely used in various studies [26,27] and whose detection range can satisfy the need of towing operations. The LiDAR computation model can manifest the vertical laser beams, spinning speed, and the laser fire frequency of real VLP-32 LiDAR. As the sea surface only serves as an interface to divide the under-water part and the above-water part of the ship model, the accurate sea-surface model only provides a small improvement compared to the single-plane sea-surface model for the proposed algorithm [28]. To simplify the problem, a single plane parallel to the sea level is adopted and the position is set to the designed draught water line of the ship model. In this case, the simulation system can only produce reflection points above the designed draught water line, which is in line with the actual situation.

Table 5. Technical information of the VLP-32 LiDAR.

Information	Value
laser beam	32
spinning speed	300 RPM
maximum range	200 m
laser fire frequency	21,700 Hz
vertical field of view	40 deg
horizontal field of view	360 deg
horizontal angular resolution	0.08 deg

The first model is a typical container ship with large board-side planes, and the dimensional information is given in Table 6. As shown in Figure 15, the test scene consists of the target container ship and an observing boat. We equip the observing boat with LiDAR to collect point clouds and use IMU to calibrate the LiDAR. The observing boat is placed 70 m from the starboard of the target container ship and can move freely. The input attitude of the container ship model is set in accordance with the output of a 4-DOF motion model of a container ship [29,30]. The movement of the container ship in the experiment is set as follows: the ship will first make a turn to trace the target course of 22°. After the course angle reaches 22°, the ship will turn to −17°and maintain its course. As shown in Table 7, the parameters of each method in this test are set consistently. The leaf size of the voxel filter is 1 m.

Table 6. Dimensional information of the container model.

Information	Value
main size	125 m
molded breadth	25 m
molded depth	13 m
designed draught	8 m
board-side inclination angle	0.05 deg

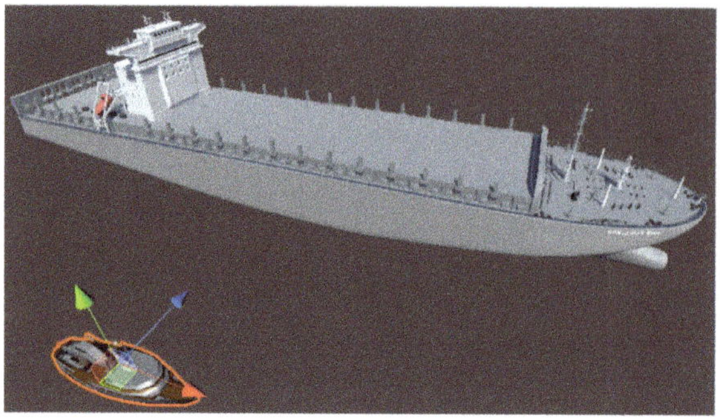

Figure 15. The test scene of the container ship motion simulation. The coordinate arrows show the installation position of LiDAR on the observing boat.

Table 7. The parameter settings of each method in the container ship motion simulation.

Method	Parameter	Meaning	Value
Improved RANSAC with VF	ε_a	Max distance of neighbor points	10 m
	ξ_a	Threshold of normal vector angle	0.003 rad
	ϵ_a	Distance tolerance of point to plane	0.05 m
Standard RANSAC with VF	ϵ_b	Distance tolerance of point to plane	0.05 m
Improved RANSAC without VF	ε_c	Max distance of neighbor points	10 m
	ξ_c	Threshold of normal vector angle	0.003 rad
	ϵ_c	Distance tolerance of point to plane	0.05 m

The second model is a typical yacht with a smaller board-side plane than the container ship, and the dimensional information is given in Table 8. As shown in Figure 16, the test scene consists of a target yacht and an observing boat. The observing boat was placed 30 m from the starboard of the target yacht. The input attitude data of the yacht was set by a set of collected data by the GNSS/IMU system on a real yacht whose basic dimension was roughly the same as the adopted model. As shown in Table 9, the parameters in each method were the same. The leaf size of the voxel filter is 10 cm.

Table 8. Dimensional information of the yacht model.

Information	Value
main size	20 m
molded breadth	6 m
molded depth	3 m
designed draught	1 m
board-side inclination angle	0.1 deg

Figure 16. The test scene of the yacht motion simulation. The coordinate arrows show the installation position of LiDAR on the observing boat.

Table 9. The parameter settings of each method in the yacht motion simulation.

Method	Parameter	Meaning	Value
Improved RANSAC with VF	ε_a	Max distance of neighbor points	3 m
	ξ_a	Threshold of normal vector angle	0.04 rad
	ϵ_a	Distance tolerance of point to plane	0.0075 m
Standard RANSAC with VF	ϵ_b	Distance tolerance of point to plane	0.0075 m
Improved RANSAC without VF	ε_c	Max distance of neighbor points	3 m
	ξ_c	Threshold of normal vector angle	0.04 rad
	ϵ_c	Distance tolerance of point to plane	0.0075 m

3.2.2. Test Results of the Container Ship Model Motion Simulation

First, we calibrate the point cloud using the IMU data of the observing boat and extract the feature plane. As shown in Figure 17a, standard RANSAC faces interference from irrelevant points on the upper structure part of the container ship, while Figure 17b shows that an improved RANSAC can extract the board-side plane of the container.

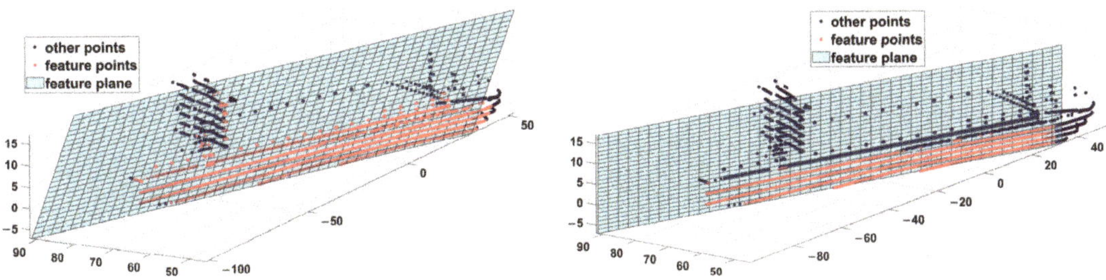

(a) (b)

Figure 17. The comparison of the feature-plane-extraction results of standard RANSAC and improved RANSAC in the container ship motion simulations: (**a**) shows the feature-plane-extraction result of standard RANSAC; (**b**) shows the feature-plane-extraction result of our improved RANSAC.

After extracting the feature planes, we then calculate the attitude states. The estimation results and the corresponding 4DOF model output (the true value in this test) are shown in Figures 18 and 19. As shown in Figure 18, the roll angle estimated by improved RANSAC can follow the trend of the true value, while the estimated roll angle of the standard RANSAC method is unsteady and far from the true value. All three methods can estimate the yaw angle in a way that is close to the true value. From the enlarged figure, we can notice that the lines of the improved RANSAC are closer to the true value line. As shown in Figure 19, the rate value lines of standard RANSAC can follow the trend of the true value, but fluctuate and contain lots of turning points, while the improved RANSAC estimation lines are steady and close to the true value line. We calculate the evaluation indexes of each method in Table 10.

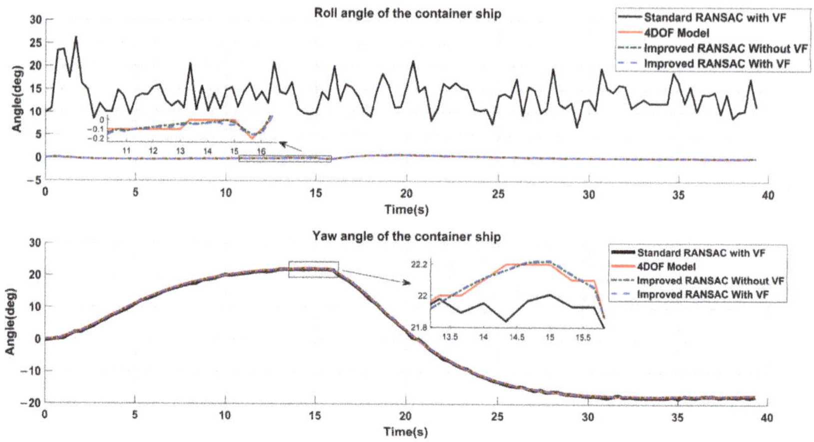

Figure 18. The roll angle and yaw angle results in the container ship simulations.

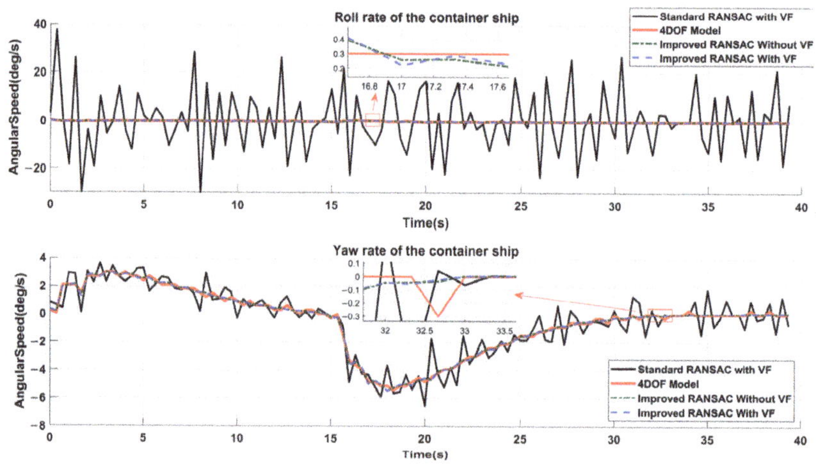

Figure 19. The roll rate and yaw rate results in the container ship simulations.

Table 10. The evaluation indexes of each method in the container ship motion simulations.

Method	Attitude State	RMSE	MAE
Improved RANSAC with VF	Roll Angle	0.0373°	0.0316°
	Yaw Angle	0.0350°	0.0294°
	Roll Rate	0.1201°/s	0.0859°/s
	Yaw Rate	0.1433°/s	0.1163°/s
Improved RANSAC without VF	Roll Angle	0.0377°	0.0307°
	Yaw Angle	0.0349°	0.0293°
	Roll Rate	0.1088°/s	0.0766°/s
	Yaw Rate	0.1430°/s	0.1162°/s
Standard RANSAC with VF	Roll Angle	13.9727°	13.5472°
	Yaw Angle	0.2845°	0.2383°
	Roll Rate	13.0475°/s	10.0682°/s
	Yaw Rate	0.8129/s	0.6323°/s

3.2.3. Test Results of the Yacht Model Motion Simulation

First, we calibrate the point cloud using the IMU data of the observing boat and extract the feature plane. As shown in Figure 20a, standard RANSAC faces interference from irrelevant points on other parts of the yacht, while Figure 20b shows that an improved RANSAC can extract the board-side plane of the yacht.

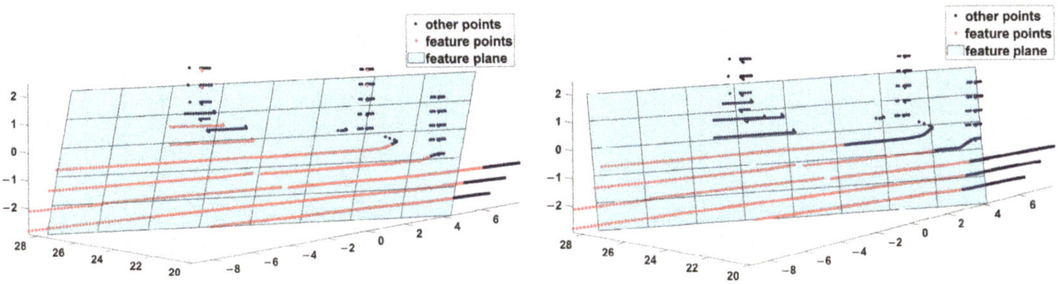

(a) (b)

Figure 20. The comparison of the feature-plane-extraction results of standard RANSAC and improved RANSAC in the yacht motion simulations: (**a**) shows the feature-plane-extraction result of standard RANSAC; (**b**) shows the feature-plane-extraction result of our improved RANSAC.

Then, we calculate the attitude state based on the extraction results. The estimation results of each method and the GNSS/IMU system data (the true value in this test) are shown in Figures 21 and 22. In Figure 21, the roll angle line of standard RANSAC is twisty and far from the true value. The yaw angle line of standard RANSAC can follow the trend of true value despite some fluctuations. Both the yaw angle lines and roll angle lines of the improved RANSAC are steady and close to the true value. In Figure 22, the roll rate and yaw rate lines of the standard RANSAC can follow the trend of the true value but suffer from large fluctuations, while the rate lines of the improved RANSAC are smooth and close to the true value. As shown in Table 11, we calculate the evaluation indexes of each method.

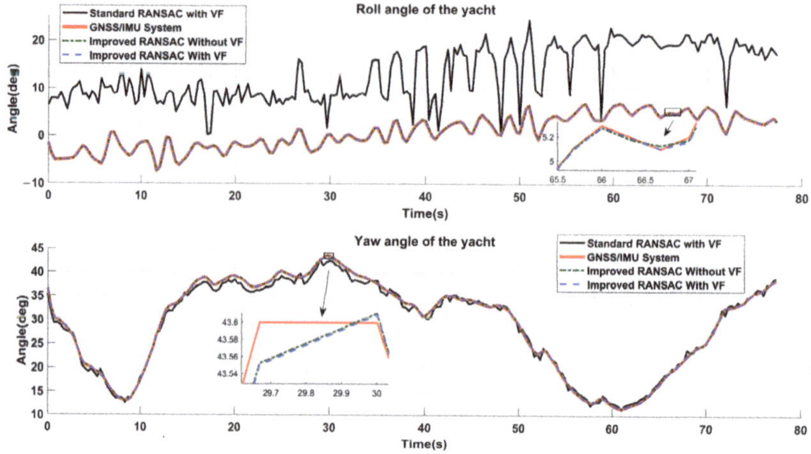

Figure 21. The roll angle and yaw angle results in the yacht simulations.

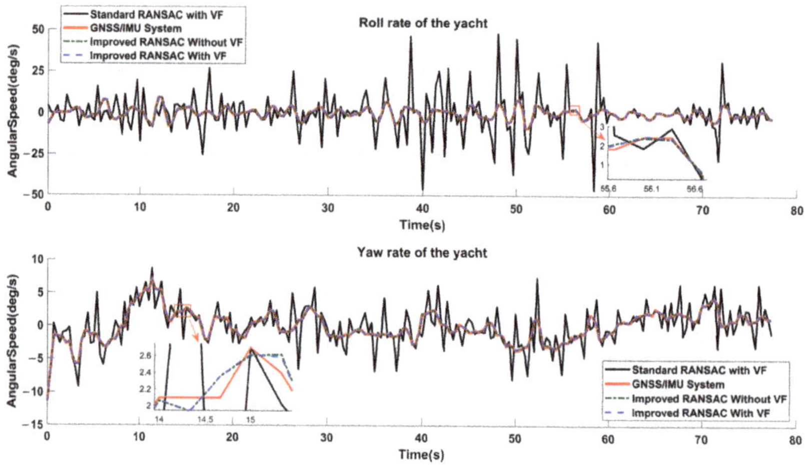

Figure 22. The roll rate and yaw rate results in the yacht simulations.

Table 11. The evaluation indexes of each method in the yacht motion simulations.

Method	Attitude State	RMSE	MAE
Improved RANSAC with VF	Roll Angle	0.0466°	0.0368°
	Yaw Angle	0.0365°	0.0295°
	Roll Rate	0.1942°/s	0.1513°/s
	Yaw Rate	0.1362°/s	0.1090°/s
Improved RANSAC without VF	Roll Angle	0.0440°	0.0346°
	Yaw Angle	0.0363°	0.0292°
	Roll Rate	0.1843°/s	0.1449°/s
	Yaw Rate	0.1354°/s	0.1083°/s
Standard RANSAC with VF	Roll Angle	12.7396°	12.2805°
	Yaw Angle	0.8079°	0.6695°
	Roll Rate	11.5327°/s	6.9814°/s
	Yaw Rate	2.1816°/s	1.5662°/s

4. Discussion

4.1. Discuss the Feature-Plane-Extraction Ability of the Methods

From the comparisons of Figures 12, 17 and 20, we can observe that the feature-plane-extraction results of improved RANSAC are more accurate than those of the standard RANSAC. We note that the noise points in the three tests mainly consist of the unknown reflection points and irrelevant points from other parts of the target ship. In Figure 11, we can see that the normal vector-based filter of the improved RANSAC can rule out both kinds of noise points sufficiently. The filtering process can reduce the possibility of false extraction of the feature plane and thus improve the extraction ability of RANSAC. Accurate feature-plane extraction can be achieved by improving RANSAC with our normal vector-based filtering algorithm.

4.2. Discuss the Attitude Estimation Accuracy of the Methods

Our estimation framework can provide an accurate target attitude state when the feature plane is properly extracted. We can infer from Tables 4, 10 and 11 that both improved RANSAC with VF and improved RANSAC without VF can achieve better estimation results than standard RANSAC regarding both RMSE and MAE. In Table 12, we calculate the average evaluation indexes of each method to further discuss the accuracy.

Table 12. The average evaluation indexes [1] of each method.

Attitude State	Method	Average RMSE	Average MAE
Roll Angle (°)	Improved RANSAC with VF	0.7070	0.5864
	Improved RANSAC without VF	0.6899	0.5651
	Standard RANSAC with VF	13.7487	13.2383
Yaw Angle (°)	Improved RANSAC with VF	0.4640	0.3894
	Improved RANSAC without VF	0.4556	0.3884
	Standard RANSAC with VF	2.9247	3.0406
Roll Rate (°/s)	Improved RANSAC with VF	6.4897	5.3755
	Improved RANSAC without VF	6.1695	4.8528
	Standard RANSAC with VF	54.9583	30.9593
Yaw Rate (°/s)	Improved RANSAC with VF	4.3042	3.0639
	Improved RANSAC without VF	4.1156	3.0290
	Standard RANSAC with VF	4.6054	3.2921

[1] Average indexes of each method are computed by averaging its indexes of three tests.

Both methods using improved RANSAC show great improvements compared with standard RANSAC in terms of roll angle, yaw angle, and roll rate, while the yaw rate estimation results are slightly improved. One possible reason for this is that the irrelevant points from other parts of the target ship might be static and have little influence on yaw rate estimation.

To show the improvement in accuracy in more detail, we calculate the angle and angular rate average evaluation indexes of improved RANSAC with VF and standard RANSAC with VF in Table 13. The angle reduction percentages of RMSE and MAE are 91.62% and 92.93%, respectively. The angular rate reduction percentages of RMSE and MAE are 71.81% and 75.36%, respectively. The above results show that our method can estimate the attitude state more accurately than standard RANSAC.

Table 13. The angle and angular rate average evaluation indexes [1] and reduction percentage.

Attitude State	Method	Average RMSE	Average MAE
Angle	Improved RANSAC with VF	0.5855°	0.4879°
	Standard RANSAC with VF	8.3367°	8.1395°
	Reduction Percentage [2]	91.62%	92.93%
Angular Rate	Improved RANSAC with VF	5.3970°/s	4.2197°/s
	Standard RANSAC with VF	29.7819°/s	17.1257°/s
	Reduction Percentage [2]	71.81%	75.36%

[1] Angle average indexes are computed by averaging the data from roll angle and yaw angle. Angular rate average indexes are computed by averaging the data from roll rate and yaw rate. [2] Reduction percentage is calculated by subtracting the division of "improved RANSAC with VF" by "standard RANSAC with VF" from one.

4.3. Discuss the Density Sensitiveness of Improved RANSAC

To investigate the density sensitiveness of our method, we applied the improved RANSAC with VF and improved RANSAC without VF to estimate the attitude. The voxel filter applied in this framework can reduce the density of the point cloud and help reduce time consumption. Given that the proposed method mainly involves the extraction of major geometric features of a ship model and that the voxel filter usually affects minor details, the drawback of the voxel filter should not be a concern. The experimental results in Tables 4, 10 and 11 indicate that the above two methods achieve similar estimation results, which supports the application of a voxel filter. We note that the adopted voxel filter and leaf size can preserve the features in the point cloud and, if the feature is clearly scanned, the density of the point cloud will have little influence on the estimation process.

4.4. Discuss the General Applicability of Improved RANSAC

Our method requires the extraction of the feature plane and estimation of the attitude. Different ship types may have different feature planes. In this paper, we used a real USV, a container ship model, and a yacht model as targets to verify our method. The above targets have varied shapes and sizes, and our method achieved satisfying results on these targets (indicated in Tables 4, 10 and 11), which demonstrates that our method can be applied to different ship types.

The drawback is that the parameters of our method for different ship targets are different. We comment that this problem can be solved by constructing a parameter database for typical ship types and applying existing classification methods [31–33] to acquire the type of target ship.

4.5. Discuss Single LiDAR and Multiple LiDARs

In this study, we applied single LiDAR to observe the target ship and develop an attitude-estimation framework. The mechanism of LiDAR makes the occlusion problem of single LiDAR inevitable. Under the decreased relative scanning angle of LiDAR, the occlusion of the ship's bow or stern will invalidate the method, which is an important issue to resolve in the future. A promising direction is combining multiple LiDARs with the attitude-estimation framework proposed in this paper to cover the blind area of a single LiDAR.

5. Conclusions

In this study, we proposed an attitude estimation framework that extracts the board-side plane through an improved RANSAC method and computes the roll angle, yaw angle, roll rate, and yaw rate of the target ship. We added a normal vector-based filter to the standard RANSAC method to rule out the noise points in a point cloud and improve the accuracy of the feature-extraction process. Specifically, we first prepared the point cloud through down-sampling and calibration transformation. Next, we estimated the normal vectors of the points and used our filter algorithm to rule out noise points. Then, the RANSAC process provided the optimal plane function, which was adopted to calculate attitude information according to geometric relation. Next, we conducted a real experiment via a USV and two numerical tests using two typical ship models (container ship and yacht) to show the accuracy and general applicability of the proposed method.

The experimental results reveal that the average mean absolute errors of the estimated angle and angular rate are 0.4879 deg and 4.2197 deg/s, respectively, which are 92.93% and 75.36% more accurate than the estimation based on standard RANSAC. Remarkably, we demonstrated the filtering ability and the practical applicability of our method in real water-pool experiments under real environmental noises. We also investigated the density sensitiveness of our method and found that the density of input point clouds has little influence if the feature is clearly scanned.

However, there are some defects in this study owing to the limitations of our experimental resources. The classification of target ships and auto-adaptation of the parameters could be achieved using more abundant data sets, such as real target ships of more types. In addition, the occlusion problem of a single LiDAR can be solved by merging multiple LiDARs, which will be the focus of our future study.

Author Contributions: Conceptualization, S.W.; methodology, S.W.; software, S.W. and X.Y.; validation, S.W. and Y.X.; formal analysis, S.W. and Y.X.; investigation, S.W.; resources, S.W., X.Y. and H.W.; data curation, S.W.; writing—original draft preparation, S.W.; writing—review and editing, Y.X. and S.W.; visualization, S.W. and Y.X.; supervision, H.W.; project administration, H.W.; funding acquisition, H.W. All authors have read and agreed to the published version of the manuscript.

Funding: This work is sponsored by the National Key R&D Program of China, grant number 2022YFE0125200; the National Natural Science Foundation of China, grant number 52271348.

Institutional Review Board Statement: Not applicable.

Informed Consent Statement: Not applicable.

Data Availability Statement: Not applicable.

Acknowledgments: Much appreciation to each reviewer for their valuable comments and suggestions to improve the quality of this note. We appreciate the work of Lei Dong, who provided the valuable data used in the experiments.

Conflicts of Interest: The authors declare no conflict of interest.

References

1. Waskito, K.T.; Sasa, K.; Chen, C.; Kitagawa, Y.; Lee, S.-W. Comparative Study of Realistic Ship Motion Simulation for Optimal Ship Routing of a Bulk Carrier in Rough Seas. *Ocean. Eng.* **2022**, *260*, 111731. [CrossRef]
2. Zhuang, J.; Zhang, L.; Zhao, S.; Cao, J.; Wang, B.; Sun, H. Radar-Based Collision Avoidance for Unmanned Surface Vehicles. *China Ocean Eng.* **2016**, *30*, 867–883. [CrossRef]
3. Zhang, L.; Du, Z.; Valdez Banda, O.A.; Goerlandt, F.; Du, L.; Li, X. Collision Prevention of Ship Towing Operation under Environmental Disturbance. *Ocean. Eng.* **2022**, *266*, 112870. [CrossRef]
4. Huang, L.-M.; Duan, W.-Y.; Han, Y.; Chen, Y.-S. A Review of Short-Term Prediction Techniques for Ship Motions in Seaway. *Chuan Bo Li Xue/J. Ship Mech.* **2014**, *18*, 1534–1542. [CrossRef]
5. He, Y.; Liu, X.; Zhang, K.; Mou, J.; Liang, Y.; Zhao, X.; Wang, B.; Huang, L. Dynamic Adaptive Intelligent Navigation Decision Making Method for Multi-Object Situation in Open Water. *Ocean. Eng.* **2022**, *253*, 111238. [CrossRef]
6. Du, Z.; Reppa, V.; Negenborn, R.R. Cooperative Control of Autonomous Tugs for Ship Towing. *IFAC-PapersOnLine* **2020**, *53*, 14470–14475. [CrossRef]
7. Naus, K.; Wąż, M.; Szymak, P.; Gucma, L.; Gucma, M. Assessment of Ship Position Estimation Accuracy Based on Radar Navigation Mark Echoes Identified in an Electronic Navigational Chart. *Measurement* **2021**, *169*, 108630. [CrossRef]
8. Wu, Y.; Chu, X.; Deng, L.; Lei, J.; He, W.; Królczyk, G.; Li, Z. A New Multi-Sensor Fusion Approach for Integrated Ship Motion Perception in Inland Waterways. *Measurement* **2022**, *200*, 111630. [CrossRef]
9. Yan, Z.; Cheng, L.; He, R.; Yang, H. Extracting Ship Stopping Information from AIS Data. *Ocean. Eng.* **2022**, *250*, 111004. [CrossRef]
10. Liu, J.; Yan, X.; Liu, C.; Fan, A.; Ma, F. Developments and Applications of Green and Intelligent Inland Vessels in China. *J. Mar. Sci. Eng.* **2023**, *11*, 318. [CrossRef]
11. Lu, X.; Li, Y. Motion Pose Estimation of Inshore Ships Based on Point Cloud. *Measurement* **2022**, *205*, 112189. [CrossRef]
12. Zhang, X.; Wang, C.; Jiang, L.; An, L.; Yang, R. Collision-Avoidance Navigation Systems for Maritime Autonomous Surface Ships: A State of the Art Survey. *Ocean. Eng.* **2021**, *235*, 109380. [CrossRef]

13. Thompson, D.; Coyle, E.; Brown, J. Efficient LiDAR-Based Object Segmentation and Mapping for Maritime Environments. *IEEE J. Ocean. Eng.* **2019**, *44*, 352–362. [CrossRef]
14. Wang, J.; Ma, F. Obstacle Recognition Method for Ship Based on 3D Lidar. In Proceedings of the 2021 6th International Conference on Transportation Information and Safety (ICTIS), Wuhan, China, 22–24 October 2021; pp. 588–593.
15. Chen, C.; Li, Y. Ship Berthing Information Extraction System Using Three-Dimensional Light Detection and Ranging Data. *J. Mar. Sci. Eng.* **2021**, *9*, 747. [CrossRef]
16. Wang, H.; Yin, Y.; Jing, Q. Comparative Analysis of 3D LiDAR Scan-Matching Methods for State Estimation of Autonomous Surface Vessel. *JMSE* **2023**, *11*, 840. [CrossRef]
17. Nocerino, A.; Opromolla, R.; Fasano, G.; Grassi, M.; Fontdegloria Balaguer, P.; John, S.; Cho, H.; Bevilacqua, R. Experimental Validation of Inertia Parameters and Attitude Estimation of Uncooperative Space Targets Using Solid State LIDAR. *Acta Astronaut.* **2023**, *210*, 428–436. [CrossRef]
18. Ma, R.; Chen, C.; Yang, B.; Li, D.; Wang, H.; Cong, Y.; Hu, Z. CG-SSD: Corner Guided Single Stage 3D Object Detection from LiDAR Point Cloud. *ISPRS J. Photogramm. Remote Sens.* **2022**, *191*, 33–48. [CrossRef]
19. Fischler, M.A.; Bolles, R.C. Random Sample Consensus: A Paradigm for Model Fitting with Applications to Image Analysis and Automated Cartography. In *Readings in Computer Vision*; Fischler, M.A., Firschein, O., Eds.; Morgan Kaufmann: San Francisco, CA, USA, 1987; pp. 726–740, ISBN 978-0-08-051581-6.
20. Grilli, E.; Menna, F.; Remondino, F. A Review of Point Clouds Segmentation and Classification Algorithms. *Int. Arch. Photogramm. Remote Sens. Spat. Inf. Sci.* **2017**, *42*, 339–344. [CrossRef]
21. Wang, B.; Lan, J.; Gao, J. LiDAR Filtering in 3D Object Detection Based on Improved RANSAC. *Remote Sens.* **2022**, *14*, 2110. [CrossRef]
22. Yang, L.; Li, Y.; Li, X.; Meng, Z.; Luo, H. Efficient Plane Extraction Using Normal Estimation and RANSAC from 3D Point Cloud. *Comput. Stand. Interfaces* **2022**, *82*, 103608. [CrossRef]
23. Dong, G.; Yan, M.; Zheng, Z.; Ma, X.; Sun, Z.; Gao, J. Experimental Investigation on the Hydrodynamic Response of a Moored Ship to Low-Frequency Harbor Oscillations. *Ocean. Eng.* **2022**, *262*, 112261. [CrossRef]
24. Wang, S.; Hu, Q.; Xiao, D.; He, L.; Liu, R.; Xiang, B.; Kong, Q. A New Point Cloud Simplification Method with Feature and Integrity Preservation by Partition Strategy. *Measurement* **2022**, *197*, 111173. [CrossRef]
25. Sun, Q.; Tang, Z.; Gao, J.; Zhang, G. Short-Term Ship Motion Attitude Prediction Based on LSTM and GPR. *Appl. Ocean. Res.* **2022**, *118*, 102927. [CrossRef]
26. Chekakta, Z.; Zenati, A.; Aouf, N.; Dubois-Matra, O. Robust Deep Learning LiDAR-Based Pose Estimation for Autonomous Space Landers. *Acta Astronaut.* **2022**, *201*, 59–74. [CrossRef]
27. Soilán, M.; González-Aguilera, D.; del-Campo-Sánchez, A.; Hernández-López, D.; Del Pozo, S. Road Marking Degradation Analysis Using 3D Point Cloud Data Acquired with a Low-Cost Mobile Mapping System. *Autom. Constr.* **2022**, *141*, 104446. [CrossRef]
28. Wijaya, A.P.; Naaijen, P.; Andonowati; van Groesen, E. Reconstruction and Future Prediction of the Sea Surface from Radar Observations. *Ocean. Eng.* **2015**, *106*, 261–270. [CrossRef]
29. Guo, H.; Zou, Z. System-Based Investigation on 4-DOF Ship Maneuvering with Hydrodynamic Derivatives Determined by RANS Simulation of Captive Model Tests. *Appl. Ocean. Res.* **2017**, *68*, 11–25. [CrossRef]
30. Zhang, G.; Zhang, X.; Pang, H. Multi-Innovation Auto-Constructed Least Squares Identification for 4 DOF Ship Manoeuvring Modelling with Full-Scale Trial Data. *ISA Trans.* **2015**, *58*, 186–195. [CrossRef]
31. Aljumaily, H.; Laefer, D.F.; Cuadra, D.; Velasco, M. Point Cloud Voxel Classification of Aerial Urban LiDAR Using Voxel Attributes and Random Forest Approach. *Int. J. Appl. Earth Obs. Geoinf.* **2023**, *118*, 103208. [CrossRef]
32. Fang, L.; You, Z.; Shen, G.; Chen, Y.; Li, J. A Joint Deep Learning Network of Point Clouds and Multiple Views for Roadside Object Classification from Lidar Point Clouds. *ISPRS J. Photogramm. Remote Sens.* **2022**, *193*, 115–136. [CrossRef]
33. Xu, Z.; Liu, K.; Chen, K.; Ding, C.; Wang, Y.; Jia, K. Classification of Single-View Object Point Clouds. *Pattern Recognit.* **2023**, *135*, 109137. [CrossRef]

Journal of *Marine Science and Engineering*

MDPI

Article

Data-Driven Parameter Estimation of Nonlinear Ship Manoeuvring Model in Shallow Water Using Truncated Least Squares Support Vector Machines

Haitong Xu * and **C. Guedes Soares**

Centre for Marine Technology and Ocean Engineering (CENTEC), Instituto Superior Técnico, Universidade de Lisboa, Av. Rovisco Pais, 1049-001 Lisboa, Portugal; c.guedes.soares@centec.tecnico.ulisboa.pt

* Correspondence: haitong.xu@centec.tecnico.ulisboa.pt

Abstract: A data-driven method, the truncated LS-SVM, is proposed for estimating the nondimensional hydrodynamic coefficients of a nonlinear manoeuvring model. Experimental data collected in a shallow water towing tank are utilized in this study. To assess the accuracy and robustness of the truncated LS-SVM method, different test data sizes are selected as the training set. The identified nondimensional hydrodynamic coefficients are presented, as well as the corresponding parameter uncertainty and confidence intervals. The validation is carried out using the reference data, and statistical measures, such as the correlation coefficient, centred RMS difference, and standard deviation are employed to quantify the similarity. The results demonstrate that the truncated LS-SVM method effectively models the hydrodynamic force prediction problems with a large training set, reducing parameter uncertainty and yielding more convincing results.

Keywords: data-driven; parameter estimation; large-scale training set; truncated LS-SVM; shallow water

Citation: Xu, H.; Guedes Soares, C. Data-Driven Parameter Estimation of Nonlinear Ship Manoeuvring Model in Shallow Water Using Truncated Least Squares Support Vector Machines. *J. Mar. Sci. Eng.* **2023**, *11*, 1865. https://doi.org/10.3390/jmse11101865

Academic Editor: Gerasimos Theotokatos

Received: 13 August 2023
Revised: 20 September 2023
Accepted: 22 September 2023
Published: 26 September 2023

1. Introduction

The numerical simulation of marine surface ships has played an increasingly important role in modern maritime engineering and design; it can be used for ship manoeuvrability prediction [1,2], safety evaluation [3], and ship operation simulators [4], which benefit from the fast development of computer technology and ship manoeuvring theory [5,6]. This innovative technique involves utilizing computational models and algorithms to replicate the complex behaviour of ships in various environmental conditions, such as waves, wind, and currents. The performance, stability, and safety of marine surface ships can be assessed easily, without the need for expensive physical prototypes and extensive sea trials. Moreover, it facilitates the study of emergency scenarios, aiding in the development of robust safety measures. For example, it can be observed that the size and number of ships are increasing in harbours, and the heavy traffic conditions require the operators to be very careful in steering the marine surface ships [7], which inevitably gives a high requirement for the prediction of the manoeuvring characteristics of ships in shallow water. Several works can be found on such topics such as [8–13] just to name a few.

The nonlinear manoeuvring models are typically used in simulators of marine ships, such as the Abkowitz model [14,15] and its revised version [16–19], the MMG model [20–22], and the vectorial model [23–25]. The manoeuvring model is based on a set of equations that consider various factors affecting a ship's turning performance. These factors include the ship's hull form, propulsion system, rudder characteristics, and environmental conditions such as wind and current. Those manoeuvring models are developed by approximating forces and moments using a set of specific hydrodynamic terms, which may vary between different ships. They are usually determined by the ship hull characters, speed, and environmental conditions and have different values for specific ships. Consequently, when dealing with a manoeuvring model, the primary objective is to determine the values of the hydrodynamic coefficients associated with these terms.

The most reliable way is to directly measure the values using the ship model in a towing tank [6,26–28], but those tests are usually expensive and time-consuming, and only linear terms can be directly measured. Besides the towing tests in the laboratory, there are also other ways to investigate the manoeuvrability of surface ships, such as sea trials and scaled ship model tests. The free-running ship model test is a promising alternative solution [13,29–33], which is much cheaper than full-scale tests [34–36]. Suzuki et al. [37] conducted a study on the manoeuvrability of a tank ship, utilizing ship model tests for validation. Free-running model tests were employed to assess the impact of shallow water on the manoeuvring behaviour of the very large crude carrier KVLCC2 [29]. System identification methods were used to estimate the ship manoeuvring mathematical models using simulation data and free-running tests.

System identification is a crucial process in the field of engineering, and it involves the estimation and characterization of mathematical models that represent the dynamic behaviour of complex systems. It can be used to extract information about a system's behaviour by observing its inputs and outputs. Therefore, system identification also plays an important role in building the mathematical model for marine vessels [38–42]. This involves conducting experiments and analysing ship manoeuvring test data to determine the relationships between inputs and outputs, typically in the form of mathematical models or transfer functions. Åström and Källström [41] applied the system identification techniques to obtain the parameters of ship steering dynamics. The Least squares is an important method and was widely used for various applications [43–45]. Wang et al. [46] proposed a hybrid recursive least squares method for online identification.

Qian et al. [47] proposed an optimized deep long short-term memory network framework (LSTM) to predict the ship trajectory of inland water, and the experimental results showed that the GA-LSTM model can effectively improve the accuracy and speed of trajectory prediction. An offline genetic algorithm was used to estimate the ship's manoeuvrability using CFD simulations of free-running model tests [48]. Xu and Guedes Soares [49] discussed the parameter error and convergence problem of the hydrodynamic coefficients estimation of a nonlinear manoeuvring model. Wang et al. [50] proposed a generalized ellipsoidal function-based fuzzy neural network (GEBF-FNN) to describe the reference model for a large tanker. The obtained models were used to simulate the typical zig-zag manoeuvres with moderate and extreme steering. Dong et al. [51] proposed a math-data integrated prediction (MDIP) model for ship manoeuvring motion, where the variable-order hydrodynamic derivatives were used. The results show that the proposed model can offer a stronger generalization, and possibly be used for the ship manoeuvring motion prediction.

Recently, a kernel-based machine learning method, support vector machine (SVM), has been used for the manoeuvring modelling of ships, considering its various advantages [15,38–40,52–54] such as relatively memory efficient, global and unique solution, and sparseness [55–59]. However, there are also some disadvantages. One of them is that SVM is not suitable for large data sets. As indicated in [60], the size of the training set for the LS-SVM should be limited to fewer than 2000 data points. A limited training set makes it not suitable for the data-driven modelling problem of complex systems, for example, the manoeuvring modelling of large container ships in shallow water, since the kinematic theory behind the shallow water effect is still blurred.

Recently, a novel version of the Support Vector Machine (SVM), known as the truncated LS-SVM, was introduced for the analysis of ship manoeuvrability based on Planar Motion Mechanism (PMM) tests, using a large-scale training dataset. The truncated LS-SVM was successfully applied to estimate hydrodynamic coefficients in various conditions, including deep water [24,61], shallow water [62,63], and free-running ship mode tests [13,64]. The size of the training dataset plays a critical role in the performance of parameter estimation methods, and the same holds for the truncated LS-SVM. Consequently, it is imperative to investigate the impact of training dataset sizes on the determination of hydrodynamic coefficients and the associated parameter uncertainty.

This paper aims to analyse data-driven parameter estimation of a nonlinear manoeuvring model using PMM test data. Additionally, it will validate the performance of the truncated LS-SVM across a range of training set sizes from 1000 to 10000. The paper will discuss parameter uncertainty arising from noise and provide confidence intervals for the identified parameters. The validation will be carried out using the statistical merits of the prediction and reference data, such as the correlation coefficient (R), centred RMS difference, and standard deviation.

2. Ship Manoeuvring Model

Surface ships in wave conditions are commonly treated as rigid bodies, with their motion described by 6 Degrees of Freedom (DOF), as shown in Figure 1. These DOFs encompass surge, sway, and yaw (manoeuvring motion), as well as roll, pitch, and heave motion (seakeeping motion). In the context of manoeuvring studies, the focus is typically on coupled motions within the horizontal plane, with an assumption of constant values for frequency-dependent added mass and potential damping [23].

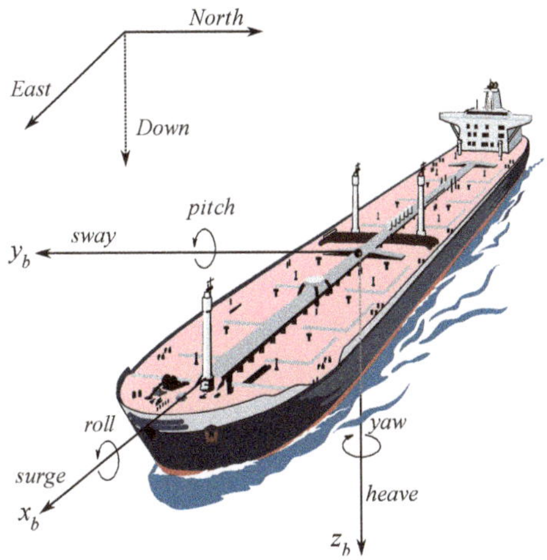

Figure 1. Coordinate systems to study the motions of ships in waves.

To describe the manoeuvring motion of the ship, an empirical manoeuvring model is presented in this section. The equations governing the ship's behaviour under the influence of hydrodynamic forces and moments are defined as

$$\begin{aligned} &(m + X_{\dot{u}})\dot{u} - mvr - mx_G r^2 = X_q + X_p \\ &(m + Y_{\dot{v}})\dot{v} + (mx_G + Y_{\dot{r}})\dot{r} + mur = Y_q \\ &(mx_G + N_{\dot{v}})\dot{v} + (I_{zz} + N_{\dot{r}})\dot{r} + mx_G ur = N_q \end{aligned} \tag{1}$$

where m and Izz are the mass and inertial moment of the ship, respectively. $X_{\dot{u}}, Y_{\dot{v}}, Y_{\dot{r}}, N_{\dot{v}}, N_{\dot{r}}$ are the added mass coefficients, x_G is the longitudinal coordinate of the centre of mass, and X_p is the surge force induced by a propeller. The quasi-steady hydrodynamic forces and moments on the ship hull and rudder are X_q, Y_q, N_q.

In this paper, only the hull forces and moment are considered because the PMM test data were carried out using the bare model ship hull. The dimensionless forces and moment are defined as multi-variate regression polynomials depending on the nondimensional velocities [65], $u' = u/U, v' = v/U, r' = rL/U$.

$$X'_q = X'_0 + X'_{uu}u'u' + X'_{vr}v'r' \tag{2}$$

$$Y'_q = Y'_0 + Y'_v v' + Y'_r r' + Y'_{vv} v' |v'| + Y'_{vr} v' |r'| + Y'_{rr} r' |r'| \quad (3)$$

$$N'_q = N'_0 + N'_v v' + N'_r r' + N'_{rv} r' |v'| + N'_{vr} v' |r'| + N'_{r|r|} r' |r'| \quad (4)$$

The above quasi-polynomial regression models are the revised version of the nonlinear manoeuvring models that were proposed by Inoue et al. [66,67]. This model gives a satisfactory agreement with the full-scale trial results and can be used for the prediction of ship manoeuvrability in the initial ship design stage. To describe the shallow water effect on the hydrodynamic forces, it is assumed that the values of the hydrodynamic coefficients in Equations (2)–(4) are related to the shallow water effect. The hydrodynamic forces and moments are nondimensionalized by using the prime system recommended by SNAME [68].

$$X'_q = X_q \Big/ 0.5\rho U^2 LT, Y'_q = Y_q \Big/ 0.5\rho U^2 LT, N'_q = N_q \Big/ 0.5\rho U^2 L^2 T \quad (5)$$

where ρ is the water density, L is the ship length, U is the ship speed over ground, and T is the draught at the midship. The hydrodynamic coefficients in Equations (2)–(4) are dimensionalized using the factors given in Table 1.

Table 1. Dimensional factors for the hydrodynamic parameters.

Coef.	Dimensional Factor	Coef.	Dimensional Factor	Coef.	Dimensional Factor
$X'_{\dot{u}}$	$0.5\rho L^2 T$	Y'_v	$0.5\rho LTU$	N'_0	$0.5\rho L^2 TU^2$
X'_0	$0.5\rho LTU^2$	Y'_r	$0.5\rho L^2 TU$	N'_v	$0.5\rho L^2 TU$
X'_{uu}	$0.5\rho LT$	$Y'_{v\|v\|}$	$0.5\rho LT$	N'_r	$0.5\rho L^3 TU$
X'_{vr}	$0.5\rho L^2 T$	$Y'_{v\|r\|}$	$0.5\rho L^2 T$	$N'_{r\|v\|}$	$0.5\rho L^3 T$
$Y'_{\dot{v}}$	$0.5\rho L^2 T$	$Y'_{r\|r\|}$	$0.5\rho L^3 T$	$N'_{v\|r\|}$	$0.5\rho L^3 T$
$Y'_{\dot{r}}$	$0.5\rho L^3 T$	$N'_{\dot{v}}$	$0.5\rho L^3 T$	$N'_{r\|r\|}$	$0.5\rho L^4 T$
Y'_0	$0.5\rho LTU^2$	$N'_{\dot{r}}$	$0.5\rho L^4 T$		

3. Duisburg Model Tested in Shallow Water

The hydrodynamic parameter estimation training dataset comprises planar motion mechanism (PMM) test data employing the Duisburg Test Case (DTC) ship model. The DTC ship model is a well-known and widely used benchmark in the field of ship hydrodynamics and manoeuvring. Many experimental tests were carried out using the DTC ship model and the results serve as a standardized test case for assessing and validating numerical simulation techniques, particularly those related to ship manoeuvring performance. It is used as a fundamental reference point for assessing and advancing the capabilities of numerical simulations in ship manoeuvring. Its standardized geometry and parameters make it an invaluable tool for improving the accuracy of ship design and performance prediction methods, ultimately benefiting the maritime industry as a whole.

The DTC model tests were carried out under the support of the SHOPERA project [69–71]. The main reason for using the PMM test is the quality of the data. The PMM test is a critical experimental method used to assess and characterize the hydrodynamic behaviour of ship models. This test involves a specialized apparatus known as a PMM that allows for controlled and precise movement of the ship model in a testing tank, simulating different types of ship motions. During the PMM ship test, the model is placed in a large water tank, and the PMM system precisely controls its movements. The model is subjected to various input motions, replicating the effects of waves, wind, and other environmental forces. The data on how the ship model responds to these simulated conditions, including its resistance, stability, and motion characteristics can be collected using data acquisition instruments.

In this paper, the PMM tests were conducted in a towing tank under shallow water conditions at Flanders Hydraulics Research (FHR), where measurements of hydrodynamic forces and moments acting on the bare hull were acquired. The quality of the test data used in this paper is very high and reliable, and to some extent, this can reduce the uncertainty due to noise.

The towing tank at FHR measures 87.5 m in length, 7 m in width, and has a maximum water depth of 0.5 m, rendering it suitable for conducting model tests in both shallow and very shallow water conditions. More detailed information can be found in [72,73]. Figure 2 displays the DTC ship model positioned on the carriage within the towing tank during testing, while Table 2 provides the key dimensions of the ship model.

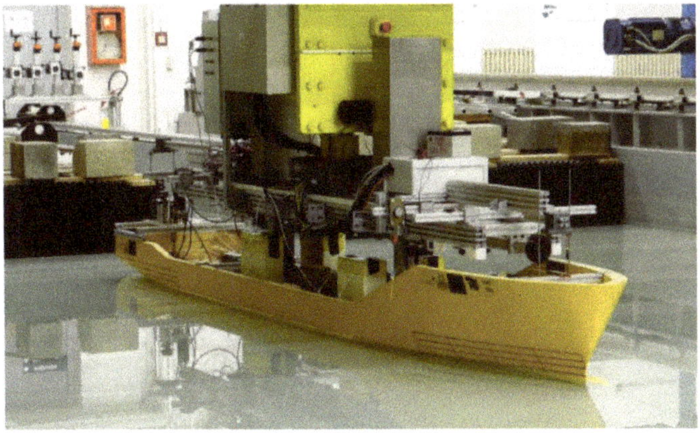

Figure 2. Planar motion mechanism tests of DTC ship model in shallow water. Reprinted from Ref. [63] with permission from Elsevier, 2023.

Table 2. The dimensions of DTC model (1:89.11).

Description	
Length between pp (L_{pp})	3.984 m
Draught (T)	0.163 m
Beam (B)	0.572 m
Block coefficient (C_b)	0.661
Mass	242.8 kg
centre of gravity in x-direction (x_G)	−0.052 m
Moment of inertia along z-axis (I_{zz})	219 kg m^2

The 60 PMM test cases were executed in the towing tank with a water depth, of 0.3254 m (the water depth to draught ratio h/T is 2). The raw results of all model tests were 40 Hz time series, and the four force gauges were installed on the towing platform. The surge, sway forces, and yaw moments were calculated based on the measured signals of the four separate force gauges. The tests included the pure drift, pure sway, and coupled sway–yaw test. To fully activate the response of the ship and obtain rich information, the velocities of towing speed and the amplitude and frequency of the oscillatory motion in tests were changed during the tests; for example, 3 different speeds, 7 drift angles, 3 amplitudes, and 2 frequencies were considered. The PMM tests are described in Figure 3. As can be observed, the pure drift was carried out using the ship model with a fixed drift angle. The pure sway test is the ship model oscillated around the y-axis with zero drift angle. During the pure yaw test, the ship model moved forward with a sinusoidal oscillation in the y-axis with zero sway speed, as presented in Figure 3c. The coupled sway–yaw test was the pure yaw test with no zero-drift angle.

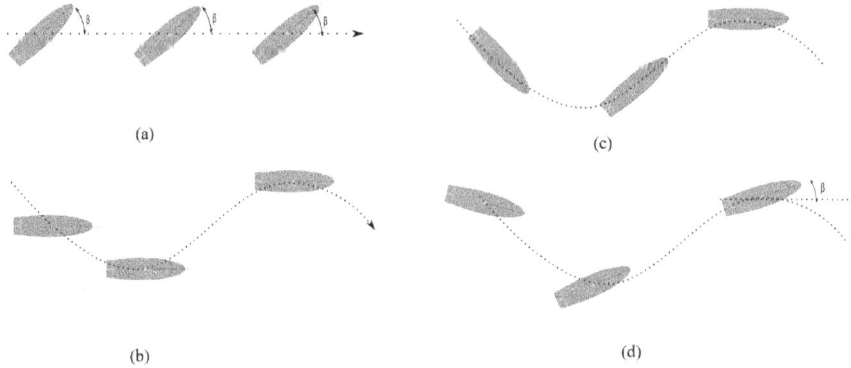

Figure 3. Ship model towing tests: (a) pure drift; (b) pure sway; (c) pure yaw; and (d) pure yaw +drift.

4. Truncated LS-SVM

Least Squares Support Vector Machines (LS-SVM) are a supervised learning algorithm that extends the original concept of SVM from classification to regression tasks. SVMs have a good performance for classification problems by finding a hyperplane that best separates two classes while maximizing the margin between them. LS-SVM adapts this idea to regression problems where the goal is to predict continuous numerical values.

The main objective of LS-SVM is to find a hyperplane that best fits the data by minimizing the regression error. It focuses on minimizing the error between the actual target values and the predicted values along with a regularization term. The regularization term helps prevent overfitting. The kernel function is also used in LS-SVMs, where it can map data into a higher-dimensional feature space, making them capable of handling non-linear relationships in the data.

This section introduces a novel iteration of the support vector machine, referred to as the truncated LS-SVM, and delves into the parameter uncertainty resulting from data noise. The classical LS-SVM was proposed by [60], and it is obtained by reformulating the minimization problem using the regression errors, as presented in [60].

One of the significant advantages of LS-SVM is that it achieves results by solving a set of linear equations, as opposed to the quadratic programming (QP) problems typically associated with classical SVMs. It can simplify the required computation, but unfortunately, the sparseness of standard SVM is lost. Therefore, the classical LS-SVM is not recommended for large-scale data applications, or more precisely, for large-scale training problems. As recommended by [60], the size of the training set is usually restricted to about N = 2000. In the following part, the truncated LS-SVM will be proposed for the manoeuvring modelling with the large-scale training set. The uncertainty of the identified parameters is also analysed. The classical LS-SVM is given as follows:

$$\underbrace{\begin{bmatrix} 0 & \vec{1} \\ \vec{1} & K(\cdot)+C^{-1}I \end{bmatrix}}_{A} \underbrace{\begin{bmatrix} b \\ \vec{\alpha} \end{bmatrix}}_{\theta} = \underbrace{\begin{bmatrix} 0 \\ \vec{Y} \end{bmatrix}}_{Y} \tag{6}$$

where I is an identity matrix of size N, $\vec{\alpha} = [\alpha_1, \cdots \alpha_N]^T$ are the Lagrange multipliers, $\vec{Y} = [y_1, \cdots, y_N]^T$ is the output vector, and $K(x_k \cdot x_i) = \varphi(x_k)^T \varphi(x_i)$, $i = 1, \cdots, N$ is the kernel function, which is positive definite and satisfies the Mercer condition [74]. To estimate the values of the hydrodynamic coefficients, the linear kernel function is chosen. As can be observed in Equation (6), the dimension of matrix A increases exponentially with the size of the training set, which will result in unstable solutions. The obtained parameters are usually sensitive to noise and drift from the true values.

In the following part, the singular values decomposition is introduced for the kernel matrix analysis, and it is given as

$$A = \sum_{i=1}^{n} u_i \sigma_i v_i^T = U \Sigma V^T \tag{7}$$

Then, substituting into Equation (6) gives

$$\theta = \left(U \Sigma V^T \right)^{-1} Y = V \Sigma^{-1} U^T Y = \sum_{i=1}^{n} \frac{v_i u_i{}^T}{\sigma_i} Y \tag{8}$$

where the matrix, U, is orthogonal with the eigenvectors chosen from AA^T, and the matrix, V, is orthogonal and its eigenvectors are chosen from $A^T A$. Σ is a diagonal matrix.

Assume that the output data, y, contains the noise, δy, then the noise will propagate to a perturbation in the solution in Equation (8):

$$\hat{\theta} \doteq \theta_{true} + \delta\theta = \left(U \Sigma V^T \right)^{-1} (y_{true} + \delta y) \tag{9}$$

Then, the perturbation in the solution due to the noise can be obtained as follows:

$$\delta\theta = V \Sigma^{-1} U^T \delta y = \sum_{i=1}^{n} \frac{v_i u_i{}^T}{\sigma_i} \delta y \tag{10}$$

As presented in Equation (10), and with the discrete Picard condition [75], the portion of the singular values can be kept when the ill-conditioned matrix is obtained from the measured data. The data noise can be magnified and potentially dominate the solutions when the singular values are small. Therefore, to diminish the error propagation due to the noise, it is preferred to neglect the smaller singular values in the matrix Σ. The matrix can be presented as

$$A_r = U_r \Sigma_r V_r{}^T \tag{11}$$

The truncated value, r, plays a trade-off role between the size of the regularized solutions and their fit to the given data, and the L-curve [76] can be used to obtain the optimal value.

To quantify how random measurement errors in the data, y, propagate to the identified parameters, the error propagation matrix can be calculated using

$$V_{\hat{\theta}} = \left[\frac{\partial \hat{\theta}}{\partial y} \right] V_y \left[\frac{\partial \hat{\theta}}{\partial y} \right]^T \tag{12}$$

The standard error of the parameters, $\sigma_{\hat{\theta}}$, is the square root of the diagonal of the error propagation matrix. Then, the confidence intervals for the identified parameters are given as follows:

$$\hat{\theta} - t_{(1-a/2)} \ \sigma_{\hat{\theta}} \leq \theta \leq \hat{\theta} + t_{(1-a/2)} \ \sigma_{\hat{\theta}} \tag{13}$$

where $1 - a$ is the desired confidence level, and t is the Student's t statistic. Typically, for the large-scale training set where the number of the measured data is much larger than the number of the estimated parameters, t is 1.96 for 95% confidence intervals and 1.28 for 80% confidence intervals.

With the identified models, it is necessary to validate the models by comparing the prediction with the new test data, which was not used in the training process. Several statistical merits are used to qualify their similarity. Given reference data, $y = [y_i y_2, \cdots, y_N]^T$

and the prediction data, $\hat{y} = [\hat{y}_i, \hat{y}_2, \cdots, \hat{y}_N]^T$, the correlation coefficients can be calculated as follows:

$$R = \frac{1}{N}\sum_{n=1}^{N}(y_i - \overline{y})(\hat{y}_i - \overline{\hat{y}})/\sigma_y\sigma_{\hat{y}} \quad (14)$$

where, $\overline{y}$ and $\overline{\hat{y}}$ are the mean values of the reference data and prediction data, respectively. σ_y and $\sigma_{\hat{y}}$ are the standard deviations, and are calculated as follows:

$$\sigma_y = \frac{1}{N}\sum_{n=1}^{N}(y_i - \overline{y}),\ \sigma_{\hat{y}} = \frac{1}{N}\sum_{n=1}^{N}(\hat{y}_i - \overline{\hat{y}}) \quad (15)$$

The centred root-mean-square (RMS) difference is given below.

$$E'^2 = \frac{1}{N}\sum_{n=1}^{N}[(y_i - \overline{y})(\hat{y}_i - \overline{\hat{y}})]^2 \quad (16)$$

5. Data-Driven Manoeuvring Modelling of DTC Model

In this section, the proposed system identification method, the truncated LS-SVM, is employed to estimate the nondimensional hydrodynamic coefficients using the PMM tests in shallow water. To test the performance of the proposed truncated LS-SVM for large-scale data-driven modelling, the different training set sizes are considered for the surge, sway, and yaw models. The training set is defined as Surge_$\{ID\}$, Sway_$\{ID\}$, and Yaw_$\{ID\}$, where ID represents the number of data points, as presented in Table 3. As suggested by [60], the data size should be restricted below 2000 because kernel matrix size grows with the number of data points. In this study, the size of the training data is set in different ranges from 1200 to 10,000, and the training data contains the pure drift, pure sway, pure yaw, and coupled sway and yaw tests, as described in Section 3.

Table 3. Training set size for the data-driven parameter estimation.

ID	1	2	3	4	5	6	7
Training set size	1200	1300	1500	1700	2000	2300	2900
ID	**8**	**9**	**10**	**11**	**12**	**13**	**14**
Training set size	4000	5000	6000	7000	8000	9000	10,000

The truncated LS-SVM is used to identify the nondimensional hydrodynamic coefficients of the ship hull, and the results are presented in Figure 4. As can be observed, the parameters converge to a constant value as the training set size grows. When the training set is around 4000, the obtained results change slightly, even while the training set size still grows. The 80% and 95% confidence intervals of the identified parameters are also presented in the figure. The confidence intervals of the identified parameters represent the theoretical long-run frequency of confidence intervals that contain the true values.

From Figure 4, the confidence intervals decrease with the training set size, which indicates that the margin of error decreases. In plain words, the large scale of the training set can provide more confidence and robust results. Since the noise in data is randomly recorded during data collection due to the device and environmental disturbance, the proposed method can diminish the noise effect, to a certain extent, by using the large-scale training set.

The truncated LS-SVM is also employed for parameter estimation of the nondimensional hydrodynamic coefficients of the nonlinear sway model, as given in Equation (3). There are eight parameters to be estimated, and the training set size was set as indicated in Table 3. The obtained values of the nondimensional coefficients are presented in Figure 5, as well as the confidence intervals of the corresponding ones.

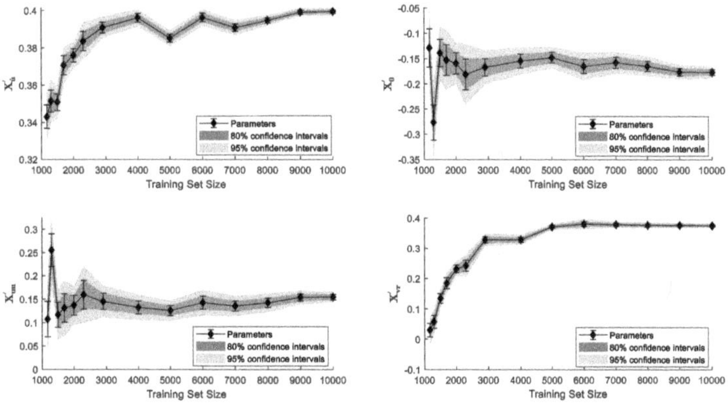

Figure 4. The identified hydrodynamic coefficients of the surge model with 80% and 95% confidence intervals.

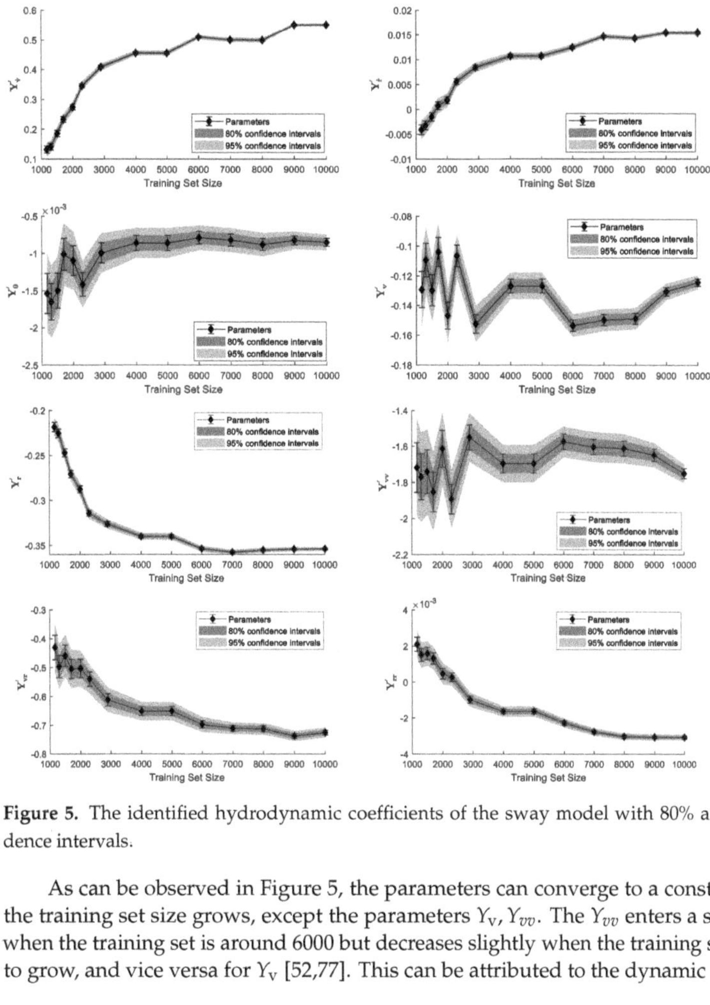

Figure 5. The identified hydrodynamic coefficients of the sway model with 80% and 95% confidence intervals.

As can be observed in Figure 5, the parameters can converge to a constant value as the training set size grows, except the parameters Y_{v}, Y_{vv}. The Y_{vv} enters a stable period when the training set is around 6000 but decreases slightly when the training set continues to grow, and vice versa for Y_{v} [52,77]. This can be attributed to the dynamic cancellation, which results from the multicollinearity of the two terms. The 80% and 95% confidence intervals of the identified parameters are also presented in the figure. From Figure 5, the confidence intervals decrease with the training set size, which also indicates that the large-scale training set can diminish the parameter uncertainty.

The nondimensional hydrodynamic coefficients of the yaw model are identified and presented in Figure 6. The results can converge to a constant value as the training set size grows, as shown in Figure 6. The confidence intervals are also given in the figure, and they decrease with the training set size, which indicates that the uncertainty has been diminished.

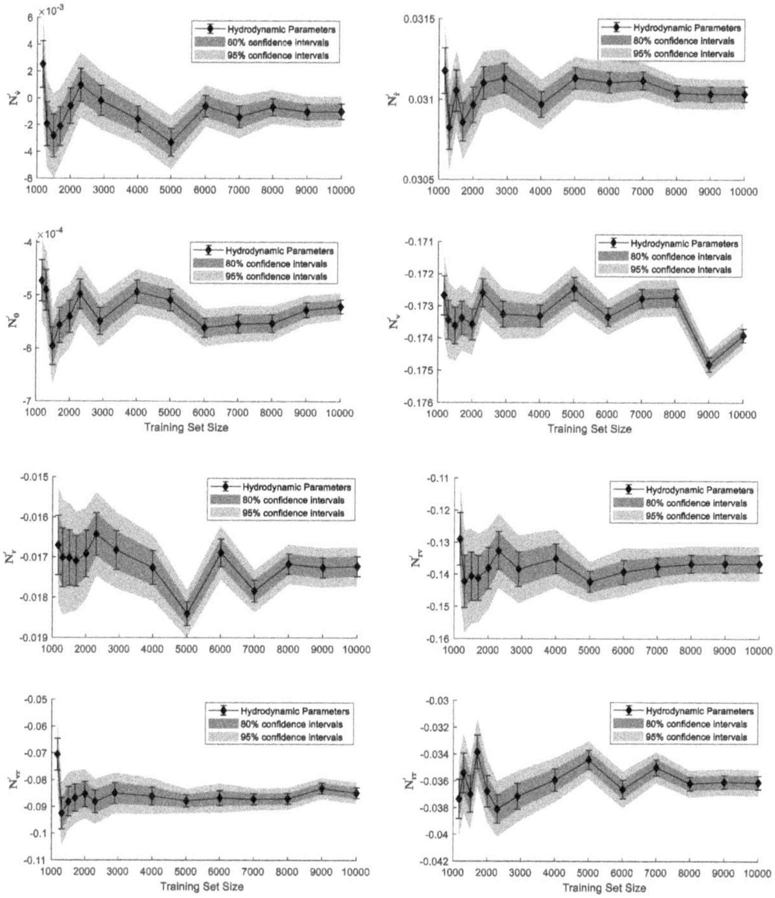

Figure 6. The identified hydrodynamic coefficients of the yaw model with 80% and 95% confidence intervals.

To validate the results, which are obtained using the different training set sizes, the models are employed to reproduce the hydrodynamic forces and moments that were measured during the tests. The harmonic yaw and sway test are selected as the test data, and the Taylor diagram [78] is used to show how closely the prediction matches the observations (experimental data). The similarity is quantified in terms of their correlation, root-mean-square differences, and amplitude of variations.

As presented in Figure 7, the statistical merits of the test data are indicated using the red line, and the values of the correlation coefficient (R), centred RMS difference, and standard deviation are presented in Table 4. From Figure 7b, the correlation coefficients of yaw models are very high, and very close to the reference data for all the cases, even when the training set is small. It indicates that the designed PMM tests fully activate the response of yaw motion and are suitable for parameter estimation of yaw motion.

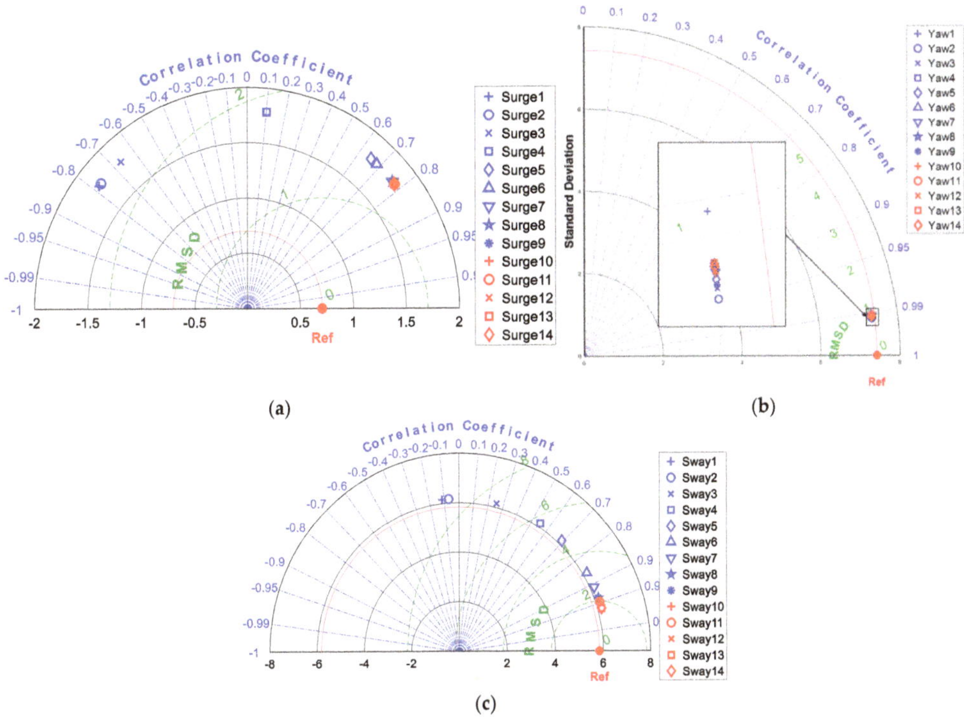

Figure 7. Taylor diagram showing the validation performance of the obtained models: (**a**) surge model; (**b**): yaw model; and (**c**): Sway model.

Table 4. The statistical merits of the predictions of the obtained surge, sway, and yaw models.

ID	SURGE			SWAY			YAW		
	R	*E′*	*STD*	*R*	*E′*	*STD*	*R*	*E′*	*STD*
REF.	1.000	0.000	0.704	1.000	0.000	5.829	1.000	0.000	7.422
1	−0.783	2.232	1.785	−0.114	6.734	6.168	0.990	1.027	7.352
2	−0.772	2.133	1.785	−0.069	6.598	6.168	0.992	0.914	7.352
3	−0.668	1.919	1.785	0.254	5.984	6.168	0.992	0.943	7.352
4	0.102	1.776	1.785	0.548	5.316	6.168	0.992	0.956	7.352
5	0.654	1.648	1.785	0.693	4.778	6.168	0.992	0.944	7.352
6	0.685	1.619	1.785	0.860	3.802	6.168	0.992	0.962	7.352
7	0.768	1.412	1.785	0.909	3.032	6.168	0.992	0.961	7.352
8	0.767	1.413	1.785	0.938	2.472	6.168	0.992	0.958	7.352
9	0.779	1.322	1.785	0.938	2.472	6.168	0.992	0.950	7.352
10	0.779	1.304	1.785	0.949	2.017	6.168	0.992	0.956	7.352
11	0.779	1.309	1.785	0.947	2.083	6.168	0.992	0.962	7.352
12	0.778	1.312	1.785	0.947	2.091	6.168	0.992	0.954	7.352
13	0.778	1.312	1.785	0.961	1.719	6.168	0.992	0.964	7.352
14	0.778	1.314	1.785	0.960	1.730	6.168	0.992	0.960	7.352

For the surge and sway motion (Figure 7a,c), the correlation coefficients are negative when the training set size is small, which indicates that the obtained models are negatively related to the test data. In this case, the obtained model cannot be used to predict the surge and sway forces on the ship hull. With the training set size growing, the correlation coefficients can achieve a good level. The centred RMS difference (E') is also presented in Figure 7 by using green contours. For the surge and sway models, the centred RMS differences decrease with the training set size, and it can be found graphically that the markers are close to the Ref. in Figure 7a,c.

It is necessary to point out that the training set size does not change the standard deviation of the prediction of the obtained models, it largely depends on the structure of the nonlinear manoeuvring models. The standard deviation of the sway model agrees very well with the reference data but is greater in the surge case. It also can be confirmed by Figure 8, which presents the predicted surge, sway forces, and yaw moments during the PMM test. The manoeuvring models obtained using the training set size (ID 14 with 10,000 samples) are chosen for validation. From Figure 8, it can be observed that there are more oscillations in the predictions of the surge and sway model.

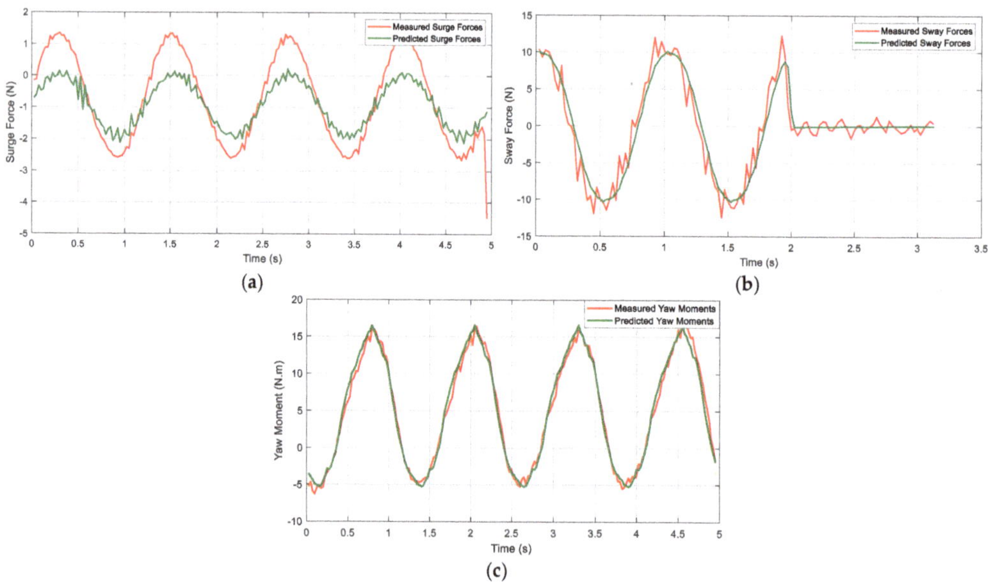

Figure 8. Reproduction of the hydrodynamic hull forces and moment: (**a**) Pure yaw with constant drift angle; (**b**): pure sway; and (**c**): harmonic yaw test in shallow water using the obtained models.

6. Conclusions

This paper investigates the data-driven parameter estimation of a nonlinear ship manoeuvring model using the truncated LS-SVM, where the PMM tests are used as the training and validation set. The truncated LS-SVM is employed to estimate the nondimensional hydrodynamic coefficients with the 14 different training set sizes, and the parameter uncertainty due to the noise is also presented, as well as the confidence intervals of the identified parameters.

The results indicate that the truncated LS-SVM is capable of the modelling problem using a large-scale training set. The obtained parameters can converge to the constant values and their uncertainty can be diminished as the training set size grows, as well as the margin of confidence intervals. Therefore, the truncated LS-SVM can diminish the parameter uncertainty and provide a robust result. The validation is also carried out using statistical measures: the correlation coefficient (R), centred RMS difference, and the standard deviation presented graphically using the Taylor diagram. It can be concluded that the PMM test can fully activate the response of yaw motions and provide rich information for the parameter estimation of the yaw model. The large-scale training set can increase the credibility of results by diminishing uncertainties. With the increase of the training set size for surge and sway models, the obtained models agree well with the reference data. This paper focuses on the prediction of the hydrodynamic forces and moments of the bare ship hull in shallow water under the assumption that the values of the hydrodynamic coefficient are directly affected by the shallow water depth. The hydrodynamic terms related to the rudder, propeller, and their interaction are neglected due to the lack of test data, which is the limitation of this paper. In a future study, it is suggested to carry out the PMM test in shallow water using the hull with rudder and propeller, and the hydrodynamic terms explicitly related to the shallow water features should also deserve more attention. The determination of the optimal values of the parameters for the truncated LS-SVM is also an interesting topic for further investigation.

Author Contributions: Conceptualization and methodology, H.X. and C.G.S.; software, validation, formal analysis, investigation, data curation, visualization, and writing—original draft preparation, H.X.; writing—review and editing, H.X. and C.G.S., supervision, project administration and funding acquisition, C.G.S. All authors have read and agreed to the published version of the manuscript.

Funding: This work was performed within the Strategic Research Plan of the Centre for Marine Technology and Ocean Engineering, financed by the Portuguese Foundation for Science and Technology (Fundação para a Ciência e Tecnologia—FCT) under contract UIDB/UIDP/00134/2020. The PMM data were collected in the experiments performed during the Project "SHOPERA-Energy Efficient Safe SHip OPERAtion", which was funded by the EU under contract 605221.

Institutional Review Board Statement: This study does not involve humans or animals.

Informed Consent Statement: Not applicable.

Data Availability Statement: Not applicable.

Conflicts of Interest: The authors declare no conflict of interest.

Nomenclature

LS-SVM	Least-squares support vector machine
RMS	Root mean square
MMG	Manoeuvring Modelling Group
LSTM	Long short-term memory network
GA	Genetic Algorithm
CFD	Computational fluid dynamics
SVM	Support vector machine
PMM	Planar Motion Mechanism
DOF	Degrees of Freedom
DTC	Duisburg Test Case
FHR	Flanders Hydraulics Research
QP	Quadratic programming
R	Correlation coefficient

References

1. Yoshimura, Y.; Sakurai, H. Mathematical Model for the Manoeuvring Ship Motion in Shallow Water (3rd Report): Manoeuvrability of a Twin-Propeller Twin Rudder Ship. *J. Kansai Soc. Nav. Archit.* **1989**, *211*, 115–126.
2. Sutulo, S.; Guedes Soares, C. On the Application of Empiric Methods for Prediction of Ship Manoeuvring Properties and Associated Uncertainties. *Ocean Eng.* **2019**, *186*, 106111. [CrossRef]
3. Perera, L.P.; Carvalho, J.P.; Guedes Soares, C. Fuzzy Logic Based Decision Making System for Collision Avoidance of Ocean Navigation under Critical Collision Conditions. *J. Mar. Sci. Technol.* **2011**, *16*, 84–99. [CrossRef]
4. Varela, J.M.; Guedes Soares, C. Interactive 3D Desktop Ship Simulator for Testing and Training Offloading Manoeuvres. *Appl. Ocean Res.* **2015**, *51*, 367–380. [CrossRef]

5. Clarke, D. A History of Ship Manoeuvrability Theory and Practice. In Proceedings of the International Conference on Marine Simulation and Ship Manoeuvrability, Newcastle upon Tyne, UK, 8–11 September 2015; pp. 1–27.
6. Sutulo, S.; Guedes Soares, C. An Algorithm for Optimized Design of Maneuvering Experiments. *J. Sh. Res.* **2002**, *46*, 214–227. [CrossRef]
7. Hinostroza, M.A.; Xu, H.; Guedes Soares, C. Path-Planning and Path-Following Control System for Autonomous Surface Vessel. In *Maritime Transportation and Harvesting of Sea Resources*; Guedes Soares, C., Teixeira, A.P., Eds.; Taylor & Francis Group: London, UK, 2018; pp. 991–998.
8. Inoue, S.; Murayama, K. Calculation of Turning Ship Derivatives in Shallow Water. *West Japan Soc. Nav. Archit.* **1969**, *37*, 73–85.
9. Sutulo, S.; Guedes Soares, C. Development of a Multifactor Regression Model of Ship Maneuvering Forces Based on Optimized Captive-Model Tests. *J. Sh. Res.* **2006**, *50*, 311–333. [CrossRef]
10. Delefortrie, G.; Eloot, K.; Lataire, E.; Van Hoydonck, W.; Vantorre, M. Captive Model Tests Based 6 DOF Shallow Water Manoeuvring Model. In Proceedings of the 4th MASHCON, Hamburg, Germany, 23–25 May 2016. [CrossRef]
11. Kaidi, S.; Smaoui, H.; Sergent, P. Modeling the Maneuvering Behavior of Container Carriers in Shallow Water. *J. Waterw. Port Coastal Ocean Eng.* **2018**, *144*, 04018017. [CrossRef]
12. Lataire, E.; Vantorre, M.; Delefortrie, G.; Candries, M. Mathematical Modelling of Forces Acting on Ships during Lightering Operations. *Ocean Eng.* **2012**, *55*, 101–115. [CrossRef]
13. Xu, H.; Hinostroza, M.A.; Wang, Z.; Guedes Soares, C. Experimental Investigation of Shallow Water Effect on Vessel Steering Model Using System Identification Method. *Ocean Eng.* **2020**, *199*, 106940. [CrossRef]
14. Abkowitz, M.A. Measurement of Hydrodynamic Characteristics from Ship Maneuvering Trials by System Identification. *SNAME Trans.* **1980**, *88*, 283–318.
15. Luo, W.; Guedes Soares, C.; Zou, Z. Parameter Identification of Ship Maneuvering Model Based on Support Vector Machines and Particle Swarm Optimization. *J. Offshore Mech. Arct. Eng.* **2016**, *138*, 031101. [CrossRef]
16. Xu, H.; Hinostroza, M.A.; Guedes Soares, C. Estimation of Hydrodynamic Coefficients of a Nonlinear Manoeuvring Mathematical Model With Free-Running Ship Model Tests. *Int. J. Marit. Eng.* **2018**, *160*, A-213–A-226. [CrossRef]
17. Moreira, L.; Fossen, T.I.; Guedes Soares, C. Path Following Control System for a Tanker Ship Model. *Ocean Eng.* **2007**, *34*, 2074–2085. [CrossRef]
18. Xu, H.; Guedes Soares, C. Waypoint-Following for a Marine Surface Ship Model Based on Vector Field Guidance Law. In *Maritime Technology and Engineering 3*; Guedes Soares, C., Santos, T.A., Eds.; Taylor & Francis Group: London, UK, 2016; Volume 1, pp. 409–418.
19. Xu, H.; Guedes Soares, C. An Optimized Energy-Efficient Path Following Algorithm for Underactuated Marine Surface Ship Model. *Int. J. Marit. Eng.* **2018**, *160*, A-411–A-421. [CrossRef]
20. Du, P.; Cheng, L.; Tang, Z.J.; Ouahsine, A.; Hu, H.B.; Hoarau, Y. Ship Maneuvering Prediction Based on Virtual Captive Model Test and System Dynamics Approaches. *J. Hydrodyn.* **2022**, *34*, 259–276. [CrossRef]
21. Miyauchi, Y.; Maki, A.; Umeda, N.; Rachman, D.M.; Akimoto, Y. System Parameter Exploration of Ship Maneuvering Model for Automatic Docking/Berthing Using CMA-ES. *J. Mar. Sci. Technol.* **2022**, *27*, 1065–1083. [CrossRef]
22. Sukas, O.F.; Kinaci, O.K.; Bal, S. System-Based Prediction of Maneuvering Performance of Twin-Propeller and Twin-Rudder Ship Using a Modular Mathematical Model. *Appl. Ocean Res.* **2019**, *84*, 145–162. [CrossRef]
23. Fossen, T.I. *Handbook of Marine Craft Hydrodynamics and Motion Control*; John Wiley & Sons, Ltd.: Chichester, UK, 2011. [CrossRef]
24. Xu, H.; Hassani, V.; Guedes Soares, C. Comparing Generic and Vectorial Nonlinear Manoeuvring Models and Parameter Estimation Using Optimal Truncated Least Square Support Vector Machine. *Appl. Ocean Res.* **2020**, *97*, 102061. [CrossRef]
25. Berge, S.P.; Fossen, T.I. On the Properties of the Nonlinear Ship Equations of Motion. *Math. Comput. Model. Dyn. Syst.* **2000**, *6*, 365–381. [CrossRef]
26. Maimun, A.; Priyanto, A.; Muhammad, A.H.; Scully, C.C.; Awal, Z.I. Manoeuvring Prediction of Pusher Barge in Deep and Shallow Water. *Ocean Eng.* **2011**, *38*, 1291–1299. [CrossRef]
27. Hassani, V.; Fathi, D.; Ross, A.; Sprenger, F.; Selvik, Ø.; Berg, T.E.T.E.; Fathi, D.; Sprenger, F.; Berg, T.E.T.E. Time Domain Simulation Model for Research Vessel Gunnerus. In Proceedings of the ASME 2015 34th International Conference on Ocean, Offshore and Arctic Engineering, St-John's, NL, Canada, 31 May–5 June 2015; ASME: St-John's, NL, Canada, 2015; Volume 7, p. V007T06A013. [CrossRef]
28. Vantorre, M.; Eloot, K. Requirements for Standard Harmonic Captive Manoeuvring Tests. *Control Eng. Pract.* **1998**, *6*, 643–652. [CrossRef]
29. Eloot, K.; Delefortrie, G.; Vantorre, M.; Quadvlieg, F. Validation of Ship Manoeuvring in Shallow Water through Free-Running Tests. In Proceedings of the ASME 2015 34th International Conference on Ocean, Offshore and Arctic Engineering, St-John's, NL, Canada, 31 May–5 June 2015; ASME: St-John's, NL, Canada, 2015; Volume 7, p. V007T06A017. [CrossRef]
30. Sutulo, S.; Guedes Soares, C. An Algorithm for Offline Identification of Ship Manoeuvring Mathematical Models from Free-Running Tests. *Ocean Eng.* **2014**, *79*, 10–25. [CrossRef]
31. Xu, H.; Hinostroza, M.A.; Guedes Soares, C. Modified Vector Field Path-Following Control System for an Underactuated Autonomous Surface Ship Model in the Presence of Static Obstacles. *J. Mar. Sci. Eng.* **2021**, *9*, 652. [CrossRef]
32. Lu, S.; Cheng, X.; Liu, J.; Li, S.; Yasukawa, H. Maneuvering Modeling of a Twin-Propeller Twin-Rudder Inland Container Vessel Based on Integrated CFD and Empirical Methods. *Appl. Ocean Res.* **2022**, *126*, 103261. [CrossRef]

33. Mucha, P.; Dettmann, T.; Ferrari, V.; el Moctar, O. Experimental Investigation of Free-Running Ship Manoeuvers under Extreme Shallow Water Conditions. *Appl. Ocean Res.* **2019**, *83*, 155–162. [CrossRef]
34. Guedes Soares, C.; Sutulo, S.; Francisco, R.A.; Santos, F.M.; Moreira, L. Full-Scale Measurements of The Manoeuvring Capabilities of A Catamaran. In Proceedings of the International Conference on Hydrodynamics of High Speed Craft, RINA, London, UK, 24–25 November 1999; pp. 1–12. [CrossRef]
35. Guedes Soares, C.; Francisco, R.A.; Moreira, L.; Laranjinha, M. Full-Scale Measurements of the Maneuvering Capabilities of Fast Patrol Vessels, Argos Class. *Mar. Technol.* **2004**, *41*, 7–16. [CrossRef]
36. Xu, H.; Hinostroza, M.A.; Hassani, V.; Guedes Soares, C. Real-Time Parameter Estimation of a Nonlinear Vessel Steering Model Using a Support Vector Machine. *J. Offshore Mech. Arct. Eng.* **2019**, *141*, 061606. [CrossRef]
37. Suzuki, R.; Tsukada, Y.; Ueno, M. Estimation of Full-Scale Ship Manoeuvrability in Adverse Weather Using Free-Running Model Test. *Ocean Eng.* **2020**, *213*, 107562. [CrossRef]
38. Wang, Z.; Xu, H.; Xia, L.; Zou, Z.; Guedes Soares, C. Kernel-Based Support Vector Regression for Nonparametric Modeling of Ship Maneuvering Motion. *Ocean Eng.* **2020**, *216*, 107994. [CrossRef]
39. Zhu, M.; Hahn, A.; Wen, Y.-Q.; Bolles, A. Identification-Based Simplified Model of Large Container Ships Using Support Vector Machines and Artificial Bee Colony Algorithm. *Appl. Ocean Res.* **2017**, *68*, 249–261. [CrossRef]
40. Luo, W.; Moreira, L.; Guedes Soares, C. Manoeuvring Simulation of Catamaran by Using Implicit Models Based on Support Vector Machines. *Ocean Eng.* **2014**, *82*, 150–159. [CrossRef]
41. Åström, K.J.; Källström, C.G. Identification of Ship Steering Dynamics. *Automatica* **1976**, *12*, 9–22. [CrossRef]
42. Wang, H.; Liu, J.; Zhi, J.; Fu, C. The Improvement of Quantum Genetic Algorithm and Its Application on Function Optimization. *Math. Probl. Eng.* **2013**, *2013*, 1–10. [CrossRef]
43. Ding, F. Least Squares Parameter Estimation and Multi-Innovation Least Squares Methods for Linear Fitting Problems from Noisy Data. *J. Comput. Appl. Math.* **2023**, *426*, 115107. [CrossRef]
44. Xu, L.; Ding, F.; Zhu, Q. Separable Synchronous Multi-Innovation Gradient-Based Iterative Signal Modeling from On-Line Measurements. *IEEE Trans. Instrum. Meas.* **2022**, *71*, 6501313. [CrossRef]
45. Xu, L. Parameter Estimation for Nonlinear Functions Related to System Responses. *Int. J. Control Autom. Syst.* **2023**, *21*, 1780–1792. [CrossRef]
46. Wang, N.; Sun, J.C.; Er, M.J.; Liu, Y.C. Hybrid Recursive Least Squares Algorithm for Online Sequential Identification Using Data Chunks. *Neurocomputing* **2016**, *174*, 651–660. [CrossRef]
47. Qian, L.; Zheng, Y.; Li, L.; Ma, Y.; Zhou, C.; Zhang, D. A New Method of Inland Water Ship Trajectory Prediction Based on Long Short-Term Memory Network Optimized by Genetic Algorithm. *Appl. Sci.* **2022**, *12*, 4073. [CrossRef]
48. Bonci, M.; Viviani, M.; Broglia, R.; Dubbioso, G. Method for Estimating Parameters of Practical Ship Manoeuvring Models Based on the Combination of RANSE Computations and System Identification. *Appl. Ocean Res.* **2015**, *52*, 274–294. [CrossRef]
49. Xu, H.; Guedes Soares, C. Convergence Analysis of Hydrodynamic Coefficients Estimation Using Regularization Filter Functions on Free-Running Ship Model Tests with Noise. *Ocean Eng.* **2022**, *250*, 111012. [CrossRef]
50. Wang, N.; Er, M.J.; Han, M. Large Tanker Motion Model Identification Using Generalized Ellipsoidal Basis Function-Based Fuzzy Neural Networks. *IEEE Trans. Cybern.* **2015**, *45*, 2732–2743. [CrossRef]
51. Dong, Q.; Wang, N.; Song, J.; Hao, L.; Liu, S.; Han, B.; Qu, K. Math-Data Integrated Prediction Model for Ship Maneuvering Motion. *Ocean Eng.* **2023**, *285*, 115255. [CrossRef]
52. Luo, W.; Li, X. Measures to Diminish the Parameter Drift in the Modeling of Ship Manoeuvring Using System Identification. *Appl. Ocean Res.* **2017**, *67*, 9–20. [CrossRef]
53. Zhang, X.-G.; Zou, Z.-J. Estimation of the Hydrodynamic Coefficients from Captive Model Test Results by Using Support Vector Machines. *Ocean Eng.* **2013**, *73*, 25–31. [CrossRef]
54. Wang, Z.; Zou, Z.; Guedes Soares, C. Identification of Ship Manoeuvring Motion Based on Nu-Support Vector Machine. *Ocean Eng.* **2019**, *183*, 270–281. [CrossRef]
55. Suykens, J.A.K.; Vandewalle, J. Least Squares Support Vector Machine Classifiers. *Neural Process. Lett.* **1999**, *9*, 293–300. [CrossRef]
56. Cortes, C.; Vapnik, V. Support-Vector Networks. *Mach. Learn.* **1995**, *20*, 273–297. [CrossRef]
57. Chen, L.; Zhou, S. Sparse Algorithm for Robust LSSVM in Primal Space. *Neurocomputing* **2018**, *275*, 2880–2891. [CrossRef]
58. Ma, H.; Ding, F.; Wang, Y. A Novel Multi-Innovation Gradient Support Vector Machine Regression Method. *ISA Trans.* **2022**, *130*, 343–359. [CrossRef]
59. Ma, H.; Ding, F.; Wang, Y. Multi-Innovation Newton Recursive Methods for Solving the Support Vector Machine Regression Problems. *Int. J. Robust Nonlinear Control* **2021**, *31*, 7239–7260. [CrossRef]
60. Suykens, J.A.K.; Van Gestel, T.; De Brabanter, J.; De Moor, B.; Vandewalle, J. *Least Squares Support Vector Machines*; World Scientific: Singapore, 2002. [CrossRef]
61. Xu, H.; Hassani, V.; Guedes Soares, C. Truncated Least Square Support Vector Machine for Parameter Estimation of a Nonlinear Manoeuvring Model Based on PMM Tests. *Appl. Ocean Res.* **2020**, *97*, 102076. [CrossRef]
62. Xu, H.; Guedes Soares, C. Hydrodynamic Coefficient Estimation for Ship Manoeuvring in Shallow Water Using an Optimal Truncated LS-SVM. *Ocean Eng.* **2019**, *191*, 106488. [CrossRef]

63. Xu, H.; Guedes Soares, C. Manoeuvring Modelling of a Containership in Shallow Water Based on Optimal Truncated Nonlinear Kernel-Based Least Square Support Vector Machine and Quantum-Inspired Evolutionary Algorithm. *Ocean Eng.* **2020**, *195*, 106676. [CrossRef]
64. Hinostroza, M.; Xu, H.; Guedes Soares, C. Manoeuvring Test for a Self-Running Ship Model in Various Water Depth Conditions. In *Sustainable Development and Innovations in Marine Technologies*; Georgiev, P., Guedes Soares, C., Eds.; Taylor & Francis Group: Abingdon, UK, 2019; pp. 187–196.
65. Sutulo, S.; Guedes Soares, C. Development of a Core Mathematical Model for Arbitrary Manoeuvres of a Shuttle Tanker. *Appl. Ocean Res.* **2015**, *51*, 293–308. [CrossRef]
66. Inoue, S.; Hirano, M.; Kijima, K.; Takashina, J. A Practical Calculation Method of Ship Maneuvering Motion. *Int. Shipbuild. Prog.* **1981**, *28*, 207–222. [CrossRef]
67. Inoue, S.; Hirano, M.; Kijima, K. Hydrodynamic Derivatives on Ship Manoeuvring. *Int. Shipbuild. Prog.* **1981**, *28*, 112–125. [CrossRef]
68. SNAME. *Nomenclature for Treating the Motion of a Submerged Body through a Fluid*; Society of Naval Architects and Marine Engineers: New York, NY, USA, 1950.
69. Sprenger, F.; Maron, A.; Delefortrie, G.; van Zwijnsvoorde, T.; Cura-Hochbaum, A.; Lengwinat, A.; Papanikolaou, A. Experimental Studies on Seakeeping and Maneuverability of Ships in Adverse Weather Conditions. *J. Sh. Res.* **2017**, *61*, 131–152. [CrossRef]
70. Eloot, K.; Vantorre, M.; Delefortrie, G.; Lataire, E. Running Sinkage and Trim of the DTC Container Carrier in Harmonic Sway and Yaw Motion: Open Model Test Data for Validation Purposes. In Proceedings of the 4th International Conference on Ship Manoeuvring in Shallow and Confined Water (MASHCON): Ship Bottom Interaction, Hamburg, Germany, 23–25 May 2016; Uliczka, K., Böttner, C.-U., Kastens, M., Eloot, K., Delefortrie, G., Vantorre, M., Candries, M., Lataire, E., Eds.; Bundesanstalt für Wasserbau: Hamburg, Germany, 2016; pp. 251–261. [CrossRef]
71. el Moctar, O.; Shigunov, V.; Zorn, T. Duisburg Test Case: Post-Panamax Container Ship for Benchmarking. *Sh. Technol. Res.* **2012**, *59*, 50–64. [CrossRef]
72. Van Kerkhove, G.; Vantorre, M.; Delefortrie, G. Advanced Model Testing Techniques for Ship Behaviour in Shallow and Confined Water. In Proceedings of the AMT '09: 1st International Conference on Advanced Model Measurement Technology for the EU Maritime Industry, Nantes, France, 1–2 September 2009; pp. 158–172.
73. Papanikolaou, A.; Zaraphonitis, G.; Bitner-Gregersen, E.; Shigunov, V.; El Moctar, O.; Guedes Soares, C.; Reddy, D.N.; Sprenger, F. Energy Efficient Safe SHip Operation (SHOPERA). *Transp. Res. Procedia* **2016**, *14*, 820–829. [CrossRef]
74. Vapnik, V.N. The Nature of Statistical Learning Theory. *IEEE Trans. Neural Networks* **1995**, *8*, 187. [CrossRef]
75. Hansen, P.C. The Discrete Picard Condition for Discrete Ill-Posed Problems. *BIT Numer. Math.* **1990**, *30*, 658–672. [CrossRef]
76. Hansen, P.C.; O'Leary, D.P. The Use of the L-Curve in the Regularization of Discrete Ill-Posed Problems. *SIAM J. Sci. Comput.* **1993**, *14*, 1487–1503. [CrossRef]
77. Hwang, W. Application of System Identification to Ship Maneuvering. Ph.D. Thesis, Massachusetts Institute of Technology, Cambridge, MA, USA, 1980.
78. Taylor, K.E. Summarizing Multiple Aspects of Model Performance in a Single Diagram. *J. Geophys. Res. Atmos.* **2001**, *106*, 7183–7192. [CrossRef]

MDPI AG
Grosspeteranlage 5
4052 Basel
Switzerland
Tel.: +41 61 683 77 34

Journal of Marine Science and Engineering Editorial Office
E-mail: jmse@mdpi.com
www.mdpi.com/journal/jmse

www.ingramcontent.com/pod-product-compliance
Lightning Source LLC
Chambersburg PA
CBHW040829250926
51850CB00007BA/1298
9783725815326